OPTIMIZATION AND LOGISTICS CHALLENGES IN THE ENTERPRISE

Springer Optimization and Its Applications

VOLUME 30

Aims and Scope
Optimization has been expanding in all directions at an astonishing rate during the last few decades. New algorithmic and theoretical techniques have been developed, the diffusion into other disciplines has proceeded at a rapid pace, and our knowledge of all aspects of the field has grown even more profound. At the same time, one of the most striking trends in optimization is the constantly increasing emphasis on the interdisciplinary nature of the field. Optimization has been a basic tool in all areas of applied mathematics, engineering, medicine, economics and other sciences.

The series *Springer Optimization and Its Applications* publishes undergraduate and graduate textbooks, monographs and state-of-the-art expository works that focus on algorithms for solving optimization problems and also study applications involving such problems. Some of the topics covered include nonlinear optimization (convex and nonconvex), network flow problems, stochastic optimization, optimal control, discrete optimization, multiobjective programming, description of software packages, approximation techniques and heuristic approaches.

OPTIMIZATION AND LOGISTICS CHALLENGES IN THE ENTERPRISE

Edited By

WANPRACHA CHAOVALITWONGSE
Rutgers University, Piscataway, NJ, USA

KEVIN C. FURMAN
ExxonMobil Research and Engineering, Annandale
NJ, USA

PANOS M. PARDALOS
University of Florida, Gainesville, FL, USA

 Springer

Editors

Wanpracha Chaovalitwongse
Department of Industrial and
  Systems Engineering
Rutgers State University of
  New Jersey
96 Frelinghuysen Rd.
Piscataway NJ 08854
USA
wchaoval@rci.rutgers.edu

Panos M. Pardalos
Department of Industrial and
  Systems Engineering
University of Florida
303 Weil Hall P.O.Box
116595
Gainesville FL 32611-6595
USA
pardalos@ise.ufl.edu

Kevin C. Furman
ExxonMobil Research and Engineering
Corporate Strategic Research
1545 Route 22 East
Annandale NJ 08801
USA
kevin.c.furman@exxonmobil.com

ISSN 1931-6828
ISBN 978-1-4899-8348-0   ISBN 978-0-387-88617-6 (eBook)
DOI 10.1007/978-0-387-88617-6
Springer Dordrecht Heidelberg London New York

Mathematics Subject Classification (2000): 90B06, 90C06, 90C59

Cover illustration: Picture taken by Elias Tyligadas

Printed on acid-free paper

Springer is part of Springer Science+Business Media (www.springer.com)

Dedicated to our loving families

Preface

This book represents a collection of computational challenges and recent advances in supply chain, logistics, and optimization research that practically applies to a collaborative and integrative environment in the enterprise. This book has been designed in response to an explosion of interest by academic researchers and industrial practitioners in highly effective coordination between supply chain partners, dynamic collaborative and strategic alliance relationships, and efficient logistics and supply chain network designs that commonly arise in a wide variety of industries. Rather than concentrating on just methodology or techniques (such as mathematical programming or simulation) or specific application areas (such as production, inventory or transportation), we have chosen to present the reader with a collection of topics, which bridge the gap between the operations research and mathematical optimization research from the academic arena with industrial practice. It will also be of value to investigators and practitioners in academic institutions and industry who become involved in supply chain and logistics in enterprise operations as an aid in translating computational optimization techniques to their colleagues in management levels. This book will be very appealing to graduate (and advanced undergraduate) students, researchers and practitioners across a wide range of industries (e.g., pharmaceutical, chemical, transportation, shipping, etc.), who require a detailed overview of the practical aspects of the design, conduct, and the analysis of supply chain and logistics problems arising in real life. For this reason, our audience is assumed to be very diverse and heterogeneous, including:

(a) researchers in operations research from engineering, computer science, statistics and mathematics domains as well as practitioners in industry (e.g., strategic planning directors, operation advisors, senior managers);

(b) researchers from engineering and business domains as well as supply chain and logistics practitioners in industry (e.g., management systems directors, supply chain managers, site supervisors);

(c) researchers in systems engineering and chemical and manufacturing process operations fields as well as investigators and practitioners in the industry who become involved in some way in systems operations (e.g., operations supervisors, process engineers, systems analysts).

There are four major research themes in this book: Process Industry, Supply Chain and Logistics Design, Supply Chain Operation, and Networking and Transportation. Each theme addresses the answer to a classic, yet extremely important, question from industry, "How do we go from the mathematical modeling and optimization techniques to the practical solutions to the enterprise's operations?"

The first theme includes four chapters focused on optimization and logistics challenges in the process industry. The first chapter, by Grossmann and Furman, lays down the platform of this book by discussing the integration of optimization systems in the process industry throughout an entire enterprise. Enterprise-wide optimization involves the coordinated optimization of research, development, supply, manufacturing, and distribution operations across business functions, organizations and the hierarchy of strategic, tactical and operational decision making. In the second chapter, Zyngier and Kelly consider novel optimization models of inventory for logistics problems in the process industry. These ideas can be applied to process production and distribution planning and scheduling models. The third chapter, by Ierapetritou and Li, presents a review of the methodologies developed to address uncertainty in chemical process planning and scheduling. Recent progress in the areas of sensitivity analysis and parametric programming are highlighted in their application to planning and scheduling in the chemical process industry. In the fourth chapter, Assavapokee et al. address decision making under uncertainty by developing a relative robust optimization algorithm. This work has an impact on supply chain network infrastructure design problems.

The second theme includes four chapters that provides reviews and challenges in supply chain models and logistics design. The fifth chapter, by Mulvey and Erkan, illustrates a supply chain risk management model with a global production problem involving movement of currency. The design of the supply chain includes uncertainty in production and as well as the risks embedded in global financial markets. In the sixth chapter, Miller provides a historical perspective and recommendation on the methods and approaches in the use of optimization technology for decision support by firms at the strategic level down through operations. The seventh chapter, by He et al., presents mathematical models for hub location problems as well as recent advances in optimization used to solve the problems. These problems are primarily of interest for supply chain modelers, especially in warehouse location design. In the eight chapter, Chen et al. present a review of the well-known Nested Partitions method for the solution of discrete optimization problems. A hybrid framework combining mathematical programming and Nest Partitions is developed and demonstrated on the intermodel hub location class of problems.

The third theme includes three chapters that address issues in supply chain operation. In the ninth chapter, Benli presents a new modeling scheme for scheduling problems that arise in supply chain optimization. This novel framework is illustrated via the lot streaming problem in the production planning area. In the tenth chapter, Metan and Thiele propose a dynamic and data-driven approach to inventory management that incorporates both historical information and addresses seasonality. This approach is demonstrated through extensive computational results with the news vendor problem. In the 11th chapter, Gong *et al.* consider a problem in task scheduling for service restoration planning. They apply a combined mathematical programming and constraint programming approach to this problem modeled with multiple objective functions.

The last theme includes four chapters that present recent advances in mathematical programming and algorithms developed for logistics networking and transportation problems. In the 12th chapter, Liang and Chaovalitwongse propose a new network model for the aircraft maintenance routing problem. The new model utilizes the idea of using bidirectional flows of aircrafts. The resulting model is very compact and scalable and has been applied to real-life problems. In the 13th chapter, Shen *et al.* develop a chance constraint model and tabu search solution procedure to look at vehicle routing in which one wishes to minimize unmet demand while addressing uncertainty in both demand and travel times. This application is important in the area of supply chain distribution during disaster scenarios. In the 14th chapter, Agarwal *et al.* study both carrier alliances and shipper collaborations as they apply to sea, air and trucking cargo. Game theoretic models are developed to analyze the benefits and sustainability issues surrounding these forms of collaboration. In the last chapter, Arulselvan *et al.* consider wireless agents in a mobile ad hoc network to determine the routing that maximizes connectivity. New formulations and heuristic algorithms are presented to address this problem which can arise in several military applications.

In order to complete this volume, we have dealt with the authors and anonymous referees over the past few years. The experience has been challenging, yet extremely rewarding. We truly hope that the reader will find the fundamental research and applications chapters presented here as stimulating and valuable as we did. We want to thank Prof. Altannar Chinchuluun from the University of Florida for proofreading the final volume. Last but not least we cannot thank the authors and anonymous referees enough for their time, efforts and dedication to make this volume successful.

New Jersey

September 2008

Wanpracha Art Chaovalitwongse

Kevin C. Furman

Panos M. Pardalos

Contents

Part I Process Industry

Part II Supply Chain and Logistics Design

Part III Supply Chain Operation

Part IV Networking and Transportation

List of Contributors

Richa Agarwal
Department of Industrial and
Systems Engineering
Georgia Institute of Technology
Atlanta, Georgia, 30332-0100
richaa@amazon.com

Jane C. Ammons
Department of Industrial and
Systems Engineering
Georgia Institute of Technology
Atlanta, Georgia, 30332-0100
jane.ammons@isye.gatech.edu

Ashwin Arulselvan
Department of Industrial and
Systems Engineering
University of Florida
Gainesville, FL 32611
ashwin@ufl.edu

Tiravat Assavapokee
Department of Industrial Engineering
University of Houston
Houston, Texas 77201-4008
tiravat.assavapokee@mail.uh.edu

Omer S. Benli
Department of Industrial Systems
California State University
Long Beach, CA 90840
obenli@csulb.edu

W. Art Chaovalitwongse
Department of Industrial and
Systems Engineering
Rutgers University
Piscataway, NJ 08854
wchaoval@rci.rutgers.edu

Anthony Chen
Department of Civil Engineering
Utah State University
Logan, UT 84322-4110
achen@engineering.usu.edu

Weiwei Chen
Department of Industrial and
Systems Engineering
University of Wisconsin-Madison
Madison, WI 53706
wchen26@wisc.edu

Clayton W. Commander
Air Force Research Laboratory
Munitions Directorate
Eglin AFB, FL 32542
clayton.commander@eglin.af.mil

Maged M. Dessouky
Department of Industrial and
Systems Engineering
University of Southern California
Los Angeles, CA 90089
maged@usc.edu

Özlem Ergun
Department of Industrial and
Systems Engineering
Georgia Institute of Technology
Atlanta, Georgia, 30332-0100
oergun@isye.gatech.edu

Hafize G. Erkan
Department of Operations Research
and Financial Engineering
Princeton University
Princeton NJ 08544
herkan@alumni.princeton.edu

Kevin C. Furman
Corporate Strategic Research
ExxonMobil Research and
Engineering
Annandale, NJ 08801
kevin.c.furman@exxonmobil.com

Jing Gong
Department of Decision Sciences and
Engineering Systems
Rensselaer Polytechnic Institute
Troy, NY 12180
gongj@rpi.edu

Ignacio E. Grossmann
Department of Chemical Engineering
Carnegie Mellon University
Pittsburgh, PA 15213
grossmann@cmu.edu

Xiaozheng He
Department of Civil Engineering
University of Minnesota
Minneapolis, MN 55455
hexxx069@umn.edu

Michael J. Hirsch
Network Centric Systems
Raytheon Inc.
St. Petersburg, FL 33710
michael_j_hirsch@Raytheon.com

Lori Houghtalen
Department of Industrial and
Systems Engineering
Georgia Institute of Technology
Atlanta, Georgia, 30332-0100
lhoughtalen@babson.edu

Marianthi Ierapetritou
Department of Chemical and
Biochemical Engineering
Rutgers University
Piscataway, NJ 08854-8058
marianth@soemail.rutgers.edu

Jeffrey D. Kelly
Honeywell Process Solutions
85 Enterprise Blvd., Suite 100
Markham, ON, L6G0B5, Canada
Jeff.Kelly@honeywell.com

Earl E. Lee
Department of Civil and
Environmental Engineering
University of Delaware
Newark, DE 19702
elee@udel.edu

Zukui Li
Department of Chemical and
Biochemical Engineering
Rutgers University
Piscataway, NJ 08854-8058
zukui@eden.rutgers.edu

Zhe Liang
Department of Industrial and
Systems Engineering
Rutgers University
Piscataway, NJ 08854
liangzhe@eden.rutgers.edu

Henry Liu
Department of Civil Engineering
University of Minnesota
Minneapolis, MN 55455
henryliu@umn.edu

Gokhan Metan
American Airlines
Fort Worth, TX 78155
gom204@lehigh.edu

Tan Miller
College of Business Administration
Rider University
Lawrenceville, NJ 08648-3001
tanjean@verizon.net

John E. Mitchell
Department of Mathematical
Sciences
Rensselaer Polytechnic Institute
Troy, NY 12180
mitchj@rpi.edu

John M. Mulvey
Department of Operations Research
and Financial Engineering
Princeton University
Princeton NJ 08544
mulvey@princeton.edu

Fernando Ordóñez
Department of Industrial and
Systems Engineering
University of Southern California
Los Angeles, CA 90089
fordon@usc.edu

Okan Orsan Ozener
Department of Industrial and
Systems Engineering
Georgia Institute of Technology
Atlanta, Georgia, 30332-0100
oozener@isye.gatech.edu

Panos M. Pardalos
Department of Industrial and
Systems Engineering
University of Florida
Gainesville, FL 32611
pardalos@ufl.edu

Liang Pi
Department of Industrial and
Systems Engineering
University of Wisconsin-Madison
Madison, WI 53706
lpi@wisc.edu

Matthew J. Realff
Department of Chemical and
Biomolecular Engineering
Georgia Institute of Technology
Atlanta, Georgia, 30332-0100
matthew.realff@chbe.gatech.edu

Zhihong Shen
Department of Industrial and
Systems Engineering
University of Southern California
Los Angeles, CA 90089
shenz@usc.edu

Leyuan Shi
Department of Industrial and
Systems Engineering
University of Wisconsin-Madison
Madison, WI 53706
leyuan@engr.wisc.edu

Aurélie Thiele
Department of Industrial and
Systems Engineering
Lehigh University
Bethlehem, PA 18015
aut204@lehigh.edu

William A. Wallace
Department of Decision Sciences and
Engineering Systems
Rensselaer Polytechnic Institute
Troy, NY 12180
wallaw@rpi.edu

Danielle Zyngier
Honeywell Process Solutions
85 Enterprise Blvd., Suite 100
Markham, ON, L6G0B5, Canada
Danielle.Zyngier@honeywell.com

Process Industry

Challenges in Enterprise-wide Optimization for the Process Industries

Ignacio E. Grossmann[1] and Kevin C. Furman[2]

[1] Department of Chemical Engineering
Carnegie Mellon University, Pittsburgh, PA 15213
grossmann@cmu.edu
[2] Corporate Strategic Research
ExxonMobil Research & Engineering, Annandale, NJ 08801
kevin.c.furman@exxonmobil.com

Summary Enterprise-wide optimization (EWO) is a new emerging area that lies at the interface of chemical engineering and operations research, and has become a major goal in the process industries due to the increasing pressures for remaining competitive in the global marketplace. EWO involves optimizing the operations of supply, manufacturing, and distribution activities of a company to reduce costs and inventories. A major focus in EWO is the optimal operation of manufacturing facilities, which often requires the use of nonlinear process models. Major operational items include planning, scheduling, real-time optimization, and inventory control. This chapter provides an overview of major challenges in the development of deterministic and stochastic linear/nonlinear optimization models and algorithms for the optimization of entire supply chains that are involved in EWO problems. We specifically review three major challenges: (a) modeling of planning and scheduling, (b) multiscale optimization, and (c) handling of uncertainties. Finally, we also discuss briefly the algorithmic methods and tools that are required for tackling these problems, and we conclude with future research needs in this area.

1 Introduction

The goal of this chapter is to provide a general overview of enterprise-wide optimization (EWO) as applied to the process industry. This chapter is largely an integration of the ideas, concepts, and reviews that have appeared in previous papers by Grossmann [74], Mendez et al. [146], Mendez et al. [147], Erdirik-Dogan and Grossmann [53], and Neiro and Pinto [155].

The chapter also has been motivated by our experiences in organizing the 2003 Conference Foundations on Computer-Aided Process Operations (see the web site at http://www.cheme.cmu.edu/focapo/), and in the Enterprise-wide Optimization project of the Center of Advanced Process Decision making at Carnegie Mellon (see the web site at http://egon.cheme.cmu.edu/ewocp/),

W. Chaovalitwongse et al. (eds.), *Optimization and Logistics Challenges in the Enterprise*, Springer Optimization and Its Applications 30, DOI 10.1007/978-0-387-88617-6_1, © Springer Science+Business Media, LLC 2009

which involves faculty from Carnegie Mellon, Lehigh, and University of Pittsburgh in collaboration with ABB, Air Products, BP, Dow Chemical, ExxonMobil, and Nova Chemicals.

The process industry is a key industrial sector in the United States. For instance, the U.S. chemical industry is the major producer in the world (25% of world production) with shipments reaching $506 billion and a record $109.3 billion in exports in 2004. EWO has become a major goal in this industry due to the increasing pressure for remaining competitive in the global marketplace. EWO involves simultaneously optimizing the operations of R&D, supply, manufacturing, and distribution activities of a company to reduce costs and inventories. A major focus in EWO is the scheduling of manufacturing facilities, as well as their modeling at the proper level of detail, often requiring nonlinear process models. Major operational items include planning, scheduling, real-time optimization, and inventory control. One of the key features in EWO is integration of the information and decision-making among the various functions that comprise the supply chain of the company. This is being achieved with modern information technology (IT) tools, which together with the Internet have promoted e-commerce. To fully realize the potential of transactional IT tools, the development of sophisticated deterministic and stochastic linear and nonlinear optimization models and algorithms (analytical IT tools) is needed to explore and analyze alternatives of the supply chain to yield overall optimum economic performance, as well as high levels of customer satisfaction. An additional challenge is the integrated and coordinated decision making across the various functions in a company (purchasing, manufacturing, distribution, sales), across various geographically distributed organizations (vendors, facilities and markets), and across the hierarchy of various levels of decision making (strategic, tactical, and operational).

As an example of a large-scale EWO problem consider in Figure 1 the supply chain in the petroleum industry, which comprises many intermediate steps starting from the exploration phase at the wellhead, going through trading and transportation, before reaching the refinery, and finally the primary and secondary distribution and delivery of its products, some at the retail level (e.g., gasoline). In this case it is clear that the effective coordination of the various stages is essential to accomplish the goal of EWO. As another example,

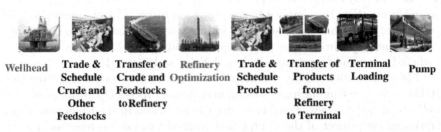

Wellhead | Trade & Schedule Crude and Other Feedstocks | Transfer of Crude and Feedstocks to Refinery | Refinery Optimization | Trade & Schedule Products | Transfer of Products from Refinery to Terminal | Terminal Loading | Pump

Fig. 1. Supply chain in the petroleum industry (courtesy ExxonMobil)

Discovery   **Market**

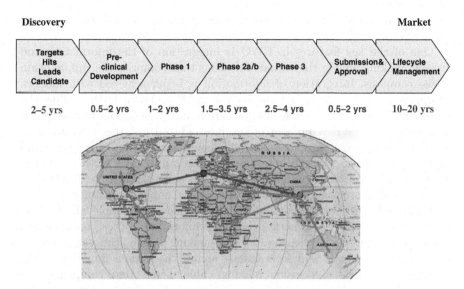

Targets Hits Leads Candidate	Pre-clinical Development	Phase 1	Phase 2a/b	Phase 3	Submission& Approval	Lifecycle Management
2–5 yrs	0.5–2 yrs	1–2 yrs	1.5–3.5 yrs	2.5–4 yrs	0.5–2 yrs	10–20 yrs

Fig. 2. Supply chain in the pharmaceutical industry with R&D coupled with manufacturing and distribution

Figure 2 shows the supply chain in the pharmaceutical industry that starts with the R&D phase for the testing of new drugs in the pharmaceutical industry, a major bottleneck, that must be eventually coupled with manufacturing and global distribution for drugs that achieve certification

From the two examples we can define EWO as optimizing the operations of R&D, material supply, manufacturing, distribution, and financial activities of a company to reduce costs and inventories, and to maximize profits, asset utilization, responsiveness, and customer satisfaction. The goal of achieving EWO in the two examples is clearly still elusive and motivates the research challenges outlined in the chapter.

As described above, EWO involves optimizing the operations of R&D, supply, manufacturing (batch or continuous), and distribution in a company. The major operational activities include planning, scheduling, real-time optimization, and inventory control. Supply chain management might be considered an equivalent term for describing EWO (see Shapiro [200]). While there is a significant overlap between the two terms, an important distinction is that supply chain management is aimed at a broader set of real-world applications with an emphasis on logistics and distribution, which usually involve linear models, traditionally the domain of operations research. In contrast, in EWO, the emphasis is on the manufacturing facilities with a major focus being their planning, scheduling, and control, which often requires the use of nonlinear process models, and hence knowledge of chemical engineering. We should also note that many process companies are adopting the term enterprise-wide optimization to reflect both the importance of manufacturing within their supply chain, as well as the drive to reduce costs through optimization.

One of the key features in EWO is integration of the information and decision making among the various functions that comprise the supply chain of the company. Integration of information is being achieved with modern information technology (IT) tools such as SAP and Oracle that allow the sharing and instantaneous flow of information along the various organizations in a company. The development of the internet and fast speed communication also has helped to promote through e-commerce the implementation and deployment of these tools. While these systems still require further developments to fully realize the vision of creating an agile platform for EWO (i.e., transactional information), it is clear that we are not too far from it.

While software vendors provide IT tools that in principle allow many groups in an enterprise to access the same information, these tools do not generally provide comprehensive decision-making capabilities that account for complex trade-offs and interactions across the various functions, subsystems, and levels of decision making. This means that companies are faced with the problem of deciding as to whether to develop their own inhouse tools for integration or else to make use of commercial software from vendors.

There is great economic potential in EWO and some progress has been made toward the goal of developing some of the basic building blocks. However, major barriers are the lack of computational optimization models and tools that will allow the full and comprehensive application of EWO throughout the process industry. This will require a new generation of tools that allow the full integration and large-scale solution of the optimization models, as well as the incorporation of accurate models for the manufacturing facilities.

2 Challenges in Enterprise-wide Optimization

In order to realize the full potential of transactional IT tools, the development of sophisticated optimization and decision-support tools (analytical IT tools) is needed to help explore and analyze alternatives and to predict actions for the operation of the supply chain so as to yield overall optimum economic performance, as well as high levels of customer satisfaction. A major challenge that is involved in EWO of process industries is the integrated and coordinated decision making across the various functions in a company (purchasing, manufacturing, distribution, sales), across various geographically distributed organizations (vendors, facilities and markets), and across various levels of decision making (strategic, tactical and operational) as seen in Figure 3 (Shapiro [200]). The first two items conceptually deal with issues related to *spatial integration* in that they involve coordinating the activities of the various subsystems of an enterprise. The third item deals with issues related to *temporal integration* in that they involve coordinating decisions

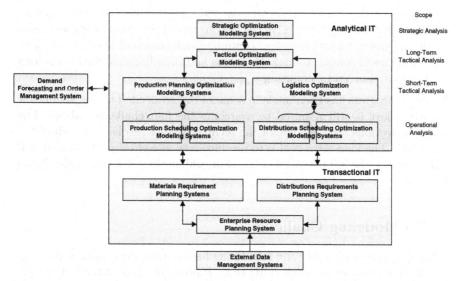

Fig. 3. Transactional and analytical IT (Tayur et al. [211])

across different timescales. Addressing these spatial and temporal integration problems is important because they provide a basis to optimize the decision making in an enterprise through the IT infrastructure.

In order to achieve EWO throughout the process industry, this goal will require a new generation of computational tools for which the following major challenges must be addressed:

a) *The modeling challenge:* What type of production planning and scheduling models should be developed for the various components of the supply chain, including *nonlinear* manufacturing processes, that through integration can ultimately achieve enterprise-wide optimization? Major issues here are the development of novel mathematical programming and logic-based models that can be effectively integrated to capture the complexity of the various operations.

b) *The multiscale challenge:* How to coordinate the optimization of these models over a given time horizon (from weeks to years) and how to coordinate the long-term strategic decisions (years) related to sourcing and investment, with the medium-term decisions (months) related to tactical decisions of production planning and material flow, and with the short-term operational decisions (weeks, days) related to scheduling and control? Major issues here involve novel decomposition procedures that can effectively work across large spatial and temporal scales.

c) *The uncertainty challenge:* How to account for stochastic variations in order to effectively handle the effect of uncertainties (e.g., demands, equipment breakdown)? Major issues here are the development of novel, meaningful, and effective stochastic programming tools.

d) *The algorithmic and computational challenge:* Given the three points above, how to effectively solve the various models in terms of efficient algorithms and in terms of modern computer architectures? Major issues here include novel computational algorithms and their implementation through distributed or grid computing.

Although progress has been made in some of the areas cited above, significant research effort is still required to overcome the four challenges above. The following sections discuss some of the technical issues involved in each of the challenges. The long-term goal is to produce new *analytical IT* tools that will help to realize the full potential of EWO in conjunction with the *transactional IT* tools.

3 The Modeling Challenge

While the area of planning and scheduling has seen the development of many models in operations research (OR) (e.g., Pinedo [167]), over the last decade a significant number of planning and scheduling models have been proposed specifically for process applications (for recent reviews, see Mendez et al. [146] and Kallrath [102]). In contrast to general OR scheduling models, the process-oriented models may require the use of material flows and balances and very often network topologies that are quite different from the more traditional serial and multistage systems (see Figures 4 and 5). Furthermore, they address both batch and continuous processes and may require the use of detailed nonlinear process models. Based on the work by Mendez et al. [146] we discuss in this section classifications of batch problems and their optimization models. We also discuss the handling of process models in the context of refinery scheduling problems.

3.1 Classification of Batch Scheduling Problems

There are a great variety of aspects that need to be considered when developing scheduling models for batch processes. First, the process layout and its topological implications have a significant influence on problem complexity. In practice many batch processes are sequential, single, or multiple stage processes, where one or several units may be working in parallel in each stage. Each batch needs to be processed following a sequence of stages defined through the product/batch recipe. However, increasingly as applications become more complex, networks with arbitrary topology must be handled. Complex product recipes involving mixing and splitting operations and material recycles need to be considered in these cases. Closely related to topology considerations are requirements/constraints on equipment in terms of its assignment and connectivity, ranging from fixed to flexible arrangements. Limited interconnections between equipment impose hard constraints on unit allocation decisions.

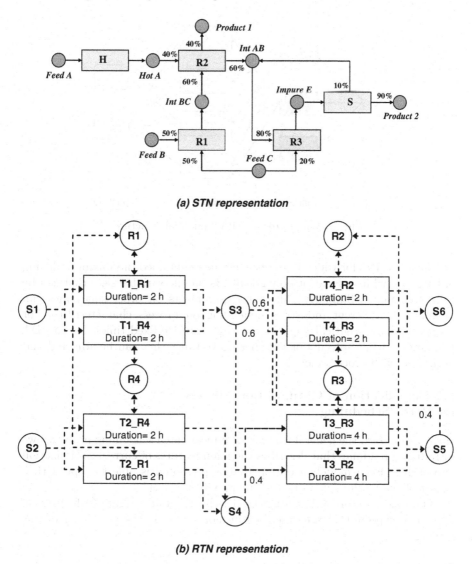

(a) STN representation

(b) RTN representation

Fig. 4. Batch process with complex network structure

Another important aspect of process flow requirements is reflected in inventory policies. These often involve finite and dedicated storage, although frequent cases include shared tanks as well as zero-wait, nonintermediate and unlimited storage policies. Material transfer is often assumed instantaneous, but in some cases like in pipeless plants it is significant and must be accounted for in corresponding modeling approaches. Another major factor is the handling of batch size requirements. For instance, pharmaceutical plants usually

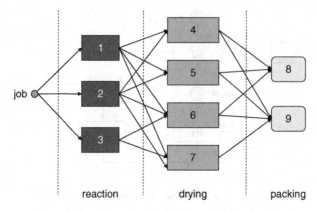

reaction   drying   packing

Fig. 5. Batch process with sequential structure

handle fixed sizes for which integrity must be maintained (no mixing/splitting of batches), while solvent or polymer plants handle variable sizes that can be split and mixed. Similarly, different requirements on processing times can be found in different industries depending on process characteristics. For example, pharmaceutical applications might involve fixed times due to FDA regulations, while solvents or polymers have times that can be adjusted and optimized with process models.

3.2 Classification of Optimization Models for Batch Scheduling

Having discussed the general features of typical batch scheduling problems we introduce a roadmap that describes the main features of current optimization approaches. Each modeling option that is presented is able to cope with a subset of the features described in the previous section.

The roadmap for optimization model classification (Figure 6) focuses on four main aspects that are described in more detail in the remainder of this section.

Time Representation

The most important issue is the time representation. Depending on whether the events of the schedule can take place only at some predefined time points or can occur at any moment during the time horizon of interest, optimization approaches can be classified into discrete and continuous time formulations. Discrete time models are based on: (i) dividing the scheduling horizon into a finite number of time intervals with predefined duration and (ii) allowing the events such as the beginning or ending of tasks to happen only at the boundaries of these time periods. Therefore, scheduling constraints have to be monitored only at known time points, which reduces the problem complexity

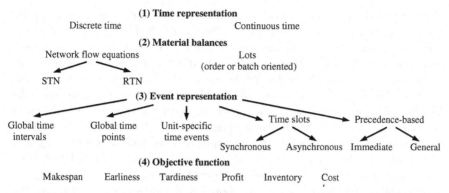

Fig. 6. Roadmap for optimization models for short-term scheduling of batch plants

and makes the model structure simpler and easier to solve, particularly when resource and inventory limitations are taken into account. On the other hand, this type of problem simplification has the disadvantage that the size of the model as well as its computational efficiency depend on the number of time intervals postulated, which is defined as a function of the problem data and the desired accuracy of the solution. Furthermore, suboptimal or infeasible schedules may be generated because of the reduction to the domain of finite timing decisions. Despite being a simplified version of the original scheduling problem, discrete formulations have proven to be very efficient, adaptable, and convenient for a wide variety of industrial applications. Also, with progress in mixed-integer linear model (MILP) solution methods it is increasingly possible to solve problems with much larger numbers of intervals (e.g., see Castro and Grossmann [33]).

Recently, a wide variety of optimization approaches have been proposed using a continuous time representation. In these formulations, timing decisions are explicitly represented as a set of continuous variables defining the exact times at which the events take place. In the general case, a variable time handling allows obtaining a significant reduction of the number of variables of the model, and at the same time more flexible solutions in terms of time can be generated. However, the modeling of variable processing times and resource and inventory limitations usually requires the definition of more complicated constraints, often involving many big-M terms, which tends to increase the model complexity and the integrality gap and may negatively impact on the capabilities of the method.

Material Balances

The handling of batches and batch sizes gives rise to two types of optimization models. The first one refers to monolithic approaches, which simultaneously deal with the optimal set of batches (number and size), the allocation and sequencing of manufacturing resources, and the timing of processing tasks.

These methods are able to deal with arbitrary network processes involving complex product recipes. Their generality usually implies large model sizes and consequently their application is currently restricted to processes involving a small number of processing tasks and rather narrow scheduling horizons. These models employ the state-task network (STN) or the resource-task network (RTN) concept to represent the problem. As shown in Figure 4(a), the STN-based models represent the problem assuming that processing tasks produce and consume states (materials). A special treatment is given to manufacturing resources aside from equipment. The STN is a directed graph that consists of three key elements: (i) *state nodes* representing feeds, intermediates, and final products; (ii) *task nodes* representing the process operations that transform material from one or more input states into one or more output states; and (iii) *arcs* that link states and tasks indicating the flow of materials. State and task nodes are denoted by circles and rectangles, respectively. In contrast, the RTN-based formulations employ a uniform treatment and representation framework for all available resources through the idea that processing and storage tasks consume and release resources at their beginning and ending times, respectively (see Figure 4(b)). In this particular case, circles represent not only states but also other resources required in the batch process such as processing units and vessels.

The second type of model assumes that the number of batches of each size is known in advance. These solution algorithms indeed can be regarded as one of the modules of a solution approach for detailed production scheduling widely used in industry, which decomposes the whole problem into two stages, batching and batch scheduling. The batching problem converts the primary requirements of products into individual batches aiming at optimizing some criterion like the plant workload. The available manufacturing resources are allocated to the batches over time in a second stage. This approach can address much larger problems than monolithic methods, but is restricted to processes with sequential recipes.

Event Representation

In addition to the time representation and material balances, scheduling models are based on different concepts to arrange the events of the schedule over time with the main purpose of guaranteeing that the maximum capacity of the shared resources is never exceeded. As can be seen in Figure 7, these concepts are classified into five different types of event representations. Figure 7 depicts a schematic representation of the same schedule obtained by using the alternative concepts. The small example given involves five batches (a, b, c, d, e) allocated to two units (J1, J2). To represent this solution, the different alternatives require: (a) 10 fixed time intervals, (b) 5 variable global time points, (c) 3 unit-specific time events, (d) 3 asynchronous times slots for each unit, and (e) 3 immediate precedence relationships or 4 general precedence relationships. Although some event representations are more general than others,

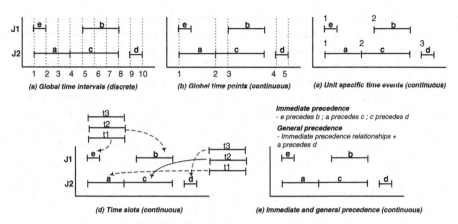

Fig. 7. Different time representations for scheduling problems

they are usually oriented toward the solution of either arbitrary network processes requiring network flow equations or sequential batch processes assuming a batch-oriented approach. Table 1 from Mendez et al. [146] summarizes the most relevant modeling characteristics and problem features related to the alternative event representations. Critical modeling issues refer to those

Table 1. General characteristics of current optimization models

Time representation	DISCRETE	CONTINUOUS					
Event representation	Global time intervals	Global time points	Unit-specific time events	Time slots*	Unit-specific immediate precedence*	Immediate precedence*	General precedence*
Main decisions	------------------ Lot-sizing, allocation, sequencing, timing ------------------				-------- Allocation, sequencing, timing ---------		
Key discrete variables	W_{ijt} defines if task I starts in unit j at the beginning of time interval t.	Ws_{in}/Wf_{in} define if task i starts/ends at time point n. $W_{inn'}$ defines if task i starts at time point n and ends at time point n'.	$Ws_{in}/Wf_{in}/$ Wf_{in} define if task i starts/is active/ends at event point n.	W_{ijk} define if unit j starts task i at the beginning of time slot k.	$X_{ii'j}$ defines if batch i is processed right before batch i' in unit j. XF_{ij} defines if batch i starts the processing sequence of unit j.	$X_{ii'}$ defines if batch i is processed right before batch i'. XF_{ij} defines if batch i starts/is assigned to unit j.	$X'_{ii'}$ defines if batch i is processed before or after batch i'. W_{ij} defines if batch i is assigned to unit j.
Type of process	------------------------------ General network ------------------------------				--------------------- Sequential ------- -------------		
Material balances	Network flow equations (STN or RTN)	Network flow equations (STN or RTN)	--- Network flow equations --- (STN)		-------------------- Batch-oriented ------------------		
Critical modeling issues	Time interval duration, scheduling period (data dependent)	Number of time points (iteratively estimated)	Number of time events (iteratively estimated)	Number of time slots (estimated)	Number of batch tasks sharing units (lot-sizing) and units	Number of batch tasks sharing units (lot-sizing)	Number of batch tasks sharing resources (lot-sizing)
Critical problem features	Variable processing times, sequence-dependent changeovers	Intermediate due dates and raw-material supplies	Intermediate due dates and raw-material supplies	Resource limitations	Inventory, resource limitations	Inventory, resource limitations	Inventory

* Batch-oriented formulations assume that the overall problem is decomposed into the lot-sizing and the short-term scheduling issues. The lot-sizing or "batching" problem is solved first in order to determine the number and size of "batches" to be scheduled.

aspects that may seriously compromise the model size, and hence the computational effort. In turn, critical problem features remark certain problem aspects that may be awkward to consider through specific basic concepts.

For discrete time formulations, the definition of global time intervals is the only option for general network and sequential processes. In this case, a common and fixed time grid valid for all shared resources is predefined and batch tasks are enforced to begin and finish exactly at a point of the grid. Consequently, the scheduling problem is reduced to an allocation problem where the main model decisions define the assignment of the time interval at which every batch task begins, which is modeled through the discrete variable W_{ijt} as shown in Table 1. A significant advantage of using a fixed time grid is that time-dependent problem aspects can be modeled in a relatively simple way without compromising the linearity of the model.

Continuous time formulations involve several alternative event representations that are focused on different types of batch processes. For instance, for general network processes global time points and unit-specific time events can be used, whereas in the case of sequential processes the alternatives involve the use of time slots and different batch precedence-based approaches. The global time point representation corresponds to a generalization of global time intervals where the timing of time intervals is treated as a new model variable. In this case, a common and variable time grid is defined for all shared resources. The beginning and the finishing times of the set of batch tasks are linked to specific time points through the key discrete variables reported in Table 1. Both for continuous STN- and RTN-based models, limited capacities of resources need to be monitored at a small number of variable time points in order to guarantee the feasibility of the solution. Models following this direction are relatively simple to implement even for general scheduling problems. In contrast to global time points, the idea of unit-specific time events defines a different variable time grid for each shared resource, allowing different tasks to start at different moments for the same event point. These models make use of the STN representation. Because of the heterogeneous locations of the event points, the number of events required is usually smaller than in the case of global time points. However, the lack of reference points for checking the limited availability of shared resources makes the formulation much more complicated and sometimes prone to errors. Special constraints and additional variables need to be defined for dealing with resource-constrained problems.

The computational efficiency of the formulations based on global time points or unit-dependent time events strongly depends on the number of time or event points predefined. Since this number is unknown a priori, a practical criterion is to determine it through an iterative procedure where the number of variable points or events is increased by 1 until there is no improvement in the objective function. This means that a significant number of instances of the model need to be solved for each scheduling problem, which may lead to a high total CPU time.

Other continuous-time formulations were initially focused on a wide variety of sequential processes, although some of them recently have been extended to also consider general batch processes. One of the first developments was based on the concept of time slots, which stands for a set of predefined time intervals with unknown durations. The main idea is to postulate an appropriate number of time slots for each processing unit in order to allocate them to the batch tasks to be performed. The selection of the number of time slots required is not a trivial decision and represents an important trade-off between optimality and computational performance. Slot-based representations can be classified into two types: synchronous and asynchronous. The synchronous representation, which is similar to the idea of global time points, defines identical or common slots across all units in such a way that the shared resources involved in network batch processes are more natural and easier to handle. Alternatively, the asynchronous representation allows the postulated slots to differ from one unit to another, which for a given number of slots provides more flexibility in terms of timing decisions than its synchronous counterpart. This representation is similar to the idea of unit-specific time events and results more appropriate for dealing with sequential batch processes.

Another alternative approach for sequential processes uses the concept of batch precedence. Model variables and constraints enforcing the sequential use of shared resources are explicitly employed in these formulations. As a result, sequence-dependent changeover times can be treated in a straightforward manner. To determine the optimal processing sequence in each unit, the concept of batch precedence can be applied to either the immediate or any batch predecessor. The immediate predecessor of a particular batch i is the batch i' that is processed right before in the same processing unit, whereas the general precedence notion extends the immediate precedence concept not only to consider the immediate predecessor but also all batches processed before in the same processing sequence. There are three different types of precedence-based mathematical formulations which are reported in Table 1. When the immediate precedence concept is applied, sequencing decisions in each processing unit can be easily determined through a unique set of model variables $X_{ii'j}$. However, in order to reduce the model size and consequently the computational effort, allocation and sequencing decisions are frequently decoupled in two different sets of model variables W_{ij} and $X_{ii'}$, as described in Table 1. The general precedence concept needs the definition of a single sequencing variable for each pair of batch tasks that can be allocated to the same shared resource. In this way, the formulation is simpler and smaller than those based on the immediate predecessor. In addition, this approach can handle the use of different types of renewable shared resources such as processing units, storage tanks, utilities, and manpower through a single set of sequencing variables without compromising the optimality of the solution. A common weakness of precedence-based formulations is the number of sequencing variables scales with the number batches to be scheduled.

Objective Function

Different measures of the quality of the solution can be used for scheduling problems, e.g., profit maximization, make-span minimization, tardiness minimization, etc. However, the criteria selected for the optimization usually has a direct effect on the model computational performance. In addition, some objective functions can be very difficult to implement for some event representations, requiring additional variables and complex constraints.

3.3 Overview of Scheduling Models

Having introduced classifications of problems and models for batch scheduling, we review in this section the models developed for the different types of event representations shown in Table 1.

The most relevant contribution for discrete time models is the state task network representation proposed by Kondili et al. [113] / Shah et al. [198] (see also Rodrigues et al.[180]). The STN model covers all the features that are included in the column on discrete time in Table 1. The MILP model consists of allocation constraints, capacity limitations, material balances, and resource balances. Changeover constraints can in principle also be added, although they are somewhat awkward and they require finer discretizations of time. Kelly and Zyngier [110] have recently worked to improve the handling of sequence-dependent changeovers. It is interesting to note how the capability for solving MILPs, such as the discrete time STN, has evolved over time. For instance, consider the classic problem shown in Fig. 4(a) (Kondili et al. [113]) with the STN MILP over a horizon of ten time units. In 1988, Kondili [112] solved this problem using a weaker form of the constraints in (1) in 908 sec and 1466 nodes on a VAX-8600 using her own LP-based branch and bound code with MINOS. In 1992, Shah [193] solved this problem in 119 sec and 149 nodes on a SUN-SPARC using the strong form of the inequality in (1) and also with his own branch and bound method. In 2003, one of the authors (Grossmann) solved the same model on his laptop IBM-T40 using CPLEX 7.5, which required only 0.45 sec and 22 nodes! Thus it is clear that a combination of better models, faster computers, and faster MILP solvers is greatly increasing the capability for solving optimization models for scheduling.

A simpler and general discrete time scheduling formulation can also be derived by means of the resource task network concept proposed by Pantelides [159]. The major advantage of the RTN formulation over the STN counterpart arises in problems involving identical equipment. Here, the RTN formulation introduces a single binary variable instead of the multiple variables used by the STN model. The RTN-based model also covers all the features in the column on discrete time in Table 1. The RTN model is very compact as it consists of resource balances and operational constraints (e.g., batch size). Sequence-dependent changeovers can also be handled but by introducing new cleaning tasks that can greatly increase the problem size.

STN and RTN models are quite general and effective in monitoring the level of limited resources at the fixed times; their major weakness is the handling of long-time horizons and relatively small processing and changeover times. Regarding the objective function, these models can easily handle profit maximization (cost minimization) for a fixed time horizon. Other objectives such as make-span minimization are more complex to implement since the time horizon and, in consequence, the number of time intervals required are unknown a priori (see Maravelias and Grossmann [139]). Recently, alternative process modeling frameworks have been proposed [106, 107, 237].

A wide variety of continuous-time formulations based both on the STN representation and the definition of global time points have been developed in the last years (see Figure 7(b)). Some of the work falling into this category is represented by the approaches proposed by Schilling and Pantelides [190], Zhang and Sargent [234], Mockus and Reklaitis [151, 152], Lee et al. [120], Giannelos and Georgiadis [66], and Maravelias and Grossmann [140].

The formulation by Maravelias and Grossmann [140], which is based on global time points and the STN representation, is able to handle most of the aspects found in standard batch processes (see first column for continuous models in Table 1). This approach is based on the definition of a common time grid that is variable and valid for all shared resources. This definition involves time points n occurring at unknown time Tn, $n = 1, 2 \ldots |N|$, when N is the set of time points. To guarantee the feasibility of the material balances at any time during the time horizon of interest, the model imposes that all tasks starting at a time point n must occur at the same time Tn. However, in order to have more flexibility in terms of timing decisions, the ending time of tasks does not necessarily have to coincide with the occurrence of a time point n, except for those tasks that need to transfer the material with a zero wait policy (ZW). For other storage policies it is assumed that the equipment can be used to store the material until the occurrence of next time point. Given that the model assumes that each task can be performed in just one processing unit, task duplication is required to handle alternative equipment and unit-dependent processing times.

The continuous-time formulation based on the RTN concept was initially proposed by Pantelides [159]. The work developed by Castro et al. [31] which was then improved in Castro et al. [32] falls into this category and is the most recent development. Major assumptions of this approach are: (i) processing units are considered individually, i.e., one resource is defined for each available unit, and (ii) only one task can be performed in any given equipment resource at any time (unary resource). These assumptions increase the number of tasks and resources to be defined, but at the same time allow reducing the model complexity. This model also covers all the features given at the column on continuous time and global time points in Table 1.

In order to gain more flexibility in timing decisions without increasing the number of time points to be defined, an original concept of event points was introduced by Ierapetritou and Floudas [88], which relaxes the global time point representation by allowing different tasks to start at different moments in different units for the same event point (see Figure 7(c)). Subsequently, the original idea was implemented in the work presented by Vin and Ierapetritou [221] and Lin et al. [128] and recently extended by Janak et al. [96]. The work by Janak et al. [96] represents the most general STN-based formulation that makes use of this type of event representation and covers all the features reported at the corresponding column in Table 1.

One of the first contributions focused on batch-oriented processes is based on the concept of time slots, which stand for a set of predefined time intervals with unknown durations [169]. A set of time slots is postulated for each processing unit in order to allocate them to the batches to be processed. Relevant work on this area is represented by the formulations developed by Pinto and Grossmann [169, 170], Chen et al. [37], and Lim and Karimi [125]. More recently, a new STN-based formulation that relies on the definition of synchronous time slots and a novel idea of several balances was developed to also deal with network batch processes (Sundaramoorthy and Karimi [207]).

The concept of batch precedence can be applied to the immediate or the general batch predecessor, which generates three different types of basic mathematical formulations. A representative example is the formulation of Cerdá et al. [34] where a single-stage batch plant with multiple equipment working in parallel is assumed. An alternative formulation is the one based on the concept of immediate batch precedence. Here allocation and sequencing decisions are divided into two different sets of binary variables as is, for instance, the work by Méndez et al. [148], where a single-stage batch plant with multiple equipment in parallel is assumed. Relevant work following this direction can also be found in Gupta and Karimi [78].

The generalized precedence notion extends the immediate precedence concept to not only consider the immediate predecessor, but also all batches processed before in the same processing sequence. In this way, the precedence concept is completely generalized which simplifies the model and reduces by half the number of sequencing variables when compared to the immediate precedence model. This reduction is obtained by defining just one sequencing variable for each pair of batch tasks that can be allocated to the same resource. Additionally, a major strength of this approach is that sequencing decisions can be easily extrapolated to different types of renewable shared resources. In this way, the use of processing units, storage tanks, utilities, and manpower can be efficiently handled through the same set of sequencing variables without compromising the optimality of the solution. Part of the work falling into this category is represented by the approaches developed by Méndez et al. [149] and Méndez and Cerdá [143, 144, 145].

3.4 Scheduling Models with Performance Models

The batch processes reviewed in the previous section assume that processing time and material balance coefficients are given. In a number of cases, however, the process is continuous as is, for instance, the case of polymer production plants. Examples of this model that use production rates for each product in each unit as a basic parameter are the scheduling models by Sahinidis and Grossmann [188], Pinto and Grossmann [168], and Erdirik-Dogan and Grossmann [53]. These models give rise to mixed integer nonlinear programming (MINLP) or MILP models depending on whether cyclic production schedules or fixed horizons are used, respectively. The more complex case of scheduling models for continuous processes involves performance models which introduces nonlinearities. A typical example is the case of refinery scheduling problems that involve blending, which we briefly review below to give some appreciation of the challenges involved in this area.

The main objective in oil refining is to convert a wide variety of crude oils into valuable final products such as gasoline, jet fuel, and diesel. Short-term blending and scheduling are critical aspects in this large and complex process. The economic and operability benefits associated with obtaining better-quality and less expensive blends, and at the same time making a more effective use of the available resources over time, are numerous and significant. A number of mathematical programming techniques have been extensively used for long-term planning as well as the short-term scheduling of refinery operations.

For planning problems, most of the computational tools have been based on successive linear programming models, such as RPMS from Honeywell, Hi-Spec Solutions (formerly Bonner and Moore [29]) and PIMS from Aspen Technology (formerly Bechtel Corp. [20]). On the other hand, scheduling problems have been addressed through linear and nonlinear mathematical approaches that make use of binary variables (MILP and MINLP codes) to explicitly model the discrete decisions to be made [195, 75].

Short-term scheduling problems have been mainly studied for batch plants. Much less work has been devoted to continuous plants. Lee et al. [118] addressed the short-term scheduling problem for the crude-oil inventory management problem. Nonlinearities of mixing tasks were reduced to linear inequalities with which the original MINLP model was relaxed to a MILP formulation that can be solved to global optimality. This linearization was possible because only mixing operations were considered (see Quesada and Grossmann [176]). However, it was later pointed out by Wenkai and Hui [228] that the proposed reformulation-linearization technique (RLT) may lead to composition discrepancy (the amounts of individual crudes delivered from a tank to crude distillation unit (CDU) are not proportional to the crude composition in the tank). The objective function is the minimization of the total operating cost, which comprises waiting time cost of each vessel in the sea, unloading cost for crude vessels, inventory cost, and changeover cost. Several examples were solved to highlight the computational performance of the proposed model. Moro et al. [154] developed a mixed-integer nonlinear programming planning model for refinery production. The model assumes that a general refinery is composed of a number of processing units producing a variety of input/output streams with different properties, which can be blended to satisfy different specifications of diesel oil demands. Each unit belonging to the refinery is defined as a continuous processing element that transforms the input streams into several products. The general model of a typical unit is represented by a set of variables such as feed flow rates, feed properties, operating variables, product flow rates, and product properties. The main objective is to maximize the total profit of the refinery, taking into consideration sales revenue, feed costs, and the total operating cost. Wenkai and Hui [228] proposed a solution algorithm that iteratively solves two mixed-integer linear programming (MIP) models and a nonlinear programming (NLP) model, resulting in better quality, stability, and efficiency than solving the MINLP model directly. Kelly and Mann [108, 109] highlight the importance of optimizing the scheduling of an oil-refinery's crude-oil feedstock from the receipt to the charging of the pipe stills. The use of successive linear programming (SLP) was proposed for solving the quality issue in this problem. More recently, Kelly [105] analyzed the underlying mathematical modeling of complex nonlinear formulations for planning models of semicontinuous facilities where the optimal operation of petroleum refineries and petrochemical plants was mainly addressed.

In addition, the off-line blending problem, also known as blend planning, has been addressed through several optimization tools. The main purpose here is to find the best way of mixing different intermediate products from the refinery and some additives in order to minimize the blending cost subject to meeting the quality and demand requirements of the final products. The term "quality" refers to meeting given product specifications. Rigby et al. [179] discussed successful implementation of decision support systems for off-line multiperiod blending problems at Texaco. Commercial applications such as Aspen Blend^{TM} and Aspen PIMS-MBO^{TM} from AspenTech are also available for dealing with online and off-line blending optimization problems. Since these software packages are restricted to solving the blending problem, resource and temporal decisions must be made a priori either manually or by using a special method. Honeywell's BLEND (formerly Bonner and Moore) and Honeywell's Production Scheduler may be used for off-line, open-loop, and short-term planning and scheduling of hydrocarbon blend shops.

To solve both subproblems simultaneously, Glismann and Gruhn [67] proposed a two-level optimization approach where a nonlinear model is used for the recipe optimization whereas a mixed-integer linear model is utilized for the scheduling problem. The proposed decomposition technique for the entire optimization problem is based on solving first the nonlinear model aiming at generating the optimal solution of the blending problem, which is then incorporated into the MILP scheduling model as fixed decisions for optimizing only resource and temporal aspects. In this way, the solution of a large MINLP model is replaced by sequential NLP and MILP models. Jia and Ierapetritou [98] proposed a solution strategy based on decomposing the overall refinery problem in three subsystems: (a) the crude-oil unloading and blending, (b) the production unit operations, and (c) the product blending and lifting

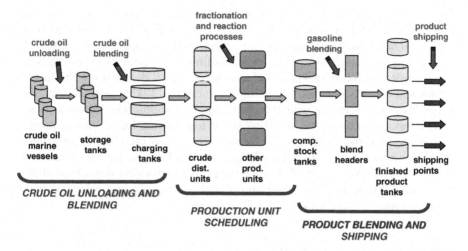

Fig. 8. Illustration of a generic refinery system

(see Figure 8). The first subproblem involves the crude oil unloading from vessels, its transfer to storage tanks, and the charging schedule for each crude oil mixture to the distillation units. The second subproblem consists of the production unit scheduling, which includes both fractionation and reaction processes. Reaction sections alter the molecular structure of hydrocarbons, in general to improve octane number, whereas fractionation sections separate the reactor effluent into streams of different properties and values. Lastly, the third subproblem is related to the scheduling, blending, storage, and lifting of final products. In order to solve each one of these subproblems in the most efficient way, a set of MILPs were developed, which take into account the main features and difficulties of each case. In particular, fixed product recipes were assumed in the third subproblem, which means that blending decisions were not incorporated into this model. The MILP formulation was based on a continuous time representation and on the notion of event points. The mathematical formulation proposed to solve each subproblem involves material balance constraints, capacity constraints, sequence constraints, assignment constraints, demand constraints, and a specific objective function. Continuous variables are defined to represent flow rates as well as starting and ending times of processing tasks. Binary variables are principally related to assignment decisions of tasks to event points or to some specific aspect of each subproblem.

In order to reduce the inherent problem difficulty, most of them rely on special assumptions that generally make the solution inefficient or unrealistic for real-world cases. Some of the common assumptions are: (a) fixed recipes for different product grades are predefined, (b) component and product flow rates are known and constant, and (c) all product properties are assumed to be linear. On the other hand, more general mixed-integer nonlinear programming formulations are capable of considering the majority of the problem features. However, as pointed out by several authors, solving logistics and quality aspects for large-scale problems is not possible in a reasonable time with current MINLP codes and global optimization techniques [108, 109, 100].

The major issue here is related to nonlinear and nonconvex constraints with which the computational performance strongly depends on the initial values and bounds assigned to the model variables. Taking into account the major weaknesses of the available mathematical approaches, Mendez et al. [147] developed a novel and iterative MILP formulation for the simultaneous gasoline short-term blending and scheduling problem of oil refinery operations. Nonlinear property specifications based on variable and preferred product recipes are effectively handled through the proposed iterative linear procedure, which allows the model to generate optimal or near-optimal solutions with modest computational effort.

4 The Multiscale Challenge

Planning and scheduling of process systems are closely linked activities. Both planning and scheduling deal with the allocation of available resources over time to perform a collection of tasks required to manufacture one or several products [27, 102, 195]. The aim in planning is to determine high-level decisions such as production levels and product inventories for given marketing forecasts and demands over a long time horizon (e.g., months to years). Scheduling, on the other hand, is defined over a short time horizon (e.g., days to weeks) and involves lower-level decisions such as the sequence and detailed timing in which various products should be processed at each equipment in order to meet the production goals set by the planning problem. Integration and coordination of planning and scheduling are key components in EWO [12, 196, 203].

Conceptually, the simplest alternative for solving planning and scheduling problems is to formulate a single simultaneous planning and scheduling model that spans the entire planning horizon of interest. However, the limitation of this approach is that when typical planning horizons are considered, the size of this detailed model becomes intractable due to the potential exponential increase in the computation. The traditional strategy for solving planning and scheduling problems is to follow a hierarchical approach in which the planning problem is solved first to define the production targets. The scheduling problem is solved next to meet these targets [27, 200]. The problem of this approach, however, is that a solution determined at the planning level does not necessarily lead to feasible schedules. These infeasibilities may arise because the effects of changeovers are neglected at the planning level, thereby producing optimistic targets that cannot be met at the scheduling level. Therefore, there is a need to develop methods and approaches that can more effectively integrate planning and scheduling [74]. We consider below two major problems related to multiscale optimization: (a) integration of planning and scheduling where the dominant aspect is the temporal scale, and (b) supply chain optimization where the dominant aspect is the spatial scale.

4.1 Integration of Production Planning and Scheduling

The fundamental issue in this area is the integration of models across very different timescales [195]. Typically, the planning model is a linear and simplified representation that is used to predict production targets and material flow over several months (up to one year). Also at this level effects of changeovers and daily inventories are neglected which tends to produce optimistic estimates that cannot be realized at the scheduling level. Scheduling models on the other hand tend to be more detailed in nature, but assume that key decisions have been taken (e.g., production targets, due dates). Two major approaches that have been investigated for integrating planning and scheduling are the following:

i) Simultaneous planning and scheduling over a common time grid. The idea here is to effectively "elevate" the scheduling model to the planning level, which leads to a very large-scale multiperiod optimization problem, since it is defined over long time horizons with a fine time discretization (e.g., intervals of one day). A good example is the use of the state-task-network for multisite planning (e.g., Wilkinson et al., [230]). To overcome the problem of having to solve a very large-scale problem, strategies based on aggregation and decomposition can be considered [13, 24](see Bassett et al. [13] Birewar and Grossmann [24] Wilkinson [229]). The former typically involve aggregating later time periods within the specified time horizon in order to reduce the dimensionality of the problem.

ii) Decomposition techniques for integrating planning and scheduling are usually based on a two-level decomposition procedure where the upper level problem (planning problem) is an aggregation of the lower level problem (scheduling). The challenge lies in developing an aggregated planning model that yields tight bounds to reduce the number of upper and lower level problems [28, 160]. Another solution approach relies on using a rolling horizon approach where the planning problem is solved by treating the first few periods in detail, while the later periods are aggregated recursively [46].

Most of the work that has been reported on integrating planning and scheduling has focused on batch processes and is based on two-level decomposition schemes [73]. Bassett et al. [13] proposed a decomposition scheme for multipurpose batch plants, where an aggregate planning problem is solved in the upper level and detailed scheduling problems are independently solved for each planning period in the lower level. Heuristic techniques that make use of shifting of operations were proposed to overcome the infeasibilities that arise in the scheduling problem. Bassett et al. [13] introduced slack variables for capacity or inventory shortfalls to remove the infeasibilities. These authors also proposed a recursive backward-rolling horizon strategy where only one interval is solved in detail at every stage of the recursion. Changeover costs and times were not considered in the scheduling model. Subrahmanyam et al. [206] proposed a hierarchical decomposition algorithm for batch plants, where the planning problem is updated at each iteration by disaggregating the aggregate constraints for all infeasible scheduling subproblems within the planning problem. The drawback of this procedure is that it may require exploring all levels of decomposition. Birewar and Grossmann [24] proposed a multiperiod linear programming (LP) formulation for simultaneous planning and scheduling of multiproduct batch plants with flow shop structure. In this formulation, batches belonging to the same products are aggregated and sequencing considerations for scheduling are accounted at the planning level by approximating the make span with the cycle time. Production shortfalls are treated through penalties. Wilkinson et al. [231] used a constraint aggregation approach for obtaining approximate solutions to the large-scale production and distribution planning problems for multiple production sites that are represented with the state-task network [113]. In their work, an upper-level aggregate model is solved to set production targets yielding a strict upper bound to the original problem, after which detailed scheduling is individually optimized for each site with fixed targets, thus decreasing the computational effort. Zhu and Majozi [235] proposed a two-level decomposition strategy for multipurpose batch plants. In the first level, the planning model is solved for the optimal allocation of raw materials to individual processes, and in the second level the raw material targets obtained at the planning model are incorporated into the scheduling models for individual processes and then solved independently. If the scheduling targets corresponding to the raw material inputs do not match the production targets predicted by the planning model, the latter is revised with more realistic targets predicted by the scheduling model.

Edririk-Dogan and Grossmann [53] have proposed a novel bilevel decomposition procedure that allows rigorous integration and optimization of planning and scheduling of continuous multiproduct plants consisting of a single processing unit. The proposed integration scheme ensures consistency and optimality within a specified tolerance while significantly reducing the computational effort. The authors developed an MILP scheduling model where sequence-dependent transition times, transition costs, and inventory costs are readily accounted for. In order to avoid nonlinearities in the objective function that are due to the inventory costs, an overestimation was developed that can be expressed in linear form. Since the MILP model becomes very expensive to solve when a large number of products and long planning horizons are considered, a bilevel decomposition procedure was proposed that allows rigorous integration and optimization of planning and scheduling. The decomposition method involves superset and subset cuts, as well as capacity cuts that eliminate many solutions from the upper-level aggregated model. The application of the algorithm was illustrated with eight examples for five products, ranging from 4 to 24 weeks and with high and low values for the lower bounds for the demands. The results showed that the proposed method is significantly faster than the full-space method, although convergence with finite tolerance is required for reasonable computational times in the larger problems. It should be noted that Erdirik-Dogan and Grossmann [54] have recently extended this solution approach to the planning and scheduling of batch plants consisting of one- and two-stage processing with intermediate products.

Sung and Maravelias [208] have recently developed an alternative approach for integrating planning and scheduling that is based on the idea of generating an attainable production region by projecting the higher-dimensional space of a scheduling model (e.g., discrete time STN) into a lower-dimensional planning space. This is performed by generating an approximation of that feasible space with procedures based on finding convex hulls.

4.2 Optimization of Supply Chains

While in the previous section the integration dealt with integration across different temporal scales, we focus in this section on supply chains where integration must take place across spatial scales. Specifically, when considering a specific decision level (strategic, tactical, or operational), it is often desired to consider the entire supply chain of a given enterprise [52, 55, 155]. A review of the various mathematical programming techniques to design and operate process supply chain is given in Kok and Graves [111] and Shah [197]. Wilkinson et al. [230] proposed an approach that integrates production and distribution in multisite facilities using the resource-task network framework of Pantelides [159]. Timpe and Kallrath [212] proposed an MILP model based on time-indexed formulation for the optimal planning of large multisite multiproduct production networks in the chemical industry. Bok et al. [28] also proposed a multiperiod supply chain optimization model for continuous flexible process networks. Tsiakis et al. [213] formulated the design of multiproduct, multiechelon supply chain networks as a MILP optimization problem, but without considering the changes in demand over the long term.

Here again problem size can become a major issue as we have to handle models across many length scales. Two major approaches are to either consider a simultaneous large-scale optimization model or else to use decomposition either in spatial or in temporal forms [115], usually using Lagrangean decomposition [73, 79]. In the case of spatial decomposition the idea is to sever the links between subsystems (e.g., manufacturing, distribution, and retail) by dualizing the corresponding interconnection constraints, which then requires the multiperiod optimization of each system. In the case of temporal decomposition the idea is to dualize the inventory constraints in order to decouple the problem by time periods. The advantage of this decomposition scheme is that consistency is maintained over every time period [92]. See also Daskin et al.[42] for combining location and inventory models.

Simultaneous optimization approaches for the integration of entire supply chains naturally lead to the definition of centralized systems. In practice, however, the operation tends to takes place as if the supply chain were a decentralized system. What is needed are coordination procedures that can maintain a certain degree of independence of subsystems [157], while at the same time aiming at objectives that are aimed at the integrated optimization of the overall system (see Perea et al. [164]).

Other recent developments in the optimization of supply chains include the incorporation of financial flows [70] as part of the optimization and the incorporation of responsiveness through the use of safety stocks [233].

Petroleum Supply Chain (from Neiro and Pinto [155])

As a specific application we focus in this section on the petroleum industry which can be characterized as a typical supply chain (see Neiro and Pinto [155] for an excellent review). All levels of decisions arise in such a supply chain, namely, strategic, tactical, and operational. In spite of the complexity involved in the decision-making process at each level, much of their management is currently still based on heuristics or on simple linear models. According to Forrest and Oettli [61], most of the oil industry still operates its planning, central engineering, upstream operations, refining, and supply and transportation groups as complete separate entities. Therefore, systematic methods for efficiently managing the petroleum supply chain must be exploited.

The petroleum supply chain is illustrated in Figure 9. Petroleum exploration is at the highest level of the chain. Decisions regarding petroleum exploration include design and planning of oil field infrastructure. Petroleum may

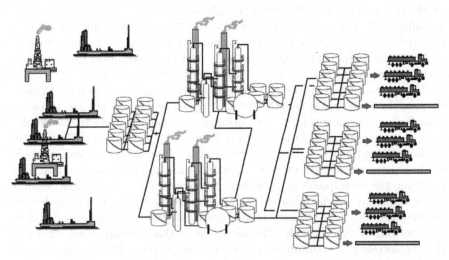

Fig. 9. General petroleum supply chain (from Neiro and Pinto [155])

be also supplied from international sources. Oil tankers transport petroleum to oil terminals, which are connected to refineries through a pipeline network. Decisions at this level incorporate transportation modes and supply planning and scheduling. Crude oil is converted to products at refineries, which can be connected to each other in order to take advantage of each refinery design within the complex. Products generated at the refineries are then sent to distribution centers. Crude oil and products up to this level are often transported through pipelines. From this level on, products can be transported either through pipelines or trucks, depending on consumer demands. In some cases, products are also transported through vessels or by train.

In general, production planning includes decisions such as individual production levels for each product as well as operating conditions for each refinery in the network, whereas product transportation focuses on scheduling and inventory management of the distribution network. Products at the last level presented in Figure 9 are actually raw materials for a variety of processes. This fact indicates that the petroleum supply chain could be further extended.

Sear [192] was probably the first to address the supply chain management in the context of an oil company. The author developed a linear programming network model for planning the logistics of a downstream oil company. The model involves crude oil purchase and transportation, processing of products and transportation, and depot operation. Escudero et al. [56] proposed an LP model that handles the supply, transformation, and distribution of an oil company that accounts for uncertainties in supply costs, demands, and product prices. Dempster et al. [45] applied a stochastic programming approach to planning problems for a consortium of oil companies. First, a deterministic multiperiod linear programming model is developed for supply, production, and distribution. The deterministic model is then used as a basis for implementing a stochastic programming formulation with uncertainty in product demands and spot supply costs. More recently, Lasschuit and Thijssen [116] point out how the petrochemical supply chain is organized and stress important issues that must be taken into account when formulating a model for the oil and chemical industry.

Important developments of subsystems of the petroleum supply chain can be found in literature. Planning of field production with multiperiod optimization models has been the most active research subject in gas and oil optimization. The well scheduling problem has been addressed in production operations optimization by developing heuristic rules obtained from real data analysis [9, 23, 131]. Mathematical programming techniques have also been applied. Using simplified planning models in petroleum production systems tend to generate LP formulations [5, 57, 65, 224]. Egg and Herring [50] improved the accuracy of LP using an iterative scheme where nonlinear simulation and LP are combined. The need of nonlinear optimization tools was pointed out by Dutta-Roy and Kattapuram [49]. Genetic algorithms have also been applied to optimize production systems but are computationally intensive [162].

Iyer et al. [91] developed a multiperiod MILP for planning and scheduling of offshore oil field infrastructure investments and operations. The nonlinear reservoir behavior is handled with piecewise linear approximation functions. A sequential decomposition technique is applied. Van den Heever and Grossmann [215] presented a nonlinear model for oil field infrastructure that involves design and planning decisions. The authors consider nonlinear reservoir behavior. A logic-based model is proposed that is solved with a bilevel decomposition technique. This technique aggregates time periods for the design problem and subsequently disaggregates them for the planning subproblem. Van den Heever et al. [216] also addressed the design and planning of offshore oil field infrastructure focusing on business rules. A disjunctive model capable of dealing with the increased order of magnitude due to the business rules is proposed. Ierapetritou et al. [89] studied the optimal location of vertical wells for a given reservoir property map. The problem is formulated as a large-scale MILP and solved by a decomposition technique that relies on quality cut constraints. Kosmidis et al. [114] described an MILP formulation for the well allocation and operation of integrated gas-oil systems, whereas Barnes et al. [10] focused on the production design of offshore platforms.

Cheng and Duran [38] focused on the worldwide transportation of crude oil based on the statement that this element of the petroleum supply chain is the central logistics that links the upstream and downstream functions, playing a crucial role in the global supply chain management in the oil industry.

At another level of the supply chain, Lee et al. [118] concentrated on the short-term scheduling of crude oil supply for a single refinery for which they proposed an MILP model. A somewhat similar model has also been developed by Shah [194] for crude oil scheduling where the scheduling time horizon is discretized into intervals of equal duration, where the requirement is that the operations must start and end at the boundaries of the intervals. This approach is more restricted as compared to that of Lee et al. [118] since the front end of the refinery is decomposed into two parts – downstream and upstream – and the models corresponding to these are solved sequentially. Más and Pinto [142] and Magalhães and Shah [135] focus on the crude oil supply scheduling. The former developed a detailed MILP formulation composed of tankers, piers, storage tanks, substations, and refineries, whereas the latter addresses a scheduling problem composed of a terminal, a pipeline, a refinery crude storage area, and its crude units. Pinto et al. [173] and Pinto and Moro [172] focused on the refinery operations. The former work focuses on production scheduling for several specific areas in a refinery such as crude oil, fuel oil, asphalt, and liquefied petroleum gas (LPG), whereas the latter addresses a nonlinear production planning. Jia and Ierapetritou [98] concentrate on the short-term scheduling of refinery operations. Crude oil unloading and blending, production unit operations, and product blending and delivery are first solved as independent problems. Each subsystem is modeled based on a continuous time formulation. Integration of the three subsystems is then accomplished by applying heuristic-based Lagrangean decomposition. Wenkai and Hui [228] studied a similar problem to that addressed by Jia and Ierapetritou [98] and proposed a new modeling technique and solution strategy to schedule crude oil unloading and storage. At the refinery level, units such as crude distillation unit and fluidized-bed catalytic cracking were modeled and a new analytical method was proposed to provide additional information for intermediate streams inside the refinery. A rigorous extension of this model can be found in Furman et al. [63], where the authors use a continuous time event formulation to schedule fluid transfer between tanks and to model the problem as an MINLP. In this work, the main idea is to allow both inputs and outputs for a tank in a single transfer event in order to reduce the number of binary variables. Recently Karuppiah et al. [104] have proposed an improved solution method for the nonconvex MINLP model for this problem involving bilinearities. The model is solved to global optimality using Lagrangian-based cuts that strengthen the lower bound of the relaxation problem in the spatial branch and bound search.

Ponnambalam et al. [174] developed an approach that combines the simplex method for linear programming with an interior point method for solving a multiperiod planning model in the oil refinery industry. Still at the production planning level, Liu and Sahinidis [133] presented a fuzzy programming approach for solving a petrochemical complex problem involving uncertainty in model parameters. Bok et al. [28] addressed the problem of long-range capacity expansion planning for a petrochemical industry.

Ross [182] formulated a planning supply network model on the petroleum distribution downstream segment. Resource allocation such as distribution centers (new and existing) and vehicles is managed in order to maximize profit. Delivery cost is determined depending on the geographic zone, trip cost, order frequency, and travel distance for each customer. Iakovou [87] proposed a model that focuses on the maritime transportation of petroleum products considering a set of transport modalities. One of the main objectives of this work was to take into account the risks of oil spill incidents. Magatão et al. [136] propose an MILP approach to aid the decision-making process for scheduling commodities on pipeline systems. On the product storage level, Stebel et al. [204] present a model involving the decision-making process on storage operations of LPG.

5 The Uncertainty Challenge

In practice it is typically necessary to make decisions before the details of uncertain parameters are known. Thus, ideally one would like to consider this uncertainty in an optimization model. Such a paradigm allows for producing robust solutions that can remain feasible over the uncertainty space, as well as making the trade-off between optimality and the randomness in the input data.

Uncertainty is a critical issue in supply chain operations. Furthermore, it is complicated by the fact that the nature of the uncertainties can be quite different in the various levels of the decision making (e.g., strategic planning vs. short-term scheduling). Most of the research in the process industry thus far has focused on operational uncertainty, such as quality, inventory management, and handling uncertain processing time (e.g., Zipkin [236], Montgomery [153], Balasubramanian and Grossmann [6]). Much less work has focused on uncertainty at the tactical level, for instance, production planning with uncertain demand [7, 80]. The reason for this is that even the corresponding deterministic optimization problems are difficult to solve; therefore, the resulting optimization problems under uncertainty are extremely complex. The review article of Li and Ierapetritou [122] presents an overview of scheduling under uncertainty for the chemical process systems area. Several methods for addressing optimization under uncertainty have been considered in the chemical process systems literature [187].

5.1 Stochastic Programming

In stochastic programming [25, 184], mathematical programs are solved over a number of stages. The fundamental idea behind stochastic programming is the concept of recourse. Recourse is the ability to take corrective action after a random event has taken place. Between each stage, some uncertainty is resolved, and the decision maker must choose an action that optimizes the current objective plus the expectation of the future objectives. The most common stochastic programs are two-stage models that may be solved in a number of ways including decomposition methods [185] and sampling-based methods [129, 199].

$$\begin{aligned} \min \ & E_\theta \left[f(x,y;\theta) \right] \\ \text{s.t.} \ & g(x,y;\theta) \leq 0 \end{aligned} \quad \longrightarrow \quad \begin{aligned} \min \ & \textstyle\sum_s p_s f(x,y_s;\theta_s) \\ \text{s.t.} \ & g(x,y_s;\theta_s) \leq 0 \end{aligned} \quad s \in \{\text{samples/scenarios}\}$$

When the second-stage (or recourse) problem is a linear program, these problems are straightforward to solve, but the more general case is where the recourse is a MILP or a MINLP. Such problems are extremely difficult to solve since the expected recourse function is discontinuous and nonconvex [187].

As an example, consider the problem of production planning across a supply chain with uncertain demands. The first-stage problem is to create a production plan. After the demand is realized, a plant process optimization problem must be solved for each plant. The challenge is to find a production plan that minimizes the expected production cost. The hierarchical nature of supply chains lends itself naturally to stochastic programming models and in particular the decomposition principles that are used to solve them.

A two-stage model is actually a special case of a more general structure called the multistage model. In this case, decision variables and constraints are divided into groups of corresponding temporal stages. At each stage some of the uncertain quantities become known. In each stage one group of decisions needs to be fixed based on what is currently known, along with trying to compensate for what remains uncertain. The model essentially becomes a nested formulation. Although these problems are difficult to solve, there is extensive potential for applications.

$$E_{\theta_1} \left[\min_{g_1(x_1;\theta_1) \leq 0} f_1(x_1;\theta_1) + ... + E_{\theta_T} \left[\min_{g_T(x_T;\theta_T) \leq 0} f_T(x_T;\theta_T) \right] \right]$$

The application of multistage optimization under uncertainty could include long-term planning of investment, production or development in which fixed decisions occur in stages over time, or leaving opportunities to consider more definite information as time passes.

The strength of stochastic programming is that it is one of the very few technologies for optimization under uncertainty that allow models to capture recourse. The modeling approach also lends a very clear way of thinking about decision making under uncertainty. Stochastic programming, however, comes with the caveat that the uncertainty of a problem needs to be modeled or assumed to take some form. Some methods for doing so include sampling from probability distribution functions and scenario generation. More sophisticated modeling forms in addition to an expected value objective function include the addition of probabilistic and expected value constraints to the optimization model.

Planning in the chemical process industry has used stochastic programming for a number of applications [2, 3, 39, 132]. The scheduling of batch plants under demand uncertainty using stochastic programming has only recently emerged as an area of active research. Gupta et al. [81] used a chance-constrained approach in conjunction with a two-stage stochastic programming model to analyze the trade-offs between demand satisfaction and production costs for a midterm supply chain planning problem. Engell et al. [51] use a scenario decomposition method for the scheduling of a multi-product batch plant by two-stage stochastic integer programming. Balasubramanian and Grossmann [7] present an approach for approximating multistage stochastic programming to the scheduling of multiproduct batch plants under demand uncertainty.

5.2 Markov Decision Processes

Markov decision processes (MDP) [175], also known as stochastic dynamic programming, is a general approach for multiperiod decision making under uncertainty. The setting for dynamic programming under uncertainty includes a time horizon discretized into stages, and at each stage the system can be in one of several states and the decision maker has to choose a decision from an action set. Transition probabilities determine the state to which the system transitions in the next stage given the decision and some random event. A cost or revenue is incurred at each stage based on the decision, some random event, and the current state. Many planning and scheduling problems can be cast as MDP.

It is well known that discrete backward dynamic programming suffers from the "curse of dimensionality" (Bellman [15]) and is computationally intractable when the state space is large. Unfortunately, as with dynamic programming, solving an MDP to optimality incurs the so-called "curse of dimensionality." Another difficulty is that the transition probabilities often can not be expressed explicitly when the system's equations are complicated or there is not a good mathematical representation of the stochastic information.

The recent developments of neuro-dynamic programming (NDP) or approximate dynamic programming (ADP) [19] and reinforcement learning [209] were devised to mitigate this "curse of dimensionality." NDP/ADP uses function approximations and other techniques from the artificial intelligence community to avoid sweeping over the entire state space. Samples are drawn either in the batch mode or the incremental mode. In the former, the state space is discretized and a limited number of points are used in regression. In the latter, the least-squares problem is solved by the stochastic gradient descent method to find the optimal coefficients. There is very little research in applying MDP for chemical process systems engineering problems. Lee and Lee [119] apply ADP to an MDP formulation for resource-constrained process scheduling, and Wei et al. [227] apply ADP to stochastic inventory optimization for a refinery case study.

5.3 Robustness and Flexibility

Although a significant amount of work has been done to address the issue of flexibility and robustness in chemical process design, these concepts have not been extensively applied to the areas of planning and scheduling. Flexibility analysis is a method for quantifying the robustness of a solution. Rotstein et al. [183] use flexibility analysis to quantify the flexibility of a batch production schedule. Vin and Ierapetritou [222] propose metrics that take into account scheduling characteristics under demand uncertainty, and develop an iterative procedure to generate schedules with improved flexibility.

Another approach in considering robustness in the case of process scheduling under uncertainty includes the reactive modification of schedules upon realization of uncertain data or unexpected events [41, 103]. With a reactive scheduling strategy, there is the possibility that too many changes might be made to the schedule to make it useful in practice. Most of the existing approaches for reactive scheduling involve heuristics. Recently Janak et al. [95] have applied reactive scheduling to a very large-scale industrial batch plant. Floudas and Lin [60] and Mendez et al. [147] provide recent reviews of reactive scheduling for chemical processes.

Robust optimization [16, 17, 18] is a more formal mathematical approach for optimization under uncertainty for flexible solutions. The aim of robust optimization is to choose a solution that is able to cope best with the various realizations of the uncertain data. The uncertain data are assumed to be unknown but bounded, and most current research assumes convexity of the uncertainty space. The optimization problem with uncertain parameters is reformulated into a counterpart robust optimization problem, or alternatively through the use of parametric programming [121]. Unlike stochastic programming, robust optimization does not require information about the probability distribution of the uncertain data and does not optimize an expected value objective function. Robust optimization promises to essentially ensure robustness and flexibility by enforcing feasibility of an optimization problem for the entire given uncertainty space. It allows mitigation of the worse-case scenario; however, since all possible realizations of the data are considered, the solution can end up being overly pessimistic and the problem is more likely to be infeasible. Balasubramanian and Grossmann [8] consider uncertain processing times through the use of fuzzy sets to obtain bounds on the makespan. Lin et al. [127], Janak et al. [97], and Li and Ierapetritou [123] apply robust optimization to process scheduling under uncertainty in processing times, product demands, and market prices.

5.4 Sensitivity Analysis and Parametric Optimization

Sensitivity analysis and parametric optimization are used in the analysis of the effects of parameter changes on the optimal solution of a mathematical programming model in an attempt to quantify the robustness and to compare it to other solutions that may arise as the parameters of the model are perturbed. Parametric optimization serves as an analytic tool in optimization under uncertainty mapping the uncertainties in the definition of the problem to optimal alternatives. From this point of view, it is the robust mathematical solution of the uncertainty problem.

Although sensitivity analysis and parametric optimization problems have been addressed successfully in the linear programming case [1, 64] they are still the subject of ongoing research for the mathematical programs that involve discrete variables in their formulation. Efforts are still being made to handle the lack of optimality criteria in the sensitivity analysis for integer optimization problems [44, 191, 202]. Parametric optimization has thus only been infrequently applied to chemical processes problems [47, 48, 166]. Jia and Ierapetritou [99] compare parametic optimization, MILP sensitivity analysis, and robust optimization for short-term process scheduling. This work has been recently extended by Li and Ierapetritou through the use of multiparametric programming [121].

6 The Algorithmic and Computational Challenges

Realizing the vision of EWO requires the development of advanced algorithms and computational architectures in order to effectively and reliably solve the large-scale optimization models. In this section we briefly outline some of the more prominent technical focus areas that are involved in this endeavor.

The necessity to bridge the gap between the process systems engineering, operations research, and computer science communities is highly apparent.

6.1 Mixed-Integer Linear Programming

When detailed nonlinear process performance models are not used, planning and scheduling problems for EWO commonly give rise to mixed-integer linear programming problems. These optimization problems can be computationally expensive to solve since in the worst case they exhibit exponential computational complexity with problem size (NP-hard). However, in the last 10 years great progress has been made in algorithms and hardware, which has resulted in an impressive improvement of our ability to solve MILPs [26, 101] through software packages such as CPLEX and XpressMP. Capitalizing on theory developed during the last 20 years, it is now possible, using off-the-shelf LP-based branch-and-cut commercial software, to solve in a few seconds MILP instances that were unsolvable just 5 years ago. This improvement has been particularly dramatic for certain classes of problems, such as the traveling salesman problem, and for certain industries, such as the commercial airlines. In contrast, for the type of problems that arise in process industries, the available LP-based branch-and-cut software is not always capable of solving industrial-size MILP models. One reason is that nonconvex functions, such as piecewise linear functions, and combinatorial constraints, such as multiple-choice, semicontinuous, fixed-charge, and job sequencing disjunctions (i.e., either job i precedes job j or vice versa), abound in optimization problems related to process industries. For such functions and constraints, the "textbook" approach implemented in the current software is often not practical.

In the current methods, nonlinearities are often modeled by introducing a large number of auxiliary binary variables and additional constraints, which typically doubles the number of variables and increases the number of constraints by the same order of magnitude. Also, with this approach, the combinatorial structure is obscured and it is not possible to take advantage of the structure. In the case of EWO, where many of these constraints appear at the same time and the sizes of the instances are considerably larger, these issues are even more serious. Recently, an alternative method, *branch-and-cut without auxiliary binary variables*, inspired by the seminal work of Beale and Tomlin [14] on special ordered sets, has proved to be promising in dealing with such constraints [43]. It consists of enforcing the combinatorial constraints algorithmically, directly in the branch-and-bound scheme, through specialized branching and the use of cutting planes that are valid for the set of feasible solutions in the space of the original decision variables. The encouraging computational results yielded by the method on some of the aforementioned constraints provide a serious indication that it may be of great impact on EWO problems for the process industries. The use of cutting planes in an LP-based branch-and-bound approach has also proven to be of significant importance in obtaining strong bounds to reduce the required amount of enumeration (see, for example, Marchand et al. [137]). It is interesting to note that in the recent version 10.0 by CPLEX there is the capability of using indicators for conditional constraints (http://www.ilog.com/products/cplex/news/whatsnew.cfm), which avoids the introduction of big-M constraints.

6.2 Constraint Programming

The relatively new field of constraint programming has recently become the state of the art for some important kinds of scheduling problems, particularly resource-constrained scheduling problems, which occur frequently in supply-chain contexts. Constraint programming (CP) can bring advantages on both the modeling and solution sides. The models tend to be more concise and easier to debug, since logical and combinatorial conditions are much more naturally expressed in a CP than in an MILP framework (e.g., Milano [150]). The solvers take advantage of logical inference (constraint propagation) methods that are well suited to the combinatorial constraints that characterize scheduling problems. In particular, the sequencing aspect of many scheduling problems—the task of determining in what order to schedule activities—can present difficulties to MILP because it is difficult to model and it gives rise to weak continuous relaxations. By contrast, a CP model readily formulates sequencing problems and offers specialized propagation algorithms that exploit their structure. Furthermore, heuristics can readily be accommodated in CP.

The greatest promise, however, lies in the integration of CP and MILP methods, which is currently a very active area of research [82, 84]. Several recent systems take some steps toward integration, such as ECLiPSe [226], OPL Studio [217], the Mosel language [40], and COMET [218]. Integration allows one to attack problems in which some of the constraints are better suited to an MILP-like approach (perhaps because they have good continuous relaxations) and others are better suited for a CP approach (because they "propagate well"). This is particularly true of supply-chain problems, in which constraints relating to resource allocation, lot sizing, routing, and inventory management may relax well, while constraints related to sequencing, scheduling, and other logical or combinatorial conditions may propagate well. In the context of scheduling problems, these models perform the assignment of jobs to machines with mixed-integer programming constraints, while the sequencing of jobs is performed with constraint programming. The motivation behind the former is to remove "big-M" constraints and exploit the optimization capability of mixed-integer programming. The motivation behind using the latter is to exploit the capability of constraint programming for effectively handling feasibility subproblems, as well as sequencing constraints. Hybrid methods have shown in some problems outstanding synergies that lead to order magnitude reductions in computation [84, 85, 86, 94, 141].

6.3 Nonlinear Programming

In order to develop real-time optimization models as part of the EWO models for process industries (energy, chemicals, and materials) high-fidelity simulation models are required that provide accurate descriptions of the manufacturing process. Most of these models consist of large sets of nonlinear equality and inequality constraints, which relate manufacturing performance to designed equipment capacities, plant operating conditions, product quality constraints, and operating costs. The sensitivity of these degrees of freedom to higher-level decisions can also be exploited by an integrated optimization formulation. The development and application of optimization tools for many of these nonlinear programming (NLP) models [158] has only recently been considered (see Biegler et al. [22]).

An important goal in EWO is the integration of these nonlinear performance models to determine optimal results from IT tools. This research task is essential because these performance models for real-time optimization ensure the feasibility of higher-level decisions (e.g., logistics and planning) for manufacturing operations. Also, these models accurately represent operating degrees of freedom and capacity expansions in the manufacturing process. As a result, incorporation of these models leads to significantly superior results than typical linear approximations to these models. Several studies have demonstrated the importance of including NLP and MINLP optimization capabilities [21, 92, 93] and the significant gains that can be made in planning and scheduling operations. On the other hand, the research challenge is that nonlinear models are more difficult to incorporate and to handle as nonlinear optimization problems because they introduce nonmonotonic behavior, nonconvexities, and local solutions. In addition the treatment of local degeneracies and ill-conditioning is more difficult and more computationally intensive optimization algorithms are required. The recent introduction of interior point (or barrier) methods for NLP [30, 220, 223] have shown significant improvements over conventional algorithms with active set strategies. Also, more recent convergence criteria have been improved with the introduction of *filter* methods [58, 223], which rapidly eliminate undesirable search regions and promote convergence from arbitrary starting points.

6.4 Mixed-integer Nonlinear Programming and Disjunctive Optimization

Developing the full range of models for EWO as given by problem (P) requires that nonlinear process models be developed for planning and scheduling of manufacturing facilities. This gives rise to MINLP problems since they involve discrete variables to model assignment and sequencing decisions, continuous variables to model flows, and amounts to be produced, and operating conditions (e.g., temperatures, yields). While MINLP optimization is still largely a rather specialized capability, it has been receiving increasing attention over the last decade. A recent review can be found in Grossmann [76]. A number of methods such as outer-approximation, extended cutting planes, and branch and bound have proved to be effective, but are still largely limited to moderate-sized problems. In addition there are several difficulties that must be faced in solving these problems. For instance in NLP subproblems with fixed values of the binary variables the problems contain a significant number of redundant equations and variables that are often set to zero, which in turn often lead to singularities and poor numerical performance. There is also the possibility of getting trapped in suboptimal solutions when nonconvex functions are involved. Finally, there is the added complication when the number of 0-1 variables is large, which is quite common in planning and scheduling problems.

To circumvent some of these difficulties, the modeling and global optimization of generalized disjunctive programs (GDP) [76] seems to hold good promise for EWO problems. The GDP problem is expressed in terms of Boolean and continuous variables that are involved in constraints in the form of equations, disjunctions, and logic propositions [177]. One motivation for investigating these problems is that they correspond to a special case of hybrid models in which all the equations and symbolic relations are given in explicit form. An important challenge is related to the development of cutting planes that provide similar quality in the relaxations as the convex hull formulation without the need of explicitly including the corresponding equations [189]. The other challenge is that global optimization algorithms [59, 186, 210] can in principle be decomposed into discrete and continuous parts, which is advantageous as the latter often represents the major bottleneck in the computations (e.g., through spatial branch-and-bound schemes; see Lee and Grossmann [124]). Finally, the extension to dynamics of these models (e.g., Barton and Lee [11] and Chachuat et al.[35]) should provide computational capabilities that are required to model real-time problems.

6.5 Computational Grid

Solving the large-scale EWO models will require significant computational effort. To achieve the goal of integrating planning across the enterprise, advances in algorithms and modeling must go hand in hand with advances in toolkits that enable algorithms to harness computational resources. One promising approach that has emerged over the last decade is to deliver computational resources in the form of a *computational grid*, which is a collection of loosely coupled, (potentially) geographically distributed, heterogeneous computing resources. The idle CPU time on these collections is an inexpensive platform that can provide significant computing power over long time periods. For example, consider the project SETI@home (http://setiathome.ssl. berkeley.edu/), which since its inception in the mid 1990s has delivered over 18,000 *centuries* of CPU time to a signal-processing effort. A computational grid is similar to a power grid in that the provided resource is ubiquitous and grid users need not know the source of the provided resource. An introduction to computational grids is given by Foster and Kessleman [62]. An advantage of computational grids over traditional parallel processing architectures is that a grid is the most natural and cost-effective manner for users of models and algorithms to obtain the required computational resource to solve EWO problems.

To allow a larger community of engineers and scientists to use computational grids, a number of different programming efforts have sought to provide the base services that grid-enabled applications require (e.g., Foster and Kesselman [62] and Livny et al. [134]). A promising approach would seem to use and augment the master-worker grid library MW [71]. The MW library is an abstraction of the master-worker paradigm for parallel computation. MW defines a *simple* application programming interface, through which the user can define the core tasks making up this computation and the actions that the master takes upon completion of a task. Once the tasks and actions are defined by the user, MW performs the necessary actions to enable the application to run on a computational grid (such as resource discovery and acquisition, task scheduling, fault-recovery, and interprocess communication).

MW was developed by the NSF-funded meta-NEOS project and used to solve numerical optimization problems of unprecedented complexity (e.g., Anstreicher et al. [4] and Linderoth and Wright [130]). A major research direction here would be the development and testing of decomposition-based and branch-and-bound–based algorithms for EWO models. The MW toolkit has already been used with great success to parallelize both decomposition-based algorithms (e.g., Linderoth and Wright [130]) and also spatial branch-and-bound algorithms (e.g., Goux and Leyffer [72] and Chen, et al. [36]). However, for EWO the current functionality in the MW toolkit is not sufficient. The simple master-worker paradigm must be augmented with features that improve its scalability and information-sharing capabilities to be able to solve the EWO models we propose.

7 Illustrative Examples

In this section we present five examples from the literature that illustrate the four challenges cited in this chapter on problems encountered in the area of EWO. Example 1 deals with a short-term scheduling problem that makes use of a hybrid model that combines mixed-integer linear programming and constraint programming. This example illustrates the challenge for developing new algorithms. Example 2 considers the simultaneous scheduling and blending of the products of a refinery. This example illustrates the potential advantages of continuous-time representation from the perspective of the modeling challenge. Example 3 deals with a multisite planning and distribution problem that incorporates nonlinear process models, illustrating the modeling challenge. Example 4 describes the simultaneous optimization of the scheduling of testing for new product development and the design of batch manufacturing facilities. This example illustrates the challenge of multi-scale modeling given the dissimilar nature of the activities and the need of combining a detailed scheduling model with a high-level design model. Finally, example 5 illustrates the third challenge with the design and planning of offshore gas field facilities under uncertainty. We should note that while the examples presented are rather modest in size compared to what ideally one would like to strive for in EWO, examples 2 and 3 correspond to real-world industrial problems.

7.1 Example 1

This example deals with the scheduling of a batch process shown in Figure 10 using the state-task network representation in which circles represent material nodes with various storage options (finite, unlimited, zero-wait, no storage) and the rectangles represent operational tasks that must be performed (e.g., mixing, reaction, separation). This batch process produces four different products, P1, P2, P3, and P4. Note that eight units are assumed to be available for performing the operations of the various tasks. Of course not all units can perform all tasks, but only a subset of them. Given data on processing times for each task as well as on the mass balance, the problem consists of determining a schedule that can produce 5 tons of the four products and that minimizes the make span (completion time). If this problem is formulated with a continuous time approach such as the one by Maravelias and Grossmann [140] in order to accommodate arbitrary processing times, the corresponding MILP cannot be solved after 10 hours of CPU time. This can be qualitatively explained by the fact that scheduling and MILP problems are NP-hard. To

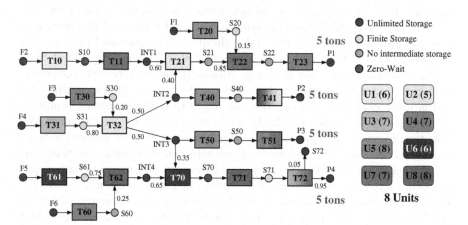

Fig. 10. State-task network for batch process manufacturing products P1, P2, P3, P4

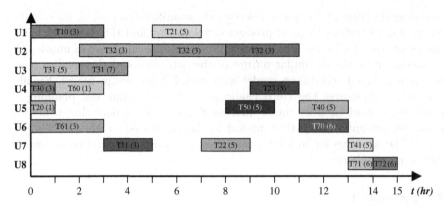

Fig. 11. Optimal schedule with make span of 15 hours

address this difficulty, however, Maravelias and Grossmann [141] developed a novel hybrid solution method that combines MILP with constraint programming. Using such a technique the problem was solved to rigorous optimality in only 5 seconds! The schedule is shown in Figure 11. This example shows the importance of special solution methods that effectively exploit the structure of scheduling problems.

7.2 Example 2

We consider in this example the simultaneous scheduling and blending of the products of a refinery. From the 12 product specifications, 4 properties are nonlinear, while the rest are linear. The basis of the example comprises nine intermediate products or components from the refinery, which can be blended in different ways to satisfy multiple demands of three gasoline grades with different specifications over a 8-day scheduling horizon, divided into six consecutive time intervals, where intervals 1, 3, 4, and 6 have a 1-day duration, whereas intervals 2 and 5 have a 2-day duration. The proposed discrete and continuous time MILP models by Mendez et al. [147] were capable of finding the solution in just one iteration. Although the discrete and continuous time representations obtained the same profit in terms of component cost and product value ($1,611,210), the continuous time representation is able to find a schedule that operates the blenders at full capacity for 2.67 days less than the discrete time representation, which can significantly reduce the total operating cost. Product schedules based on a discrete and continuous time representation in terms of Gantt charts and inventory evolution of components for both discrete and continuous time representations are shown in Figure 12. The discrete time formulation involved 679 constraints, 9 binary variables, and 757 continuous variables. The continuous time formulation comprises 832 constraints, 9 binary variables, and 841 continuous variables. Both models were solved in less than 1 second.

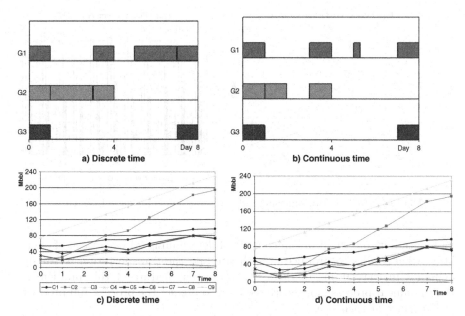

Fig. 12. Gantt charts and evolution of component stocks (example 2)

7.3 Example 3

This example deals with the production planning of a multisite production facility that must serve global markets (see Figure 13). The sites can produce 25 grades of different polymers. Given forecasts of demands over a 6- to

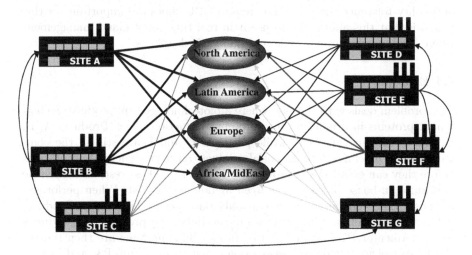

Fig. 13. Multisite planning for polymer production

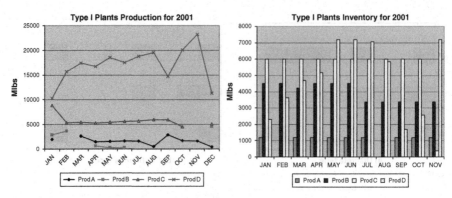

Fig. 14. Predicted production and inventory plans for one of the sites

12-month horizon, the problem consists of determining for each week of operation what grades to produce in each site and the transportation to satisfy demands in the various markets. An important feature of this problem is that nonlinear process models are required to predict the process and product performance at each site.

Neglecting effects of changeovers, the problem of optimizing the total profit can be formulated as a multiperiod NLP problem. The difficulty is that the size of the problem can become very large. For instance, a 12-month problem involves 34,381 variables and 28,317 constraints. To circumvent this problem Jackson and Grossmann [92] developed a temporal decomposition scheme based on Lagrangean relaxation. The authors showed that much better results could be obtained compared to a spatial decomposition (see Figure 14), and that the CPU times could be reduced by one or two orders of magnitude for optimality tolerances of 2-3%. The reason CPU times are important for this model is that this allows one to use it in real time for demand management when deciding what orders to accept and their deadlines.

7.4 Example 4

This problem deals with the case where a biotechnology firm produces recombinant proteins in a multipurpose protein production plant. Products A, B, D, and E are currently sold while products C and F are still in the company's R&D pipeline. Both potential products must pass successfully 10 tests before they can gain FDA approval (see Figure 15). These tests can either be performed in-house or else outsourced at double the cost. When performed in-house, they can be conducted in only one specific laboratory. Products A-C are extracellular, while D-F are intracellular. All proteins are produced in the fermentor P1 (see Figure 16). Intracellular proteins are then sent to the homogenizer P2 for cell suspension, then to extractor P3, and last to the chromatographic column P4 where selective binding is used to further

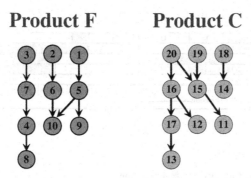

Product F   Product C

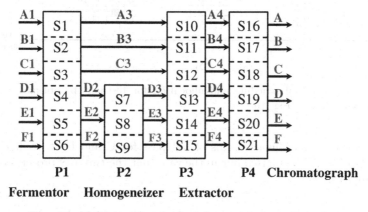

Fig. 15. Precedence tests for new proteins F and C

Fig. 16. Multistage batch plant for protein manufacturing

separate the product of interest from other proteins. Extracellular proteins after the fermentor P1 are sent directly to the extractor P3 and then to the chromatograph P4.

The problem consists of determining simultaneously the optimal schedule of tests and their allocation to labs, while at the same time deciding on the batch plant design to accommodate the new proteins. Here two major options are considered. One is to build a new plant for products C and D (assuming both pass the tests) and the other is to expand the capacity of the existing plant. This problem, formulated as an MILP problem by Maravelias and Grossmann [138], involves 612 0-1 variables, 32,184 continuous variables and 30,903 constraints. Here again one option is to solve simultaneously the full-size MILP, while the other is to decompose the problem into the scheduling and design functions using a Lagrangian relaxation technique similar to the one in example 3. The schedule predicted for the tests is shown in Figure 17(a). In Figure 17(b) it can be seen that the model selects to expand the capacity of the various units rather than building a new plant. Also, since the model

TESTING : Gantt Charts for Resources

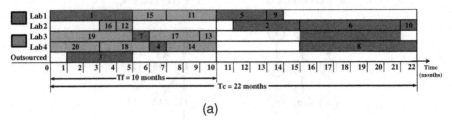

(a)

DESIGN: Expansions of Processes

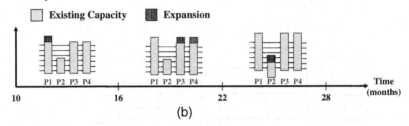

(b)

Fig. 17. (a) Optimal schedule and allocation for testing, (b) optimal capacity expansion of plant in Fig. 16

accounts for the various scenarios of fall/pass for C and F, it predicts that products D and E be phased out in the case that the two new proteins obtain FDA approval.

7.5 Example 5

This problem deals with the design and planning of an offshore facility for gas production. It is assumed that a superstructure consisting of a production platform, well platforms (one for each field), and pipelines is given (see Figure 18). It is also assumed that for some of the fields there is significant uncertainty in the size and initial deliverability (production rate) of the fields. The problem consists of determining over a given time horizon (typically 10-15 years) decisions regarding the selection and timing of the investment for the installation of the platforms, their capacities, and production profiles.

Goel and Grossmann [68] developed a mixed-integer optimization model assuming discrete probability distribution functions for the sizes and initial deliverabilities. Also, the model was simplified with a linear performance model to avoid the direct use of a reservoir simulation model. The optimization problem gives rise to a very difficult stochastic optimization problem, which has the unique feature that the scenario trees are a function of the timing of the investment decisions. Goel and Grossmann [68, 69] have developed two solution methods, a heuristic and a rigorous branch and bound search method

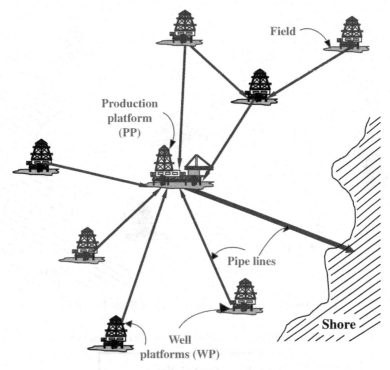

Fig. 18. Infrastructure for offshore gas production

for solving this problem. The example in Figure 19 involves six fields over 15 years, with two of the fields having uncertain sizes and deliverabilities. If one were to solve directly the deterministic equivalent problem in which all scenarios are anticipated, this would give rise to a very large multiperiod MILP model with about 16,281 0-1 variables and 2.4 million constraints which is impossible to solve directly with current solution methods. Fortunately the methods by Goel and Grossmann circumvent the solution of such a large problem. The solution is shown in Table 2, which postpones the investment in the uncertain fields to years 5 and 7, with an expected net present value (NPV) of $146 million and a risk of less than 1% that the NPV be negative. Interestingly, if one simply uses mean values for the uncertain parameters, the platforms at the uncertain fields are installed in period 1 and the financial risk increases to 8%. Obviously, in practice models like these would be solved periodically by updating them with new information on the fields.

8 Concluding Remarks

This chapter has provided an overview of the building blocks of the emerging area of EWO that is driven by needs of the chemical process industries for

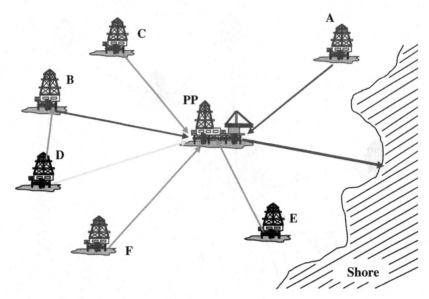

Fig. 19. Example with uncertain gas fields D and E

Table 2. Solution of stochastic model for Fig. 19

Proposed Solution	
Year 1	PP, A, B, C, F
Year 5	E
Year 7	D
ENPV	$146.32 Million

reducing costs and remaining competitive in the global marketplace. Some of the major challenges related to modeling of planning and scheduling, multi-scale optimization, and handling of uncertainties, coupled with algorithmic methods and tools have been highlighted. The ultimate goal of EWO is to be able to develop the technology to address large-scale process industry supply chain optimization problems that encounter and overcome all of the highlighted challenges. Several examples were presented to illustrate the nature of the applications and the progress that has been made in tackling some of these problems.

From this review it is clear that further research is required to expand the scope and size of planning and scheduling models that can be solved in order to achieve the goal of EWO. Of particular significance would be the development of effective decomposition schemes that have the capability of handling large-scale problems over geographically distributed sites and over wide temporal scales. There is also an increasing need for the effective global optimization of nonconvex planning and scheduling problems that incorporate process models such as blending operations. The development of effective and meaningful solution methods for stochastic programming problems for handling uncertainties continues to be of paramount importance. It is worth noting that all these developments will require a synergy of advances in basic algorithms and advances in large-scale computation that can provide the basis of a cyberinfrastructure on which EWO can be fully realized.

It is hoped that this chapter has shown that EWO offers new and exciting opportunities for research to chemical engineers addressing the challenges of the process industry supply chain. While EWO lies at the interface of chemical engineering (process systems engineering) and operations research, it is clear that chemical engineers can play a major role not only in the modeling part, but also in the algorithmic part given the strong and rich tradition that chemical engineers have built in mathematical programming. Thus, in collaboration with operations researchers, chemical engineers should be well positioned for developing novel computational models and algorithms that are to be integrated with coordination and decomposition techniques through advanced computing tools. This effort should help to expand the scope and nature of EWO models that can be effectively solved in real-world industrial problems. These models and methods have the potential of providing a new generation of analytical IT tools that can significantly increase profits and reduce costs, thereby strengthening the economic performance and competitiveness of the process industries.

Acknowledgments

The authors would like to acknowledge financial support from the Pennsylvania Infrastructure Alliance, the Center for Advanced Process Decision making at Carnegie Mellon University, ExxonMobil Corporation, and from DIMACS at Rutgers University.

References

1. Acevedo J, Pistikopoulos EN (1997) A multiparametric programming approach for linear process engineering problems under uncertainty. Ind Eng Chem Res, 36: 717
2. Ahmed S, Sahinidis NV (1998) Robust process planning under uncertainty. Ind Eng Chem Res, 37: 1883–1892
3. Ahmed S, Sahinidis NV (2000) Analytical investigations of the process planning problem. Comp Chem Eng, 23: 1605–1621

4. Anstreicher K, Brixius N, Goux J-P, Linderoth J (2002) Solving large quadratic assignment problems on computational grids. Math Prog, 91: 563–588
5. Aronofsky, JS, Williams, AC (1962) A use of linear programming and mathematical models in underground production. Man Sci, 8: 394–407
6. Balasubramanian J, Grossmann IE (2002) A novel branch and bound algorithm for scheduling flowshop plants with uncertain processing times. Comp Chem Eng, 26: 41–57
7. Balasubramanian J, Grossmann IE (2004) Approximation to multistage stochastic optimization in multiperiod batch plant scheduling under demand uncertainty. Ind Eng Chem Res, 43: 3695–3713
8. Balasubramanian J, Grossmann IE (2003) Scheduling optimization under uncertainty – an alternative approach. Comp Chem Eng, 27: 469–490
9. Barnes DA, Humphrey K, Muellenberg L (1990) A production optimization system for Western Prudhoe Bay Field, Alaska. Paper 20653-MS presented in the SPE Annual Technical Conference and Exhibition, 23-26 September, New Orleans, Louisiana
10. Barnes R, Linke P, Kokossis A (2002) Optimization of oil field development production capacity. ESCAPE-12 proceedings, The Hague, Netherlands, 631–636
11. Barton P, Lee CK (2004) Design of process operations using hybrid dynamic optimization. Comp Chem Eng, 28: 955–969
12. Bassett MH, Dave P, Doyle III FJ, Kudva GK, Pekny JF, Reklaitis GV, Subrahmanyam S, Miller DL, Zentner MG (1996) Perspectives on model based integration of process operations. Comp Chem Eng, 20: 821–844
13. Bassett MH, Pekny JF, Reklaitis GV (1996) Decompositon techniques for the solution of large-scale scheduling problems. AIChE J, 42: 3373
14. Beale EML, Tomlin JA (1970) Special facilities in a general mathematical programming system for nonconvex problems using ordered sets of variables. In: Lawrence J (ed) Proceedings of the Fifth International Conference on Operations Research Tavistock Publications, 447–454
15. Bellman, R (1957) Dynamic Programming. Princeton University Press, New Jersey
16. Ben-Tal A, Nemirovski A (1998) Robust convex optimization. Math Oper Res, 23: 769–805
17. Ben-Tal A, Nemirovski A (1999) Robust solutions to uncertain linear programs. Oper Res Let, 25: 1–13
18. Ben-Tal A, Nemirovski A (2002) Robust optimization–methodology and applications. Math Prog, 92: 453–480
19. Bertsekas DP, Tsitsiklis J (1996) Neuro-Dynamic Programming. Athena Scientific, Belmont, MA
20. Bechtel Corp (1993) PIMS (Process Industry Modeling System) User's manual, version 60 Houston, TX
21. Bhatia T, Biegler LT (1996) Dynamic optimization in the design and scheduling of multiproduct batch plants. Ind Eng Chem Res, 35: 2234
22. Biegler LT, Cervantes A, Wächter A (2002) Advances in simultaneous strategies for dynamic process optimization. Chem Eng Sci, 57: 575
23. Bieker HP, Slupphaug O, Johansen TA (2006) Real-time production optimization of offshore oil and gas production systems: a technology survey. Paper

99446-MS presented in Intelligent Energy Conference and Exhibition, 11-13 April, Amsterdam, The Netherlands

24. Birewar DB, Grossmann IE (1990) Simultaneous production planning and scheduling of multiproduct batch plants. Ind Eng Chem Res, 29: 570

25. Birge JR, Louveaux F (1997) Introduction to Stochastic Programming. Springer, New York

26. Bixby RE, Fenelon M, Gu Z, Rothberg E, Wunderling R (2002) MIP theory and practice. closing the gap. http//wwwilogcom/products/optimization/tech/research/mippdf

27. Bodington EC (1995) Planning, Scheduling and Control Integration in the Process Industries. McGraw Hill

28. Bok J, Grossmann IE, Park S (2000) Supply chain optimization in continuous flexible process networks. Ind Eng Chem Res, 39: 1279

29. Bonner and Moore Management Science (1979) RPMS (Refinery and Petrochemical Modeling System). A system description. Houston, TX

30. Byrd RH, Gilbert JC, Nocedal J (2000) An interior point algorithm for large scale nonlinear programming. Math Prog, 89: 149

31. Castro P, Barbosa-Póvoa APFD, Matos H (2001) An improved RTN continuous-time formulation for the short-term scheduling of multipurpose batch plants. Ind Eng Chem Res, 40: 2059–2068

32. Castro PM, Barbosa-Póvoa AP, Matos HA, Novais AQ (2004) Simple continuous-time formulation for short-term scheduling of batch and continuous processes. Ind Eng Chem Res, 43: 105–118

33. Castro PM, Grossmann IE (2006) An efficient MILP model for the short-term scheduling of single stage batch plants. Comp Chem Eng, 30: 1003–1018

34. Cerdá J, Henning GP, Grossmann IE (1997) A mixed-integer linear programming model for short-term scheduling of single-stage multiproduct batch plants with parallel lines. Ind Eng Chem Res, 36: 1695–1707

35. Chachuat B, Singer AB, Barton PI (2006) Global methods for dynamic optimization and mixed-integer dynamic optimization. Ind Eng Chem Res, 45: 8373–8392

36. Chen Q, Ferris M, Linderoth J (2001) FATCOP 20 advanced features in an opportunistic mixed integer programming solver. Ann Oper Res, 103: 17–32

37. Chen C, Liu C, Feng X, Shao H (2002) Optimal short-term scheduling of multiproduct single-stage batch plants with parallel lines. Ind Eng Chem Res, 41: 1249–1260

38. Cheng L, Duran MA (2003) World-wide crude transportation logistics a decision support system based on simulation and optimization. In: Grossmann, IE, McDonald, CM (eds) Proceedings of 4th International Conference on Foundations of Computers-Aided Process Operations CAChE, Coral Springs, FL, 187–201

39. Clay RL, Grossmann IE (1997) A disaggregation algorithm for the optimization of stochastic planning models. Comp Chem Eng, 21: 751–774

40. Columbani Y, Heipcke S (2002) Mosel: An overview. Dash Optimization

41. Cott BJ, Macchietto S (1989) Minimizing the effects of batch process variability using online schedule modification. Comp Chem Eng, 13: 105–113

42. Daskin MS, Coullard C, Shen Z-JM (2002) An inventory-location model formulation, solution algorithm and computational results. Ann Oper Res, 110: 83–106

43. de Farias IR (2004) Semi-continuous cuts for mixed-integer programming. In: Daniel Bienstock, D, Nemhauser, G (eds) Integer Programming and Combinatorial Optimization - 10th International IPCO Conference Proceedings New York, NY

44. Dawande MW, Hooker JN (2000) Inference-based sensitivity analysis for mixed integer/linear programming. Oper Res, 48: 623

45. Dempster MAH, Pedron NH, Medova EA, Scott JE, Sembos A (2000) Planning logistics operations in the oil industry. J Opl Res Soc, 11: 1271–1288

46. Dimitriadis AD, Shah N, Pantelides CC (1997) RTN-based rolling horizon algorithms for medium term scheduling of multipurpose plants. Comp Chem Eng, 21: S1061

47. Dua V, Pistikopoulos EN (1999) Algorithms for the solution of multiparametric mixed integer nonlinear optimization problems. Ind Eng Chem Res, 38: 3976

48. Dua V, Pistikopoulos EN (2000) Algorithms for the solution of multiparametric mixed integer linear optimization problems. Ann Oper Res, 99: 123

49. Dutta-Roy K, Kattapuram J (1997) A new approach to gas lift allocation optimization. Paper SPE 38333 presented in SPE Western Regional Meeting, Long Beach, CA

50. Egg OS, Herring T (1997) Combining linear programming and reservoir simulation to optimize asset value. SPE 37446 presented at the SPE Production Operations Symposium, March 9-11, Oklahoma City, OK

51. Engell S, Markert A, Sand G, Schultz R (2004) Aggregated scheduling of a multi-product batch plant by two-stage stochastic integer programming. Opt Eng, 5: 335–359

52. Equi L, Gallo G, Marziale S, Weintraub A (1997) A combined transportation and scheduling problem. Eur J Oper Res, 97: 94–104

53. Erdirik-Dogan M, Grossmann IE (2006) Simultaneous planning and scheduling for multiproduct continuous plants. Ind Eng Chem Res, 45: 299–315

54. Erdirik-Dogan M, Grossmann IE (2007) Planning models for parallel batch reactors with sequence-dependent changeovers. AIChE J, 53: 2284–2300

55. Erengüç S, Simpson NC, Vakharia AJ (1999) Integrated production/distribution planning in supply chains. An invited review. Eur J Oper Res, 115: 219–236

56. Escudero LF, Quintana FJ , Salmeron J (1999) CORO, a modeling and an algorithmic framework for oil supply, transformation and distribution optimization under uncertainty. Eur J Oper Res, 114: 638–656

57. Fang WY, Lo KK (1996) A generalized well-management scheme for reservoir simulation. SPE Reservoir Engineering, 5: 116–120

58. Fletcher R, Gould NIM, Leyffer S, Toint PhL, Waechter A (2002) Global convergence of a trust-region (SQP)—Filter algorithm for general nonlinear programming. SIAM J Opt, 13: 635–655

59. Floudas CA (2000) Deterministic global optimization theory, methods and applications. Kluwer Academic Publishers, Dordrecht, Netherlands

60. Floudas CA, Lin X (2004) Continuous-time versus discrete-time approaches for scheduling of chemical processes a review. Comp Chem Eng, 28: 2109–2129

61. Forrest J, Oettli M (2003) Rigorous simulation supports accurate refinery decisions In: Grossmann, IE, McDonald, CM (eds) Proceedings of 4th International Conference on Foundations of Computers-Aided Process Operations CAChE, Coral Springs, FL, 273–280

62. Foster I, Kesselman C (1999) The Grid Blueprint for a New Computing Infrastructure. Morgan-Kaufman

63. Furman KC, Jia Z, Ierapetritou MG (2007) A robust event-based continuous time formulation for tank transfer scheduling. Ind Eng Chem Res, 46: 9126–9136

64. Gal T, Nedoma J (1972) Multiparametric linear programming. Man Sci, 8: 406–422

65. Garvin WW, Crandall HW, John JB, Spellman RA (1957) Applications of linear programming in the oil industry. Man Sci, 3: 407–430

66. Giannelos NF, Georgiadis MC (2002) A simple new continuous-time formulation for short-term scheduling of multipurpose batch processes. Ind Eng Chem Res, 41: 2178–2184

67. Glismann K, Gruhn G (2001) Short-term scheduling and recipe optimization of blending processes. Comp Chem Eng, 25: 627

68. Goel V, Grossmann IE (2004) A stochastic programming approach to planning of offshore gas field developments under uncertainty in reserves. Comp Chem Eng, 28: 1409–1429

69. Goel V, Grossmann IE, El-Bakry AS, Mulkay EL (2006) A novel branch and bound algorithm for optimal development of gas fields under uncertainty in reserves. Comp Chem Eng, 30: 1076–1092

70. Guillén G, Badell M, Espuña A, Puigjaner L (2006) Simultaneous optimization of process operations and financial decisions to enhance the integrated planning/scheduling of chemical supply chains. Comp Chem Eng, 30: 421–436

71. Goux J-P, Kulkarni S, Linderoth J, Yoder M (2001) Master-worker: An enabling framework for master-worker applications on the computational grid. Cluster Comput, 4: 63–70

72. Goux J-P, Leyffer S (2003) Solving large MINLPs on computational grids. Opt Eng, 3: 327–354

73. Graves SC (1982) Using Lagrangean techniques to solve hierarchical production planning problems. Man Sci, 28: 260–275

74. Grossmann IE (2005) Enterprise-wide optimization: A new frontier in process systems engineering. AIChE J, 51: 1846–1857

75. Grossmann IE, Van den Heever SA, Harjunkoski I (2002) Discrete optimization methods and their role in the integration of planning and scheduling. AIChE Symposium Series No 326, 98: 150

76. Grossmann IE (2002) Review of nonlinear mixed-integer and disjunctive programming techniques. Opt Eng, 3: 227–252

77. Guignard M, Kim S (1987) Lagrangean decomposition: A model yielding stronger Lagrangean bounds. Math Prog, 39: 215–228

78. Gupta S, Karimi IA (2003) An improved MILP formulation for scheduling multiproduct, multistage batch plants. Ind Eng Chem Res, 42: 2365–2380

79. Gupta A, Maranas CD (1999) A hierarchical Lagrangean relaxation procedure for solving midterm planning problems. Ind Eng Chem Res, 38: 1937

80. Gupta A, Maranas CD (2003) Managing demand uncertainty in supply chain planning. Comp Chem Eng, 27: 1219–1227

81. Gupta A, Maranas CD, McDonald CM (2000) Midterm supply chain planning under demand uncertainty: Customer demand satisfaction and inventory management. Comp Chem Eng, 24: 2613
82. Hooker JN (2000) Logic-based methods for optimization combining optimization and constraint satisfaction. Wiley, New York
83. Hooker JN (2002) Logic, optimization and constraint programming. INFORMS J Comp, 14: 295–321
84. Hooker JN (2007) Integrated methods for optimization. Springer, New York
85. Hooker JN, Ottosson G (2003) Logic-based Benders' decomposition. Math Prog, 96: 33–60
86. Hooker JN, Ottosson G, Thorsteinsson E, Kim H-J (1999) On integrating constraint propagation and linear programming for combinatorial optimization. Proceedings 16th National Conference on Artificial Intelligence, MIT Press, 136–141
87. Iakovou ET (2001) An interactive multiobjective model for the strategic transportation of petroleum products risk analysis and routing. Safety Sci, 39: 19–29
88. Ierapetritou MG, Floudas CA (1998) Effective continuous-time formulation for short-term scheduling. 1 Multipurpose batch processes. Ind Eng Chem Res, 37: 4341–4359
89. Ierapetritou MG, Floudas CA, Vasantharajan S, Cullick AS (1999) Optimal location of vertical wells decomposition approach. AIChE J, 45: 844–859
90. Iyer RR, Grossmann IE (1998) A bilevel decomposition algorithm for long-range planning of process networks. Ind Eng Chem Res, 37: 474–481
91. Iyer RR, Grossmann IE, Vasantharajan S, Cullick AS (1998) Optimal planning and scheduling of offshore oil field infrastructure investment and operations. Ind Eng Chem Res, 37: 1380
92. Jackson J, Grossmann IE (2003) A temporal decomposition scheme for nonlinear multisite production planning and distribution models. Ind Eng Chem Res, 42: 3045–3055
93. Jain V, Grossmann IE (1998) Cyclic scheduling and maintenance of parallel process units with decaying performance. AIChE J, 44: 1623–1636
94. Jain V, Grossmann IE (2001) Algorithms for hybrid MILP/CP models for a class of optimization problems. INFORMS J Comp, 13: 258–276
95. Janak SL, Floudas CA, Kallrath J, Vormbrock N (2006) Production scheduling of a large-scale industrial batch plant ii reactive scheduling. Ind Eng Chem Res, 45: 8253–8269
96. Janak SL, Lin X, Floudas CA (2004) Enhanced continuous-time unit-specific event-based formulation for short-term scheduling of multipurpose batch processes: Resource constraints and mixed storage policies. Ind Eng Chem Res, 43: 2516–2533
97. Janak SL, Lin X, Floudas CA (2007) A new robust optimization approach for scheduling under uncertainty. II Uncertainty with known probability distribution. Comp Chem Eng, 31: 171–195
98. Jia Z, Iearapetritou MG (2003) Mixed-integer linear programming for gasoline blending and distribution scheduling. Ind Eng Chem Res, 42: 825–835
99. Jia Z, Ierapetritou MG (2004) Short-term scheduling under uncertainty using MILP sensitivity analysis. Ind Eng Chem Res, 43: 3782
100. Jia Z, Ierapetritou MG, Kelly JD (2003) Refinery short-term scheduling using continuous time formulation crude oil operations. Ind Eng Chem Res, 42: 3085

101. Johnson EL, Nemhauser GL, Savelsbergh MWP (2000) Progress in linear programming based branch-and-bound algorithms exposition. INFORMS J Comp, 12: 2–23

102. Kallrath J (2002) Planning and scheduling in the process industry. OR Spectrum, 24: 219–250

103. Kanakamedala KB, Reklaitis GV, Venkatasubramanian V (1994) Reactive schedule modification in multipurpose batch chemical plants. Ind Eng Chem Res, 33: 77–90

104. Karuppiah R, Furman KC, Grossmann IE (2008) Global optimization for scheduling refinery crude oil operations. Comp Chem Eng, 32: 2745–2766

105. Kelly JD (2004) Formulating production planning models. Chem Engineer Progr, January: 43–50

106. Kelly JD (2005) Modeling production-chain information. Chem Engineer Progr, February: 28–31

107. Kelly JD (2005) The unit-operation-stock superstructure (UOSS) and the quantity-logic-quality paradigm (QLQP) for production scheduling in the process industries. Proceedings of the Multidisciplinary Conference on Scheduling Theory and Applications (MISTA), 327

108. Kelly JD, Mann JL (2003) Crude-oil blend scheduling optimization: An application with multi-million dollar benefits - Part I. Hydrocarb Proc, 6: 47–53

109. Kelly JD, Mann JL (2003) Crude-oil blend scheduling optimization An application with multi-million dollar benefits - Part II. Hydrocarb Proc, 7: 72–79

110. Kelly JD, Zyngier D (2007) An improved MILP modeling of sequence-dependent switchovers for discrete-time scheduling problems. Ind Eng Chem Res 46: 4964–4973

111. Kok AG, Graves SC (2003) Supply chain management design, coordination and operation. Handbooks in Operations Research and Management Science, Elsevier, Amsterdam, The Netherlands

112. Kondili E (1988) The optimal scheduling of batch chemical processes. PhD Thesis, Imperial College, London

113. Kondili E, Pantelides CC, Sargent R (1993) A general algorithm for short-term scheduling of batch operations – I MILP formulation. Comp Chem Eng, 17: 211–227

114. Kosmidis VD, Perkins JD, Pistikopoulos EN (2002) A mixed integer optimization strategy for integrated gas/oil production. ESCAPE – 12 proceedings, The Hague, Netherlands, 697–702

115. Kulkarni RV, Mohanty RP (1996) Temporal decomposition approach for solving a multilocation plant sizing and timing problem. Prod Plan Control, 7: 27

116. Lasschuit W, Thijssen N (2003) Supporting supply chain planning and scheduling decisions in the oil & chemical industry. In: Grossmann, IE, McDonald, CM (eds), Proceedings of 4th International Conference on Foundations of Computers-Aided Process Operations CAChE, Coral Springs, FL, 37–44

117. Lasschuit W, Thijssen N (2004) Supporting supply chain planning and scheduling decisions in the oil and chemical industry. Comp Chem Eng, 28: 863–870

118. Lee H, Pinto JM, Grossmann IE, Park S (1996) Mixed-integer linear programming model for refinery short-term scheduling of crude oil unloading with inventory management. Ind Eng Chem Res, 35: 1630–1641
119. Lee JH, Lee JM (2006) Approximate dynamic programming based approach to process control and scheduling. Comp Chem Eng, 30: 1603–1618
120. Lee K, Park H-J, Lee I (2001) A novel nonuniform discrete time formulation for short-term scheduling of batch and continuous processes. Ind Eng Chem Res, 40: 4902–4911
121. Li Z, Ierapetritou MG (2007) Process scheduling under uncertainty using multiparametric programming. AIChE J 53: 3183
122. Li Z, Ierapetritou MG (2008) Process scheduling under uncertainty Review and challenges. Comp Chem Eng, 32: 715–727
123. Li Z, Ierapetritou MG (2008) Robust optimization for process scheduling under uncertainty. Ind Eng Chem Res, 47: 4148–4157
124. Lee S, Grossmann IE (2001) A global optimization algorithm for nonconvex generalized disjunctive programming and applications to process systems. Comp Chem Eng, 25: 1675–1697
125. Lim M, Karimi IA (2003) Resource-constrained scheduling of parallel production lines using asynchronous slots. Ind Eng Chem Res, 42: 6832–6842
126. Lin G, Ettl M, Buckley S, Bagchi S, Yao DD, Naccarato BL, Allan R, Kim K, Koenig L (2000) Extended-enterprise supply-chain management at IBM personal systems group and other divisions. Interfaces, 30: 7–25
127. Lin X, Janak SL, Floudas CA (2004) A new robust optimization approach for scheduling under uncertainty. I Bounded uncertainty. Comp Chem Eng, 28: 1069–1085
128. Lin X, Floudas CA, Modi S, Juhasz NM (2002) Continuous-time optimization approach for medium-range production scheduling of a multiproduct batch plant. Ind Eng Chem Res, 41: 3884–3906
129. Linderoth J, Shapiro A, Wright S (2006) The empirical behavior of sampling methods for stochastic programming. Ann Oper Res, 142: 215–241
130. Linderoth J, Wright SJ (2003) Implementing a decomposition algorithm for stochastic programming on a computational grid. Compl Opt Appl, 24: 207–250
131. Litvak ML, Hutchins LA, Skinner RC, Darlow BL, Wood RC, Kuest LJ (2002) Prudhoe Bay E-field production optimization system based on integrated reservoir and facility simulation. Paper 77643-MS presented in SPE Annual Technical Conference and Exhibition, 29 September-2 October, San Antonio, TX
132. Liu ML, Sahinidis NV (1996) Optimization in process planning under uncertainty. Ind Eng Chem Res, 35: 4154–4165
133. Liu ML, Sahinidis NV (1997) Process planning in a fuzzy environment. Eur J Oper Res, 100: 142–169
134. Livny M, Ramakrishnan R, Beyer KS, Chen G, Donjerkovic D, Lawande S, Myllymaki J, Wenger RK (1997) DEVise integrated querying and visualization of large datasets. SIGMOD Conference, 301–312
135. Magalhães MV, Shah N (2003) Crude oil scheduling. In: Grossmann IE, McDonald, CM (eds), Proceedings of 4th International Conference on Foundations of Computers-Aided Process Operations CAChE, Coral Springs, FL, 323–326

136. Magatão L, Arruda, LVR, Neves Jr F (2002) A mixed integer programming approach for scheduling commodities in a pipeline. ESCAPE – 12 proceedings, The Hague, Netherlands 715–720

137. Marchand H, Martin A, Weismantel R, Wolsey LA (2002) Cutting planes in integer and mixed integer programming. Dis Appl Math, 123: 397–446

138. Maravelias CT, Grossmann IE (2001) Simultaneous planning for new product development and batch manufacturing facilities. Ind Eng Chem Res, 40: 6147–6164

139. Maravelias CT, Grossmann IE (2003) Minimization of makespan with discrete-time state-task network formulation. Ind Eng Chem Res, 42: 6252–6257

140. Maravelias CT, Grossmann IE (2003) A new general continuous-time state task network formulation for short term, scheduling of multipurpose batch plants. Ind Eng Chem Res, 42: 3056–3074

141. Maravelias CT, Grossmann IE (2004) A hybrid MILP/CP decomposition approach for the continuous time scheduling of multipurpose batch plants. Comp Chem Eng, 28: 1921–1949

142. Más R, Pinto JM (2002) A mixed-integer optimization strategy for oil supply in distribution complexes. Opt Eng, 4: 23–64

143. Méndez CA, Cerdá J (2003) An MILP continuous-time framework for short-term scheduling of multipurpose batch processes under different operation strategies. Opt Eng, 4: 7–22

144. Méndez CA, Cerdá J (2004) An MILP framework for batch reactive scheduling with limited discrete resources. Comp Chem Eng, 28: 1059–1068

145. Méndez CA, Cerdá J (2004) Short-term scheduling of multistage batch processes subject to limited finite resources. Comp Chem Eng, 15B: 984–989

146. Méndez CA, Cerdá J, Grossmann IE, Harjunkoski I, Fahl M (2006) State-of-the-art review of optimization methods for short-term scheduling of batch processes. Comp Chem Eng, 30: 913–946

147. Méndez CA, Cerdá J, Grossmann IE, Harjunkoski I, Kabore P (2006) A simultaneous optimization approach for off-line blending and scheduling of oil-refinery operations. Comp Chem Eng, 30: 614–634

148. Méndez CA, Henning GP, Cerdá J (2000) Optimal scheduling of batch plants satisfying multiple product orders with different due-dates. Comp Chem Eng, 24: 2223–2245

149. Méndez CA, Henning GP, Cerdá J (2001) An MILP continuous-time approach to short-term scheduling of resource-constrained multistage flowshop batch facilities. Comp Chem Eng, 25: 701–711

150. Milano M (2003) Constraint and integer programming toward a unified methodology. Kluwer, Dordrecht Boston London

151. Mockus L, Reklaitis GV (1999) Continuous time representation approach to batch and continuous process scheduling. 1 - MINLP formulation. Ind Eng Chem Res, 38: 197–203

152. Mockus L, Reklaitis GV (1999) Continuous time representation approach to batch and continuous process scheduling. 2 Computational issues. Ind Eng Chem Res, 38: 204–210

153. Montgomery DC (2000) Introduction to statistical quality control. Wiley, New York

154. Moro LFL, Zanin AC, Pinto JM (1998) A planning model for refinery diesel production. Comp Chem Eng, 22: S1039–S1042

155. Neiro SMS, Pinto JM (2004) Supply chain optimization of petroleum refinery complexes. Comp Chem Eng, 28: 871–896
156. Nemhauser GL, Wolsey LA (1988) Integer and combinatorial optimization. Wiley-Interscience
157. Nishi T, Konishi M, Hasebe S, Hashimoto I (2002) Autonomous decentralized supply chain optimization system for multi-stage production processes. Proceedings of 2002 Japan-USA Symposium on Flexible Automation, 131–138
158. Nocedal J, Wright SJ (1999) Numerical optimization. Springer, New York
159. Pantelides CC (1994) Unified frameworks for the optimal process planning and scheduling. Proceedings on the Second Conference on Foundations of Computer Aided Operations, 253–274
160. Papageorgiou LG, Pantelides CC (1996) Optimal campaign planning/ scheduling of multipurpose batch/semi-continuous plants, 1 - Mathematical formulation. Ind Eng Chem Res, 35: 488
161. Papageorgiou LG, Pantelides CC (1996) Optimal campaign planning/ scheduling of multipurpose batch/semi-continuous plants, 2 - A mathematical decomposition approach. Ind Eng Chem Res, 35: 510
162. Park H-J, Lim J-S, Roh J, Kang JM, Min B-H (2006) Production-system optimization of gas fields using hybrid fuzzy-genetic approach. Paper 100179-MS, SPE Europec/EAGE Annual Conference and Exhibition, 12-15 June, Vienna, Austria
163. Pekny JF, Reklaitis GV (1998) Towards the convergence of theory and practice a technology guide for scheduling/planning methodology. AIChE Symposium Series No 94 320: 91–111
164. Perea E, Grossmann IE, Ydstie E, Tahmassebi T (2001) Dynamic modeling and decentralized control of supply chains. Ind Eng Chem Res, 40: 3369–3383
165. Perea E, Ydstie E, Grossmann IE (2003) A model predictive control strategy for supply chain optimization. Comp Chem Eng, 27: 1201–1218
166. Pertsinidis A, Grossmann IE, McRae GJ (1998) Parametric optimization of MILP programs and a framework for the parametric optimization of MINLPs. Comp Chem Eng, 22: S205
167. Pinedo M (2001) Scheduling theory, Algorithms, and systems. Prentice Hall
168. Pinto J, Grossmann IE (1994) Optimal scheduling of multistage multiproduct continuous plants. Comp Chem Eng, 9: 797–816
169. Pinto JM, Grossmann IE (1995) A continuous time mixed integer linear programming model for short-term scheduling of multistage batch plants. Ind Eng Chem Res, 34: 3037–3051
170. Pinto JM, Grossmann IE (1996) An alternate MILP model for short-term scheduling of batch plants with preordering constraints. Ind Eng Chem Res, 35: 338–342
171. Pinto J, Grossmann IE (1998) Assignment and sequencing models for the scheduling of chemical processes. Ann Oper Res, 81: 433–466
172. Pinto JM, Moro LFL (2000) A planning model for petroleum refineries. Braz J Chem Eng, 17: 575–585
173. Pinto JM, Joly M, Moro LFL (2000) Planning and scheduling models for refinery operations. Comp Chem Eng, 24: 2259–2276
174. Ponnambalam K, Vannelli A, Woo S (1992) An interior point method implementation for solving large planning problems in the oil refinery industry. Can J Chem Eng, 70: 368–374

175. Puterman ML (1994) Markov decision processes. Wiley, New York
176. Quesada I, Grossmann IE (1995) Global optimization of bilinear process networks with multicomponents flows. Comp Chem Eng, 19: 1219–1242
177. Raman R, Grossmann IE (1994) Modeling and computational techniques for logic based integer programming. Comp Chem Eng, 18: 563–578
178. Reklaitis GV (1992) Overview of scheduling and planning of batch process operations. Technical report, NATO Advanced Study Institute, Antalaya, Turkey
179. Rigby B, Lasdon LS, Waren AD (1995) The evolution of Texaco blending systems – from Omega to StarBlend. Interfaces, 2: 64
180. Rodrigues MTM, Latre LG, Rodrigues LCA (2000) Short-term planning and scheduling in multipurpose batch chemical plants A multi-level approach. Comp Chem Eng, 24: 2247–2258
181. Rogers DF, Plante RD, Wong RT, Evans JR (1991) Aggregation and disaggregation techniques and methodology in optimization. Oper Res, 39: 553
182. Ross AD (2000) Performance-based strategic resource allocation in supply networks. Int J Prod Econ, 63: 255–266
183. Rotstein GE, Lavie R, Lewin DR (1996) Syntheis of flexible and reliable short-term batch production plans. Comp Chem Eng, 20: 201
184. Ruszczynski A, Shapiro A (2003) Stochastic programming. Handbooks in operations research and management science Vol 10, Elsevier, Amsterdam, The Netherlands
185. Ruszczynski A (2003) Decomposition methods. In: Ruszczynski A, Shapiro A (eds) Stochastic programming, Handbooks in operations research and management science Vol 10, Elsevier, Amsterdam, The Netherlands
186. Sahinidis NV (1996) BARON: A general purpose global optimization software package. J Glob Opt, 8: 201–205
187. Sahinidis NV (2004) Optimization under uncertainty state of the art and opportunities. Comp Chem Eng, 28: 971–983
188. Sahinidis NV, Grossmann IE (1991) MINLP model for cyclic multiproduct scheduling on continuous parallel lines. Comp Chem Eng, 15: 85
189. Sawaya NW, Grossmann IE (2005) A cutting plane method for solving linear generalized disjunctive programming problems. Comp Chem Eng, 29: 1891–1913
190. Schilling G, Pantelides CC (1996) A simple continuous-time process scheduling formulation and a novel solution algorithm. Comp Chem Eng, 20: S1221–S1226
191. Schrage L, Wolsey L (1985) Sensitivity analysis for branch and bound integer programming. Oper Res, 33: 1008
192. Sear TN (1993) Logistics planning in the downstream oil industry. J Opl Res Soc, 44: 9–17
193. Shah N (1992) Efficient scheduling technologies for multipurpose plants. PhD Thesis, Imperial College, London
194. Shah N (1996) Mathematical programming techniques for crude oil scheduling. Comp Chem Eng, 20: S1227–S1232
195. Shah N (1998) Single- and multisite planning and scheduling current status and future challenges. AIChE Symposium Series No 94, 320: 75
196. Shah N (2004) Pharmaceutical supply chains key issues and strategies for optimization. Comp Chem Eng, 28: 929–941
197. Shah N (2005) Process industry supply chains: Advances and challenges. Comp Chem Eng, 29: 1225–1235

198. Shah N, Pantelides CC, Sargent WH (1993) A general algorithm for short-term scheduling of batch operations – II Computational issues. Comp Chem Eng, 2: 229–244
199. Shapiro A (2003) Monte Carlo sampling methods. In: Ruszczynski, A, Shapiro, A (eds) Stochastic programming. Handbooks in operations research and management science vol 10, Elsevier, Amsterdam, The Netherlands
200. Shapiro JF (2001) Modeling the supply chain. Duxbury, Pacific Grove
201. Shapiro JF (2004) Challenges of strategic supply chain planning and modeling. Comp Chem Eng, 28: 855–861
202. Skorin-Kapov J, Granot F (1987) Nonlinear integer programming. Sensitivity analysis for branch and bound. Oper Res Lett, 6: 269
203. Song JS, Yao DD (2001) Supply chain structures coordination, information, and optimization. Kluwer, Dordrecht Boston London
204. Stebel SL, Arruda, LVR, Fabro, JA, Rodrigues (2002) Modeling liquefied petroleum gas storage and distribution. ESCAPE - 12 proceedings, The Hague, Netherlands, 805–810
205. Subrahmanyam S, Bassett MH, Pekny JF, Reklaitis GV (1996) Issues in solving large scale planning, design and scheduling problems in batch chemical plants. Comp Chem Eng, 19: 577–582
206. Subrahmanyam S, Pekny JF, Reklaitis GV (1996) Decomposition approaches to batch plant design and planning. Ind Eng Chem Res, 35: 1866–1876
207. Sundaramoorthy A, Karimi IA (2005) A simpler better slot-based continuous-time formulation for short-term scheduling in multiproduct batch plants. Chem Eng Sci, 60: 2679–2702
208. Sung C, Maravelias CT (2007) An attainable region approach for effective production planning of multi-product processes. AIChE J, 53: 1298–1315
209. Sutton R, Barto A (1998) Reinforcement learning. MIT Press, Cambridge, MA
210. Tawarmalani M, Sahinidis NV (2002) Convexification and global optimization in continuous and mixed-integer nonlinear programming. Kluwer, Dordrecht Boston London
211. Tayur S, Ganeshan R, Magazine M (1999) Quantitative models for supply chain management. Kluwer Academic, Norwell, MA
212. Timpe CH, Kallrath J (2000) Optimal planning in large multi-site production networks. Eur J Oper Res, 126: 422–435
213. Tsiakis P, Shah N, Pantelides CC (2001) Design of multi-echelon supply chain networks under demand uncertainty. Ind Eng Chem Res, 40: 3585–3604
214. Turkay M, Grossmann IE (1996) Logic-based MINLP algorithms for the optimal synthesis of process networks. Comp Chem Eng, 20: 959–978
215. Van den Heever SA, Grossmann IE (2000) An iterative aggregation/disaggregation approach for the solution of a mixed-integer nonlinear oilfield infrastructure planning model. Ind Eng Chem Res, 39: 1955–1971
216. Van den Heever SA, Grossmann IE, Vasantharajan S, Edwards K (2000) Integrating complex economic objectives with the design and planning of offshore oilfield infrastructure. Comp Chem Eng, 24: 1049–1055
217. Van Hentenryck P (1999) The OPL optimization programming language. MIT Press, Boston, MA
218. Van Hentenryck P, Michel L (2005) Constraint-based local search. MIT Press, Boston, MA
219. Van Roy TJ (1983) Cross decomposition for mixed integer programming. Math Prog, 25: 46–63

220. Vanderbei RJ, Shanno, DF (1999) An interior point algorithm for nonconvex nonlinear programming. Comp Opt Appl, 13: 231–252
221. Vin JP, Ierapetritou MG (2000) A new approach for efficient rescheduling of multiproduct batch plants. Ind Eng Chem Res, 39: 4228–4238
222. Vin JP, Ierapetritou MG (2001) Robust short-term scheduling of multiproduct batch plants under demand uncertainty. Ind Eng Chem Res, 40: 4543
223. Waechter A, Biegler LT (2005) Line search filter methods for nonlinear programming motivation and global convergence. SIAM J Opt, 16: 1–31
224. Wang P, Litvak ML, Aziz K (2002) Optimization production from mature fields. Paper 77658-MS presented in SPE Annual Technical Conference and Exhibition, 29 September-2 October, San Antonio, TX
225. Wang Z, Schaefer AJ, Rajgopal J, Prokopyev O (2008) Effective distribution policies for remnant inventory supply chains. Submitted for publication
226. Wallace M, Novello S, Schimpf J (1997) ECLiPSe A platform for constraint logic programming. ICL Systems J, 12: 159–200
227. Wei J, Furman KC, Duran MA (2008) Approximate dynamic programming for stochastic inventory optimization with customer service level constraints. Submitted
228. Wenkai L, Hui CW (2003) Plant-wide scheduling and marginal value analysis for a refinery. In: Grossmann IE, McDonald CM (eds) Proceedings of 4th International Conference on Foundations of Computers-Aided Process Operations. CAChE, Coral Springs, FL, 339–342
229. Wilkinson SJ (1996) Aggregate formulations for large-scale process scheduling problems. PhD Thesis, Imperial College, London
230. Wilkinson SJ, Cortier A, Shah N, Pantelides CC (1996) Integrated production and distribution scheduling on a Europe-wide basis. Comp Chem Eng, 20: S1275–S1280
231. Wilkinson SJ, Shah N, Pantelides CC (1996) Aggregate modeling of multipurpose plant operation. Comp Chem Eng, 19: 583–588
232. Wolsey LA (2003) Strong formulations for mixed-integer programs valid inequalities and extended formulations. Math Prog, 97: 423–447
233. You, F, Grossmann, IE (2008) Design of responsive process supply chains under demand uncertainty. Comp Chem Eng, 32: 3090–3111
234. Zhang X, Sargent RWH (1996) The optimal operation of mixed production facilities – General formulation and some approaches for the solution. Comp Chem Eng, 20: 897–904
235. Zhu XX, Majozi T (2001) Novel continuous time milp formulation for multipurpose batch plants. 2 Integrated planning and scheduling. Ind Eng Chem Res, 40: 5621–5634
236. Zipkin PH (2000) Foundations of inventory management. McGraw-Hill
237. Zyngier D, Kelly JD (2008) Multi-product inventory logistics modeling in the process industries. In: Chaovalitwongse W, Furman KC, Pardalos P (eds) Computational optimization and logistics challenges in the enterprise. Springer, in preparation

Multiproduct Inventory Logistics Modeling in the Process Industries

Danielle Zyngier[1] and Jeffrey D. Kelly[2]

[1] Honeywell Process Solutions
   85 Enterprise Blvd., Suite 100, Markham, ON, L6G0B5, Canada
   `Danielle.Zyngier@honeywell.com`
[2] Honeywell Process Solutions
   85 Enterprise Blvd., Suite 100, Markham, ON, L6G0B5, Canada
   `Jeff.Kelly@honeywell.com` (Corresponding author)

Summary In this chapter the mathematical modeling of several types of inventories are detailed. The inventory types are classified as batch processes, pools, pipelines, pilelines and parcels. The key construct for all inventory models is the "fill-hold/haul-draw" fractal found in all discontinuous inventory or holdup modeling. The equipment, vessel or unit must first be "filled" unless enough product is already held in the unit. Product can then be "held" or "hauled" for a definite (fixed) or indefinite (variable) amount of time and then "drawn" out of the unit when required. Mixed-integer linear programming (MILP) modeling formulations are presented for five different types of logistics inventory models which are computationally efficient and can be readily applied to industrial decision-making problems.

1 Introduction

Inventory management is a key component in any production environment. This has been recognized not only in the chemical engineering literature but also in the operations research and industrial engineering domains. From queuing theory and Little's law [1] it is well known that variability causes congestion and congestion causes an eventual loss of capacity. In the context of production, congestion means a buildup of resources which must be properly modeled in order to accurately represent the overall capacity and capability of the system. More specifically, inventory modeling in the process industries is of vital importance when making manufacturing decisions and is applicable to batch processes, pools and pilelines. Managing inventory is also essential when distributing, transporting or chaining material from one location to another using such equipment as pipelines and parcels which may represent tank trucks, rail cars and ships.

In a batch process raw material is loaded (all at once or in several stages) and after a certain processing time elapses product is removed. A pool unit

W. Chaovalitwongse et al. (eds.), *Optimization and Logistics Challenges in the Enterprise*, Springer Optimization and Its Applications 30,
DOI 10.1007/978-0-387-88617-6_2, © Springer Science+Business Media, LLC 2009

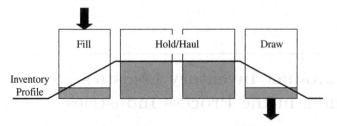

Fig. 1. Fill-hold/haul-draw (FHD) cycle

is an inventory element of the system to which indefinite amounts of material can be added and/or removed at unspecified times. Pilelines, often found for example in the mining sector, are composed of a series of piles of material that obeys a first-in, last-out inventory model.

Logistics modeling indicates that quantity, logic and logistics balances are used to approximate the system [2]. Quantity balances correspond to material balances or the conservation of matter that are very well established in the chemical engineering literature and only involve continuous variables. Logic balances consider the operating procedures, protocols and policies of the system and usually only involve discrete logic or binary variables. Logistics balances involve a combination of continuous and discrete variables and are used to link quantity and logic decisions together such as in semicontinuous constraints (Section 2.2) and in implication inequalities (Section 3.3).

Examples of intensive logic variables are process-unit start-ups, shutdowns and switch-overs-to-itself, whereas extensive logic variables relate to the timing of activities such as process-unit uptime and downtime durations. Concomitantly extensive quantity variables are the flows, flow rates and holdups while intensive quantity variables are exemplified by yields, recipes and ratios.

Inventory models obey what may be called the "fill-hold/haul-draw" (FHD) cycle (Figure 1). Continuous processes are special case inventory models that have a FHD cycle of one time period indicating that the filling and drawing occur in the same time period and that the hold is equal to zero.

From the perspective of quantity and logic (including time) of the hold stage, inventory models can be classified into increasing order of complexity starting with fixed size and fixed time (FSFT), variable size and fixed time (VSFT), fixed size and variable time (FSVT), or variable size and variable time (VSVT) (Table 1). Variable time (indefinite duration) implies a

Table 1. Classification of inventory models with respect to quantity and time of the hold/haul stage

	Batch processes	Pools	Pipelines	Pilelines	Parcels
Quantity	Fixed/Variable	Variable	Fixed	Fixed	Fixed/Variable
Time (logic)	Fixed/Variable	Variable	Variable	Variable	Fixed/Variable

Table 2. Types of inventory models with respect to stock and unit movement characteristics

	Batch processes	Pools	Pipelines	Pilelines	Parcels
Stock	Stationary	Stationary	Dynamic	Stationary	Dynamic
Unit	Stationary	Stationary	Stationary	Stationary	Dynamic
Relative movement	None	None	Exists	None	None

carry-over or intertemporal transfer of logic variables from one time period to the next and introduces degeneracy in the problem since material can enter or leave the unit at any number of different time periods, thus making the problem computationally more difficult.

Note that variable quantity and variable time are mutually exclusive in order to keep the linearity of the inventory models. A further way of classifying inventory models refers to the movement characteristics of stocks (non-renewable resources) and units (renewable resources). A summary of this classification can be found in Table 2. Batch processes, pools and pilelines are completely stationary in that both the unit and the stock remain physically at the same location. Parcels on the other hand are fully dynamic in absolute terms since the truck, rail car, ship or any other parcel unit moves the stock with the vessel but there is no relative movement between the material and the unit. A pipeline on the other hand is a unique type of unit since it exhibits relative movement between the stock and the unit, i.e., the pipeline itself is static and stock is transported through it.

In the previous table, "Relative movement" refers to the movement between the stock and the unit. An important modeling concept that can be used for formulating stock or resource balances on units is to define inlet and outlet ports (Figure 2). These elements correspond to the physical nozzles, points or "puts" through which a resource flows into or out of a unit. The idea of also referring to ports as puts is taken from the fact that the words "input" and "output" are formed by a combination of the base "put" with different prefixes ("in" and "out").

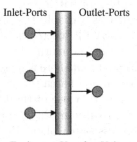

Fig. 2. Inlet and outlet ports on a unit

The ports in a system must always be associated with one or more states since the nature of the material flowing through the ports must be known. In state-task networks [3], states are referred to as being the "feedstocks, intermediate and final products." In this chapter three additional state types are defined in addition to the stock states defined in [3]. In total, there can be four different types of states: stock and utility (as in [3]), utensil and time. Stock states are the most natural ones to chemical engineers since non-renewable resources such as bulk solids, liquids, vapors and gases are included in the material balances. Utensil states take into account renewable resources such as tools (i.e., number of tools needed to support a certain unit) and labor (i.e., number of operators needed to operate a unit). Utility states such as steam, electrical power, instrument air and cooling water are a combination of stock states and utensil states in that a certain amount of utility must be used once a unit starts up (similar to utensil states) and the need for the utility may increase with the processing capacity of the unit (similar to stock states). The utilities flowing through utility states are sometimes referred to as "doubly constrained" resources since they act like both nonrenewable and renewable resources. Time states are quite unique since they may be used to build continuous-time models. The outlet time port of a batch process will have a "flow" of time which is equal to the "flow" of time on the inlet time port on the same unit plus the batch time duration. A novel projectional data model that relates the physical with the procedural elements of a system in a unit operation port state superstructure (UOPSS) can be found in [4].

The following sections explore the details of each type of inventory model commencing with batch processes and ending with parcels. After an introductory section to each unit three types of balances are presented when they exist: quantity balances, logic balances and logistics balances.

2 Batch Processes

Batch processes are well studied process units [3]. As mentioned previously continuous processes are a subset of batch processes. In the literature batch processes are usually classified as variable size and fixed time (VSFT) or fixed size and variable time (FSVT) due to the necessity of linearity when solving them using MILP. The pure quantity balance for a batch process is uninteresting in that it is assumed that there is no inventory or quantity carry-over from the previous shutdown of a batch to the start-up of the next batch. It should be noted that an important use of the FSVT batch unit operation model is to represent no or zero intermediate storage (NIS/ZIS) situations especially those found in pipeless batch plants [5]. Pipeless batch plants are those plants that do not use pipes to transfer materials but moveable vessels such as conveyors with buckets or cranes with bins.

2.1 Logic Balances

For fixed hold, batch or cycle times, a batch process can be modeled with an uptime on the physical unit. This constraint states that the unit can startup at most one mode operation per time period and if it does it must prevent all other mode operation startups until the uptime has expired. Since this constraint involves summing over all mode operations on the physical unit, the uptime constraint also implies single use in which no more than one mode operation can be active at a time in each physical unit.

$$\sum_{pl=1}^{NPL} \sum_{tt=1}^{\tau_{pl}-1} yu_{pl,t-tt} \leq 1 \quad \forall \quad t - tt \geq 0 \ and \ t = 1..NT \tag{1}$$

An interesting and alternative formulation which is less dense (i.e., contains a smaller number of nonzeros) when there are many mode operations and/or the time-period duration is small compared to the uptime can be found in [6]. In their work a network flow formulation is used with hypothetical batch-size variables denoting "idle" flow in and out of the unit and the arcs defined from $yu_{pl,t-tt}$ to $yu_{pl,t}$. In this metaphor the nodes are defined as the time periods; however, we do not present this formulation for brevity. The next set of constraints defines the dependent transition variables associated with the fixed time batch processes.

$$y_{pl,t} = \sum_{tt=0}^{\tau_{pl}-1} yu_{pl,t-tt} \quad \forall \quad pl = 1..NPL \ and \ t = 1..NT \tag{2}$$

$$yw_{pl,pl,t} = \sum_{tt=1}^{\tau_{pl}-2} yu_{pl,t-tt} \quad \forall \quad pl = 1..NPL \ and \ t = 1..NT \tag{3}$$

$$yd_{pl,t} = yu_{pl,t-\tau_{pl}+1} \quad \forall \quad pl = 1..NPL \ and \ t = 1..NT \tag{4}$$

In these equations τ_{pl} corresponds to the hold time of a logical batch unit. The setup, switch-over-to-itself and shutdown logic variables ($y_{pl,t}$, $yw_{pl,pl,t}$ and $yd_{pl,t}$, respectively) are completely dependent on the independent logic variable for the start-up of the corresponding mode operation[1] on the unit ($yu_{pl,t}$) (Figure 3) [7]. Therefore, the three dependent logic variables may be relaxed in the interval [0,1] since their integrality is completely enforced by the integrality of $yu_{pl,t}$, thus easing the computational burden of the formulation.

---

[1] This mode operation is also referred to as a task, instruction, procedure, activity or even job on the unit.

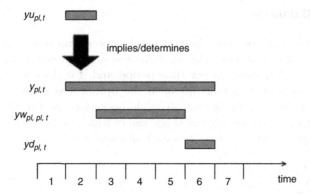

Fig. 3. Dependency between variables for fixed hold-time batch processes only.

For the case of fixed size and variable time batches (FSVT), the independent logic variable is the mode operation setup variable $y_{pl,t}$. The uptime of FSVT batch processes is modeled as follows:

$$\sum_{tt=1}^{\tau_{pl}^{\min}-1} yu_{pl,t-tt} - y_{pl,t} \leq 0 \quad \forall \quad pl = 1..NPL \ and \ t = 1..NT| \ t - tt \geq 0 \quad (5)$$

$$\sum_{tt=0}^{\tau_{pl}^{\max}} y_{pl,t+tt} - \tau_{pl}^{\max} \leq 0 \quad \forall \quad pl = 1..NPL \ and \ t = 1..NT| \ t + tt \leq NT \quad (6)$$

Additionally there must be a single-use constraint on the physical batch process when FSVT type units are modeled.

$$\sum_{pl=1}^{NPL} y_{pl,t} \leq 1 \quad \forall \quad t = 1..NT \quad (7)$$

The mode operation start-up, switch-over-to-itself and shutdown logic variables ($yu_{pl,t}$, $yw_{pl,pl,t}$ and $yd_{pl,t}$, respectively) are again dependent and may be relaxed in the interval [0,1] where their integrality is implied by the integrality of $y_{pl,t}$ and by the tight formulation of the three constraints below:

$$y_{pl,t} - y_{pl,t-1} - yu_{pl,t} + yd_{pl,t} = 0 \quad \forall \quad pl = 1..NPL \ and \ t = 1..NT \quad (8)$$

$$y_{pl,t} + y_{pl,t-1} - yu_{pl,t} - yd_{pl,t} - 2yw_{pl,pl,t} = 0 \ \forall \ pl = 1..NPL \ and \ t = 1..NT \quad (9)$$

$$yu_{pl,t} + yd_{pl,t} + yw_{pl,pl,t} \leq 1 \quad \forall \quad pl = 1..NPL \ and \ t = 1..NT \quad (10)$$

An important benefit of these constraints is that they do not require any minimization of the start-up, switch-over-to-itself and shutdown variables in the objective function of the MILP as is required in other formulations [20].

However, there is a definition change with respect to Figure 3 for these FSVT type batch processes since the shutdown is now defined for the time period after the batch has been completed and the switch-over-to-itself is extended by one time period to take its place. The trade-off is that if there is a requirement for a consecutive or "back-to-back" mode operation on the same batch unit, then a duplicate mode operation must be configured.

2.2 Logistics Balances

The logistics or quantity and logic balances for fixed time batch processes are well known in the literature [3] and are presented here for completeness. The first equation states that the amount of material or stock that is processed at any given time period t cannot be greater than the capacity of the batch unit mode operation in question. For fixed time units this semicontinuous type of logistics constraint is only enforced in the time period in which the unit mode operation starts-up as follows:

$$XH_{pl}^{\min} \cdot yu_{pl,t} \leq xh_{pl,t} \leq XH_{pl}^{\max} \cdot yu_{pl,t} \ \forall \ pl = 1..NPL \ and \ t = 1..NT \tag{11}$$

There are two factors that contribute to the flows entering and leaving batch process mode operations on the ports. The first factor is a function of a fixed yield (XFY) that only depends on the logic start-up variable of the batch process, i.e., when a unit mode operation starts up, a fixed amount of material enters and/or leaves the unit. A fixed yield that is only activated by a logic variable is needed for utensil ports in which a certain number of tools or operators is required to support a given batch and it is not necessarily a function of the size of the batch. The second factor is a function of variable yields (XVV) on the inlet and outlet ports. Variable yields account for variation in flow through the ports as a function of the batch size. The situation in which both fixed and variable yields are needed is when utility ports are used since they may require a nonzero yield of utilities such as the flow of steam or coolant when the process starts up (fixed yield) while the requirement for utilities also increases proportionally to the batch size (variable yield). The equations that model the flows through the inlet and outlet ports, respectively, are provided below.

$$xf_{i,t} \geq XFY_{i,t} \cdot yu_{pl,t} + \sum_{tt=\tau_{SFD,i}^{\min}}^{\tau_{SFD,i}^{\max}} XVV_{i,tt}^{\min} \cdot xh_{pl,t-tt}, \tag{12}$$

$$\forall \ pl = 1..NPL, i = 1..NI \ and \ t = 1..NT$$

$$xf_{i,t} \leq XFY_{i,t} \cdot yu_{pl,t} + \sum_{tt=\tau_{SFD,i}^{\min}}^{\tau_{SFD,i}^{\max}} XVV_{i,tt}^{\max} \cdot xh_{pl,t-tt}, \tag{13}$$

$$\forall \ pl = 1..NPL, i = 1..NI \ and \ t = 1..NT$$

$$xf_{j,t} \geq XFY_{j,t} \cdot yu_{pl,t} + \sum_{tt=\tau_{SDD,j}^{\min}}^{\tau_{SDD,j}^{\max}} XVY_{j,tt}^{\min} \cdot xh_{pl,t-tt}, \qquad (14)$$

$$\forall \quad pl = 1..NPL, j = 1..NJ \text{ and } t = 1..NT$$

$$xf_{j,t} \leq XFY_{j,t} \cdot yu_{pl,t} + \sum_{tt=\tau_{SDD,j}^{\min}}^{\tau_{SDD,j}^{\max}} XVY_{j,tt}^{\max} \cdot xh_{pl,t-tt}, \qquad (15)$$

$$\forall \quad pl = 1..NPL, j = 1..NJ \text{ and } t = 1..NT$$

To account for the different times at which resources enter and leave the batch process throughout the FHD cycle, two new parameters are introduced in the above constraint sets: the start-to-fill delay (SFD) and the start-to-draw delay (SDD) which apply to any type of inlet and outlet ports, respectively. Start-to-fill delay indicates by how many time periods the flow that enters the batch process through a particular port lags behind the start-up of that mode operation. Start-to-draw delay on the other hand indicates how many time periods it takes after a batch process mode operation has started up to start drawing resource out of the unit. It is interesting to point out that in a "standard" batch process unit in which resources enter the unit at the very first time period and leave the unit at the very last time period, the SFD is equal to zero and the SDD is equal to the discrete uptime minus one (τ_{pl}-1).

Different values of SFD may be used for different inlet ports on the same unit in the same mode operation. This can account for fed batch or semi-batch units where material is sequentially added to the batch process during its execution. The SDD can be interpreted in an identical way to SFD except that it is applied to the outlet ports. When different values of SDD are used in two or more ports on the same unit it indicates that resources are drawn out at different time periods during the batch cycle. The use of lower and upper bounds on SFD and SDD instead of a fixed value gives the logistics model more flexibility in terms of accurately modeling the behavior of the system. The cost of the additional flexibility is of course increased degeneracy which may make the problem computationally more intensive. Therefore, the use of dissimilar lower and upper bounds on SFD and SDD is only recommended when truly needed.

Continuous-time formulations of scheduling problems for example contain logistics balances for determining the flows coming into and out of fixed size and variable time batch process (FSVT) but these balances have not to our knowledge been previously extended to fit discrete-time formulations. The following equations consist of a novel method for modeling FSVT in discrete-time formulations.

$$xf_{i,t} \geq XFY_{i,t} \cdot yu_{pl,t} + \sum_{tt=\tau_{SFD,i}^{\min}}^{\tau_{SFD,i}^{\max}} XVY_{i,tt}^{\min} \cdot XH_{pl}^{\min} \cdot yu_{pl,t-tt}, \qquad (16)$$

$$\forall \quad pl = 1..NPL, i = 1..NI \text{ and } t = 1..NT$$

$$xf_{i,t} \leq XFY_{i,t} \cdot yu_{pl,t} + \sum_{tt=\tau_{SFD,i}^{\min}}^{\tau_{SFD,i}^{\max}} XVY_{i,tt}^{\max} \cdot XH_{pl}^{\max} \cdot yu_{pl,t-tt}, \qquad (17)$$

$$\forall \quad pl = 1..NPL, i = 1..NI \text{ and } t = 1..NT$$

$$xf_{j,t} \geq XFY_{j,t} \cdot yu_{pl,t} + \sum_{tt=\tau_{SDD,j}^{\min}}^{\tau_{SDD,j}^{\max}} XVY_{j,tt}^{\min} \cdot XH_{pl}^{\min} \cdot yd_{pl,t+tt}, \qquad (18)$$

$$\forall \quad pl = 1..NPL, j = 1..NJ \text{ and } t = 1..NT$$

$$xf_{j,t} \leq XFY_{j,t} \cdot yu_{pl,t} + \sum_{tt=\tau_{SDD,j}^{\min}}^{\tau_{SDD,j}^{\max}} XVY_{j,tt}^{\max} \cdot XH_{pl}^{\max} \cdot yd_{pl,t+tt}, \qquad (19)$$

$$\forall \quad pl = 1..NPL, j = 1..NJ \text{ and } t = 1..NT$$

When the yields on inlet and outlet ports have non-identical minimum and maximum values and the batch process must obey the law of conservation of matter, the two additional constraints are needed to enforce this balance.

$$\sum_{tt=0}^{\tau_{pl}-1} \sum_{j=1}^{NJ_{pl}} xf_{j,t+tt} - \sum_{tt=0}^{\tau_{pl}-1} \sum_{i=1}^{NI_{pl}} xf_{i,t+tt} \leq (1 - yu_{pl,t}) \cdot XVY_{pl,t}^{sum,\max} \cdot XH_{pl}^{\max},$$

$$(20)$$

$$\forall pl = 1..NPL \text{ and } t = 1..NT$$

$$\sum_{tt=0}^{\tau_{pl}-1} \sum_{j=1}^{NJ_{pl}} xf_{j,t+tt} - \sum_{tt=0}^{\tau_{pl}-1} \sum_{i=1}^{NI_{pl}} xf_{i,t+tt} \geq - (1 - yu_{pl,t}) \cdot XVY_{pl,t}^{sum,\max} \cdot XH_{pl}^{\max},$$

$$(21)$$

$$\forall pl = 1..NPL \text{ and } t = 1..NT$$

The maximum summation of variable yields on logical unit pl denoted by $XVY_{pl,t}^{sum,max}$ is determined as follows:

$$XVY_{pl,t}^{sum,max} = \max \left(\sum_{i=1}^{NI} XVY_{i,t}^{max}, \sum_{j=1}^{NJ} XVY_{j,t}^{max} \right), \tag{22}$$

$$\forall \, pl = 1..NPL \, and \, t = 1..NT$$

In specific circumstances batch processes may also be VSVT instead of either VSFT or FSVT. The approach taken in this chapter is to model the VSVT batch process as a pool with one inlet and one outlet port with constraints that will be shown in Section 3. This scenario has been addressed by [9]. The disadvantage of this approach is that multiple inlet and outlet ports cannot be supported. The poor man's alternative to handle the general VSVT is to discretize the batch sizes and to create extra mode operations for each incremental batch size. The most accurate approach but the more complex and most difficult to solve numerically is of course to linearize the term $xh_{pl,t}.yu_{pl,t}$ where there are both a variable batch size and a variable start-up event. The derivation of this is left as an exercise for the reader. Finally, another detail which we will not elaborate on is the issue of varying yields which can either increase or decrease over the relative time that unit or unit operation has been started up. In [10] and [11] informative discussions on this topic are provided with applications found in olefin plants where coking degrades the production of ethylene and in heat exchanger networks where fouling deteriorates the transfer of heat.

2.3 Solution Profile

The solution profile for a batch process with SFD of zero and SDD equal to the batch uptime minus one can be seen in Figure 4a. The entire batch process lasts for six time periods where material enters the unit in the first time period is held for four time periods and leaves the unit during the sixth time period.

A batch process with a SFD of one and a SDD equal to the batch uptime minus three can be seen in Figure 4b. It is important to recall that the

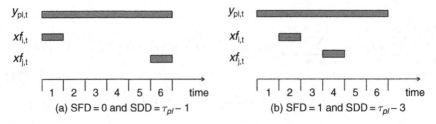

Fig. 4. Solution profiles for batch-processes

formulation presented in Section 2.2 allows for multiple inlet and outlet ports on batch processes where each inlet or outlet port can have a different value for SFD or SDD, respectively.

3 Pools

In the process industries there is always the need to store material or any other type of resource. Essentially two production protocols exist that differentiate between the methods of fulfilling customer orders in the process industries and they are make-to-stock (MTS) and make-to-order (MTO) production systems [1] of which a single plant can have both systems in place. In MTS, customer orders can be satisfied directly from inventory which implies the storage of finished products. This production notion is also known as a "push" system and it is often used when the final products are fungible goods with a relatively long shelf life such as the final products of petroleum refineries. MTO production systems on the other hand can usually not satisfy all of the customer orders from inventory directly. In this case, unfinished products are stored and must be specifically processed in order to meet the requirements of the customer orders usually requiring limited finished product inventory. Rarely in the process industries do MTS or MTO exist solely by themselves. More commonly, process industry plants are a mixture of both, with MTS being involved in the initial production stages and MTO in the final stages of production.

Pools are arguably the most flexible type of unit which can store resources indefinitely. These units may also incorporate additional details with respect to their material operations[2] such as starting-up-when-above (SUWA) a specific lot size or inventory value, shutting-down-when-above (SDWA), starting-up-when-below (SUWB) and shutting-down-when-below (SDWB). The SUWA constraint for instance ensures that the material operation on the pool unit is only allowed to start up when the lot size in the pool is above a certain upper threshold value. SUWB on the other hand only allows the pool to start up a different material-operation when the lot size is below a certain lower threshold value. Since these constraints link binary or logic variables (i.e., start-ups and shutdowns) with continuous variables (i.e., lot size) they are known as logistics implication inequalities [12]. The illustration of SUWA and SUWB for two material operations on a pool can be seen in the Figure 5.

In Figure 5 the pool is initially in material operation $y_{2,t}$. At time period 3, the lot size exceeds the SUWA threshold for logical unit 1 and the pool is thus allowed to start up material operation 1 which it only does in time period

---

[2] Note that we use mode operations for batch processes and material operations for pools. Batch processes can have one or more different types of resources or materials filling and drawing, whereas a pool has only one type of resource that can be fed or drawn, and hence the reference to it as a "pool."

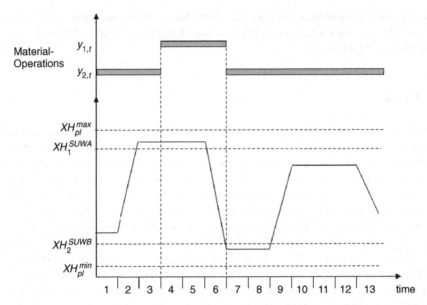

Fig. 5. Illustration of SUWA and SUWB implication constraints on pools

4 (i.e., $y_{1,4} = 1$). The pool remains in material operation 1 until time period 7 when the lot size is below the SUWB threshold for material operation 2 and then the unit is able to start up material operation 2.

In addition to the material operation logic variables, logical pool units also have a second logic variable for the suboperations of filling and drawing. If the pool is only allowed to fill or draw, it is said to be in "standing gauge" or "dead tank" policy. Conversely if the pool can fill and draw simultaneously this is referred to as "running gauge" or "live tank" policy, in which the sub-operation logic variables are not required. A more detailed requirement of standing gauge pools is the operational procedure that once a pool starts being filled it must be filled up to at least a certain upper threshold lot size before material can start to be drawn from it. There is also the corresponding requirement for the drawing suboperation. Once material starts being drawn from the pool the lot size must reach a specified lower threshold inventory amount before the pool is allowed to be filled again. These specifications entail the creation of additional pool suboperation logic constraints referred to as fill-to-full (FTF) and draw-to-empty (DTE), respectively. The illustration of FTF and DTE is shown below.

In Figure 6 the pool starts being filled in time-period 2. Until the lot size reaches the FTF threshold value of XH_{pl}^{FTF} (which happens in time period 5) material cannot be drawn from it. Once this threshold is reached any suboperation (filling or drawing) is allowed. More material is added to the pool in time period 6. In time period 7 the drawing suboperation starts.

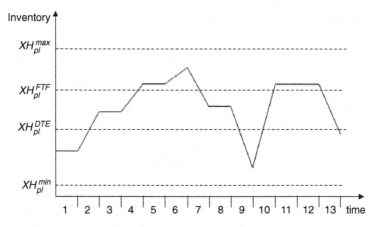

Fig. 6. Illustration of FTF and DTE suboperations on pools

Therefore the pool may only start filling again when the lot size is below the DTE threshold value of XH_{pl}^{DTE} which occurs in time period 10. The pool is filled in time period 10 to a value above XH_{pl}^{FTF}, thus enabling the drawing sub-operation which occurs in time period 13.

Standing gauge pools may also require what is known as a settling, certification or mixing time (hold time) before the material is drawn. This is known as the fill-draw delay (FDD). An example of this is when a crude oil arrives at a petroleum refinery from a ship where the crude oil may be contaminated with such unwanted components as saltwater and bulk sediment requiring settling (6 to 24-hours) before the desalting process[3]. Similarly there can also be a draw-fill delay (DFD) which specifies the time between the shutdown of the last draw from the pool and the start-up of a new fill suboperation. The fill-draw and draw-fill delays on a pool are described in Figure 7.

In the following sections the quantity, logic and logistics balances are presented for pool units with more details also found in [9].

3.1 Quantity Balances

The overall quantity balance on a pool is represented by the following equation which is known as the inventory or holdup balance constraint.

$$xh_{pp,t} = xh_{pp,t-1} + \sum_{pl=1}^{NPL}\sum_{i=1}^{NI_{pl}} xf_{i,t} - \sum_{pl=1}^{NPL}\sum_{j=1}^{NJ_{pl}} xf_{j,t} \forall \quad t = 1..NT \qquad (23)$$

---

[3] Note that we are assuming perfect mixing inside the pools. This also assumes that the material operation defines a homogeneous mixture with no component or property quality gradients.

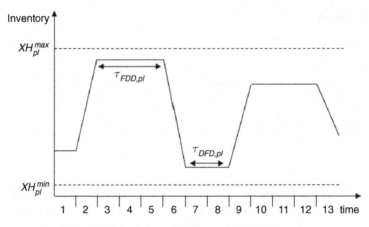

Fig. 7. Illustration of FDD and DFD on pools

This equality constraint manages the indefinite carryover or intertemporal transfer of quantity from one time period to the next. However, what makes it somewhat complicated is the fact that the physical pool unit inventory or lot size is modeled and balanced instead of the individual logical pool unit inventory. This is done in order to manage the switch over from one material operation to the other and to properly account for the heel[4] inventory during the transition. That is, when the next material operation starts up it will start with the same amount of inventory from the previous material operation and matter is neither created nor destroyed.

It should be mentioned that in equation (23) we have modeled only positive inventory (holdup) although there is also negative inventory (hold-down) which we are tacitly assuming to be zero. Positive inventory can also be likened to "forelog" in the same way negative inventory is well known to be referred to as "backlog" or "back orders" where the base word "log" implies accumulation. Actual or total inventory is then equal to the holdup minus the hold-down where both holdup and hold-down are nonnegative entities at least in terms of the MILP modeling.

3.2 Logic Balances

Pools are always assumed to be in some material service or material operation at any given time since there must always be memory of the previous material operation in order to accurately manage the inventory in any time period. This is modeled as the single-use constraint shown below:

$$\sum_{pl=1}^{NPL} y_{pl,t} = 1, \forall \quad t = 1..NT \tag{24}$$

---

[4] "Heel" inventory is commonly referred to as the residual holdup in the pool unit at time period.

The fill-draw delay (FDD) constraints relate to the timing between the last filling suboperation and the following drawing suboperation (Figure 7). There are two constraints that model FDD. The first one is a lower bound which indicates the minimum duration between the end or shutdown of the filling suboperation and the start-up of the drawing suboperation. In the following constraints i and j indicate the specific inlet and outlet ports attached to the units that are downstream and upstream of the logical pool unit, respectively. The other ipl and jpl subscripts refer to the outlet and inlet port on the logical pool unit itself.

$$y_{j,ipl,t} + y_{jpl,i,t+tt} \leq 1, \forall \quad tt = 0.. \, \tau^{\min}_{FDD,pl}, t = 1..NT | \, t + tt \leq NT \quad (25)$$

Constraint (25) stipulates that a draw cannot occur until after the lower FDD ellapses. There is also an upper bound of FDD which indicates the maximum duration between the last filling and the following drawing of resource out of the logical pool as:

$$y_{j,ipl,t-1} - y_{j,ipl,t} - \sum_{tt=1}^{\tau^{\max}_{FDD,pl}} y_{jpl,i,t+tt} \leq 0, \forall \ t = 1..NT - \tau^{\max}_{FDD,pl} | \, t + tt \leq NT$$
$$(26)$$

There is an equivalent set of constraints for the draw-fill delay (DFD). In this case the constraints refer to the time between the shutdown of the drawing suboperation and the start-up of the filling suboperation. The equations to model the lower and upper bounds on DFD can be seen below:

$$y_{jpl,i,t} + y_{j,ipl,t+tt} \leq 1, \forall \quad tt = 0.. \, \tau^{\min}_{DFD,pl}, t = 1..NT | \, t + tt \leq NT \quad (27)$$

$$y_{jpl,i,t-1} - y_{jpl,i,t} - \sum_{tt=1}^{\tau^{\max}_{DFD,pl}} y_{j,ipl,t+tt} \leq 0, \forall \ t = 1..NT - \tau^{\max}_{DFD,pl} | \, t + tt \leq NT$$
$$(28)$$

3.3 Logistics Balances

As previously mentioned the constraint start-up-when-above (SUWA) only allows a material operation on the pool to start-up when the lot size is above an upper threshold value and can be modeled as follows:

$$XH^{SUWA}_{pl} - xh_{pp,t-1} - XH^{\max}_{pl} + XH^{\max}_{pl} \cdot (y_{pl,t} - y_{pl,t-1}) \leq 0, \quad (29)$$

$$\forall \, pl = 1..NPL \, and \, t = 1..NT$$

The start-up-when-below (SUWB) constraint only allows the pool to shut down a material operation when the lot size is below a lower threshold value.

$$xh_{pp,t-1} - XH^{SUWB}_{pl} - XH^{\max}_{pl} + XH^{\max}_{pl} \cdot (y_{pl,t} - y_{pl,t-1}) \leq 0, \quad (30)$$

$$\forall \, pl = 1..NPL \, and \, t = 1..NT$$

The shut-down-when-above (SDWA) constraint only allows the pool to shut down a material operation when the lot size is above an upper threshold value.

$$XH_{pl}^{SDWA} - xh_{pp,t-1} - XH_{pl}^{\max} + XH_{pl}^{\max} \cdot (y_{pl,t-1} - y_{pl,t}) \leq 0, \qquad (31)$$

$$\forall\, pl = 1..NPL \text{ and } t = 1..NT$$

The shut-down-when-below (SDWB) constraint only allows the pool to shut down a material operation when the lot size is below a lower threshold value.

$$xh_{pp,t-1} - XH_{pl}^{SDWB} - XH_{pl}^{\max} + XH_{pl}^{\max} \cdot (y_{pl,t-1} - y_{pl,t}) \leq 0, \qquad (32)$$

$$\forall\, pl = 1..NPL \text{ and } t = 1..NT$$

In order to fulfill the requirement of fill-to-full (FTF) and draw-to-empty (DTE) on a pool additional logic variables $y_{pl,t}^{FD}$ representing the filling and drawing suboperations must be used which are equal to zero if the pool is filling and one if it is drawing[5].

$$xh_{pp,t} + \sum_{pl=1}^{NPL} \left(-XH_{pl}^{FTF} \cdot y_{pl,t} + XH_{pl}^{\max} \cdot y_{pl,t} - XH_{pl}^{\max} \cdot \left(y_{pl,t}^{FD} - y_{pl,t-1}^{FD} \right) \right) \geq 0$$

$$(33)$$

$$\forall\, t = 1..NT$$

Similarly the DTE constraint is represented by:

$$xh_{pp,t} - \sum_{pl=1}^{NPL} \left(XH_{pl}^{DTE} \cdot y_{pl,t} + XH_{pl}^{\max} \cdot y_{pl,t} - XH_{pl}^{\max} \cdot \left(y_{pl,t-1}^{FD} - y_{pl,t}^{FD} \right) \right) \leq 0$$

$$(34)$$

$$\forall\, t = 1..NT$$

3.4 Solution Profile

The solution profile for a pool is shown below. In Figure 8 the pool is filled in time period 2 and product is drawn at time period 6. Another filling suboperation occurs in time period 9 followed by a draw suboperation in time period 11. Note that both the amount of material filled and drawn to/from the pool and the duration of the hold time in the pool are variable forming two different fill-hold-draw cycles.

---

[5] Note that we avoid the use of two logic variables i.e., one for filling and one for drawing by making the logic variable represent both filling (0) and drawing (1).

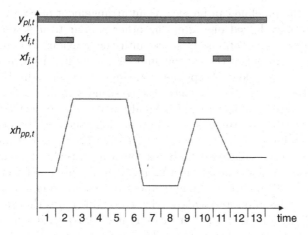

Fig. 8. Solution profile for a pool

4 Pipelines

Pipelines are multiproduct[6] transportation or distribution units that obey a first-in/first-out (FIFO) or plug-flow/hold-up pattern, i.e., the first material to be introduced into the pipeline is the first one to leave the pipeline. This behavior is illustrated in Figure 9 where the pipeline is modeled as equisized "pipes" laid out in series with the pipeline always being filled with at least one product at all times.

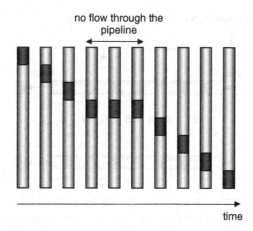

Fig. 9. First-in, first-out (FIFO) pipeline

---

[6] Multiproduct pipelines are also referred to as segregated pipelines. Pipelines that move essentially the same or only one type of material are called fungible pipelines.

In spirit, the pipeline can be seen as an arrangement of FSVT batch processes sequentially linked one after the other similar to a flow shop where each batch process maintains its stock until the previous unit in the series forces the material out of the current unit. A comprehensive overview of the state-of-the-art technology of pipeline modeling can be found in [13]. Rejowski and Pinto [14] proposed a complete discrete-time formulation for modeling pipelines and later published valid inequalities or cuts to decrease the computation time [15]. Continuous-time formulations such as the one found in [16] and also adopted by [13] may be faster than the discrete-time approach when tailored to the same case study without time-based orders, i.e., without specific release and due dates. Unfortunately continuous-time formulations are known to have difficulty in handling temporally located orders such as pipeline flow rate schedules and intermediate release and due dates that lie within the decision-making horizon but not at its extremes. Moreover in continuous-time approaches, downtime (when pipeline flow rate equals zero) on the pipeline is incorporated as additional time duration of a time slot or interval which makes it virtually impossible for the user to determine the exact time in which the downtime occurs within a time slot. Due to the practical needs of considering exogenous time-based information a discrete-time approach is preferred. In this chapter essentially the same formulation of [14] and [17] is presented where an exogenous (variable[7] but specified) pipeline flow rate schedule is provided.

A schematic of a pipeline with the notation used in this chapter can be seen below. In Figure 10 one physical pipeline pp exists and is discretized into eight pipes which are assumed to be uniform sections of the pipeline with respect to the material it holds. Ideally each pipe ppp should hold the greatest common

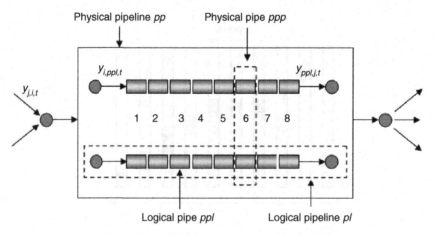

Fig. 10. Schematic of a two-product pipeline

---

[7] Variable here means a two-valued flowrate which can be either zero or some constant nonzero value.

factor (GCF) between all possible nomination amounts which will be some appropriate GCF of the minimum sizes for all products that can be carried by the pipeline. A nomination is a request from a company for transportation service on a common or third-party pipeline during a decision-making time horizon and it is a 3-tuple of pipeline, product or product type, amount, origin (source or lifting point) and destination (sink or delivery point) (cf. [16]). In the illustration below two different products can be in the same pipeline, but not at the same time in the same physical pipe, which obviously leads to the modeling of two material operations per pipe.

The dotted line box in the figure encapsulating two shaded boxes represents the physical pipe containing two logical pipes (one logical pipe for each product) where each logical pipe ppl belongs to a logical pipeline denoted as pl.

4.1 Quantity Balance

There are really no equations for quantity-only balancing of a pipeline given that the FIFO inventory is managed essentially using logic constraints and variables. The reason for this is based on the two fundamental pipeline assumptions that the pipeline is always kept full and that there is only a two-valued pipeline flow rate profile over the decision-making horizon.

4.2 Logic Balances

For pipelines, two sets of constraints are needed which model the overall setup operation of the pipeline called $y_{pp,t}$ based on the moves of material into the pipeline[8]. The logic variables $y_{i,ppl,t}$ and $y_{ppl,j,t}$ refer to the movement from the inlet port into the first pipe of the pipeline and from the last pipe of the pipeline to the outlet port, respectively, as shown in Figure 10. The following constraint states that the moves entering a particular pipeline for a particular product must be no more than one. The logic variables $y_{i,ppl,t}$ and $y_{pp,t}$ do not have to be explicitly declared as binary. However $y_{pp,t}$ must also be declared as being less than or equal to one to enforce single use on the pipeline unit. In the following equations, the notation "$ppl \in pplfirst$" indicates that the constraint refers to the first logical pipe which is connected to an inlet port. Equivalently, "$ppl \in ppllast$" indicates a constraint that refers the last logical pipe which is connected to an outlet port.

$$\sum_{i=1}^{NI} \sum_{\substack{ppl = 1 \\ ppl \in pplfirst}}^{NPPL} y_{i,ppl,t} = y_{pp,t}, \forall\, t = 1..NT \tag{35}$$

$$y_{pp,t} \leq 1 \quad , \forall\, t = 1..NT \tag{36}$$

---

[8] Note that because the pipeline is always full we do not need explicitly the flow out indication on the pipeline.

The same needs to be done for the outlet of the pipeline in order to determine the flow out of the pipeline for a particular active logical pipe (for a particular product).

$$\sum_{j=1}^{NJ} \sum_{\substack{ppl=1 \\ ppl \in ppllast}}^{NPPL} y_{ppl,j,t} = y_{pp,t}, \forall\, t = 1..NT \tag{37}$$

Note that equations (35) and (37) can be aggregated equivalently into one of the following two constraints given that when there is flow in, there is always flow out:

$$\sum_{\substack{ppl=1 \\ ppl \in pplfirst}}^{NPPL} \sum_{i=1}^{NI} y_{i,ppl,t} + \sum_{\substack{ppl=1 \\ ppl \in ppllast}}^{NPPL} \sum_{j=1}^{NJ} y_{ppl,j,t} = 2y_{pp,t} \tag{38}$$

$$\forall\, t = 1..NT$$

$$\sum_{\substack{ppl=1 \\ ppl \in pplfirst}}^{NPPL} \sum_{i=1}^{NI} y_{i,ppl,t} = \sum_{\substack{ppl=1 \\ ppl \in ppllast}}^{NPPL} \sum_{j=1}^{NJ} y_{ppl,j,t}, \tag{39}$$

$$\forall\, t = 1..NT$$

The following constraint allows the determination of whether any moves occur on an inlet port to a pipeline from anywhere in the production network. Note that there is a one-to-one mapping between the first logical pipe and an inlet port meaning that the first logical pipe can be connected to a maximum of one inlet port.

$$y_{i,ppl,t} \geq y_{j,i,t}, \forall\; t = 1..NT \text{ and } ppl \in pplfirst \tag{40}$$

In equation (40) j refers to the outlet port of any unit upstream of the pipeline and i is the inlet port of the logical pipeline. The equivalent constraint is also added for the outlet port:

$$y_{ppl,j,t} \geq y_{j,i,t}, \forall\; t = 1..NT \text{ and } ppl \in ppllast \tag{41}$$

In equation (41) j refers to the outlet port of the pipeline and i refers to the inlet port of any unit downstream of the pipeline. The following constraint states that if there is an active flow on the inlet port to the logical pipe, the logical pipe that is directly connected to the inlet port must be activated to match the material that is flowing into the pipeline.

$$y_{i,ppl,t} \leq y_{ppl,t}, \forall\; t = 1..NT \text{ and } ppl \in pplfirst \tag{42}$$

The single-use constraint below for the physical pipes states that only one logical pipe can be active at a time within the physical pipe. This is an equality constraint because the pipeline is always filled. Note that $ppl \in ppp$ implies the logical pipes ppl belong to one and only one physical pipe ppp.

$$\sum_{\substack{ppl = 1 \\ ppl \in ppp}}^{NPPL} y_{ppl,t} = 1, \forall\, t = 1..NT \tag{43}$$

The propagation or chain effect for the modeling of a pipeline with pipes is now described – this models the intertemporal and interspatial transfer of material in time and in space, respectively. If the pipeline is active, then there is transfer of material from each logical pipe to the adjacent logical pipe in the next time period for all logical pipes. When the pipeline is down, i.e., the flow rate is zero, only intertemporal transfer of quantity and logic exists; however, when the pipeline is up, both intertemporal and interspatial transfer of quantity and logic exist. Intertemporal transfer is needed due to the material hold up in the pipes.

$$y_{ppl,t-1} - y_{ppl',t} \leq 1 - y_{pp,t}, \forall\, ppl, ppl' = 1..NPPL, t = 1..NT, \tag{44}$$

$$pp = 1..NPP \text{ where } ppl \text{ and } ppl' \text{ are adjacent and connected}$$

The following constraint is required for transferring material to the outlet ports from the last logical pipe when the pipeline is active.

$$y_{ppl,t-1} - y_{ppl,j,t} \leq 1 - y_{pp,t}, \forall\ \ t = 1..NT, pp = 1..NPP, ppl \in ppllast \tag{45}$$

If the pipeline is inactive, then there must be memory of the previous logical pipe from one time period to the next; this corresponds to the intertemporal transfer of logic.

$$y_{ppl,t-1} - y_{ppl,t} \leq y_{pp,t}, \forall\ \ ppl = 1..NPPL, pp = 1..NPP, t = 1..NT \tag{46}$$

4.3 Logistics Balances

The only logistics balance constraint needed to complete the modeling of the pipeline is the calculation of the flow that leaves the pipeline. Parameter XFR_{pp} is defined as the flow rate on the physical pipeline pp per time period which can be calculated as the nominal flow rate of the pipeline times the time period duration.

$$xf_{j,t} = XFR_{pp} \cdot y_{ppl,j,t}, \ \forall\ \ j = 1..NJ, t = 1..NT, ppl \in ppllast \tag{47}$$

This equality constraint in fact models the bilinear relation: $xf_{j,t} = XFR_{pp} \cdot y_{ppl,t} \cdot y_{pp,t}$, where $y_{ppl,t}$ is the setup of the last logical pipe. If the pipeline is flowing ($y_{pp,t} = 1$) and the material in the last pipe is say product A ($y_{ppl,t} = 1$), then the flow rate $xf_{j,t}$ will equal XFR_{pp} with respect to product A flowing out. However, if any of these conditions do not hold, e.g., the pipeline is down and/or there is no material A in the last logical pipe, no flow of material A leaves the pipeline.

Constraint (47) also implies that the size of each pipe corresponds exactly to the flow rate in the pipeline over each time period. Multiple fixed and pre-determined flow rate values in the pipeline (i.e., a multivalued flow rate) can be handled using this discrete-time formulation with modification to some of the constraint sets described here. This is achieved by adding an extra index set to specific constraints where this index or subscript relates to the multi-rate. The only restriction on a multirate pipeline is that the flow rates must be a multiple of the pipe size which is fixed. Therefore the pipe size should be the greatest common factor between all flow rate values.

Sequence-dependent changeover or switchover considerations are needed in most pipeline applications due to segregation and transmixing restrictions between different materials or products. Sequence-dependent switchover constraints, however, do not need to be added to the pipeline directly. They may be applied to a continuous "sequencer" unit upstream of the pipeline where this unit contains the sequence-dependent details and can be modeled using the techniques found in for example [3], [18], [19] and [20].

For some pipelines minimum nominations (i.e., batch or block sizes) must be respected. When the smallest nomination is greater than the pipe size but is also a multiple of it (e.g., smallest nomination is 10,000 cubic meters and the pipe size = 1,000 cubic meters), then it is the responsibility of movement of material upstream of the pipeline to push the minimum nomination size to the pipeline. Once at the pipeline, the FIFO model described in this section will properly account for the product when it leaves the pipeline at an indefinite amount of time into the future[9].

It should be mentioned that special-ordered-sets one (SOS1) may be used in equations (35) and (43) for example where a higher priority or directive may be applied on the SOS1 rather than the individual members of the set. This approach works even if the members of these equations are not explicitly declared as binary given that SOS1 ensures that only one of the members is non-zero and not necessarily unity [21]. This strategy increases the solution speed at the expense of adding more global entities to the MILP problem. Decisions or logic variables relating to shorter-term time periods should have a higher priority than the remaining variables due to their impact on longer-term decisions.

In addition some pipelines may require a minimum downtime. This means that if the pipeline is shut down or inoperable, then it must remain in this way for a minimum number of time periods. Including this consideration has been observed to significantly reduce solution time given that some degeneracy

---

[9] The indefiniteness here is caused by the fact that the pipeline can have a zero flow rate during some time periods.

is removed from the problem. A related but opposite notion is the minimum number of time periods the pipeline should be in operation (uptime) to satisfy product demands. See [20] for details of how to model downtime in discrete time. A simple technique to aid in the determination of a potential amount of downtime is to model a scenario in which the pipeline only has one pipe and must always be in operation ($y_{pp,t} = 1$ for all time periods) using a simple time delay to determine the FIFO product pattern leaving the pipeline at any time. This scenario is solved with quantity penalties (i.e., artificial or elastic variables) on all of the up- and downstream material balances in and out of the pipeline. Any penalties at the solution are added over the time horizon and divided by the flow per time period. The number of time periods obtained from this calculation may be used either as (1) a theoretical number of time periods during which the pipeline should be shut down and can be specified as a lower bound on the downtime or (2) to limit the number of time periods in which $y_{pp,t}$ can be active.

It is important to note that one of the key features of the pipeline scheduling problem modeling is the handling of the interface between different materials since incompatible fluids should not be injected into the pipeline so that they are adjacent to each other. An improved handling of sequence-dependent switch-over constraints in discrete-time scheduling problems is presented by [18] and can be used to address this issue.

4.4 Solution Profile

The solution profile for a pipeline with eight pipes is shown below. In Figure 11 it is possible to see the flow in and out of the pipeline ($xf_{i,t}$ and $xf_{j,t}$, respectively) for the two different products (gray and black) and the material operations of each one of the eight pipes ($y_{ppp,t}$).

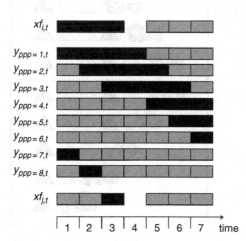

Fig. 11. Solution profile for a pipeline containing product 1 (gray) and product 2 (black)

As product 2 enters the pipeline in time periods 1, 2 and 3, two pipes of product 1 and one pipe of product 2 are displaced out of the pipeline. At time period 4 there is no flow entering or leaving the pipeline and consequently there are no changes in the material operations of the pipes for time periods 3 and 4. Pipeline activity resumes in time period 5 starting from the previous pipe material operation in time period 4 and so on.

5 PileLines

In some process industries solid product is piled in large geographical areas; for example, in the storage of ore in the mining industry. Once two or more piles have been built or filled, it is not physically possible to reclaim or draw the pile that was built first before removing the pile that was built last. This constitutes the last-in/first-out (LIFO) inventory protocol which is illustrated in Figure 12. An example of a LIFO system is a pileline which is a combination of two or more piles which can store different products but not at the same time. This chapter presents a novel MILP formulation for LIFO systems.

The structure of a pileline is somewhat similar to the pipeline although there are significant differences in the equations that represent each one. The structure of a pileline can be seen in Figure 13. In this example the pileline can store a maximum of five physical piles *ppp* (corresponding to the five columns of piles in Figure 13). There are two logical piles *ppl* in each physical

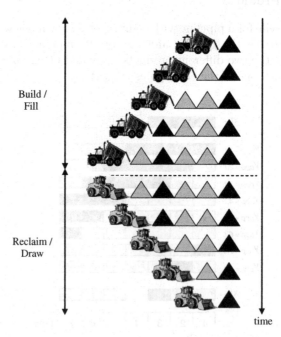

Fig. 12. Last-in, first-out (LIFO) pileline

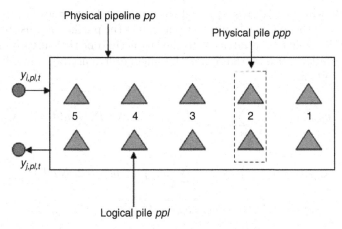

Fig. 13. Schematic of a pileline

pile *ppp* (corresponding to the two rows of piles in Figure 13 where each row represents a different product) and there is only one inlet and one outlet port to and from the pileline both located at the front of pileline.

Unlike the pipeline, the pileline does not have to be full at every time period. Building, storing (holding) and reclaiming piles can take place in any order that is consistent with the LIFO arrangement. In Figure 13 if there is product in pile 4, all of the preceding piles (1, 2 and 3) must contain product. Also, product can only be reclaimed from pile 3 if all the succeeding piles (4 and 5) are empty. Since the pileline is also assumed to have a fixed flow rate of one pile per time period, the flows into and out of the pileline are determined by the movement logic variables $y_{i,pl,t}$ and $y_{j,pl,t}$, respectively, and consequently there are really no explicit quantity balances.

5.1 Logic Balances

If there is flow into or out of the pileline, the material operations of the logical piles must change accordingly depending on what type of material was filled or drawn. If there is no flow, then there must be temporal propagation or carryover of the material operations on the logical piles similar to logical pipes. Because the setup of the physical piles $y_{pp,t}$ are independent binary variables, integrality of the setup of the logical piles $y_{ppl,t}$ is implied and therefore these variables do not need to be explicitly declared as binary.

$$y_{ppl,t} - y_{ppl,t-1} \leq y_{i,pl,t} \, , \tag{48}$$

$\forall \, ppl$ connected to $pl, ppl = 1..NPPL$ and $t = 1..NT$

$$y_{ppl,t-1} - y_{ppl,t} \leq y_{j,pl,t} \, , \tag{49}$$

$\forall \, ppl$ connected to $pl, ppl = 1..NPPL$ and $t = 1..NT$

The following constraints represent the sequence dependency or precedence restrictions on the pileline. If any material enters the pileline, it must be placed in the physical pile that is adjacent to the last active one that already contains product. Similarly if material leaves the pileline, it must be drawn from the last active pile first.

$$y_{ppp,t} - y_{ppp-1,t} \leq 1 - \sum_{pl=1}^{NPL} y_{i,pl,t}, \forall\, ppp = 2..NPPP \text{ and } t = 1..NT \qquad (50)$$

$$y_{ppp',t} - y_{ppp'-1,t} \leq 1 - \sum_{pl=1}^{NPL} y_{j,pl,t}, \forall\, ppp = 2..NPPP, \qquad (51)$$

$$ppp' = NPPP - ppp, (ppp' - 1) \geq 1 \text{ and } t = 1..NT$$

Besides associating the physical piles with one or more logical piles, the following constraint also enforces single use on the piles since no more than one material operation or product can exist in a physical pile at any time period given that the setup variable of the physical piles $y_{ppp,t}$ is a binary variable.

$$\sum_{ppl=1}^{NPPL} y_{ppl,t} = y_{ppp,t}, \qquad (52)$$

$$\forall\, ppl \text{ connected to } ppp, \ ppp = 1..NPPP \text{ and } t = 1..NT$$

5.2 Logistics Balances

The overall quantity balance on a pileline at any given time period is calculated as the amount of material in the pileline in the previous time period plus the amount of material that comes in minus the amount of material that is removed from the pileline.

$$xh_{pp,t} = xh_{pp,t-1} + XFR_{pp} \cdot \left(\sum_{pl=1}^{NPL} y_{i,pl,t} - \sum_{pl=1}^{NPL} y_{j,pl,t} \right), \qquad (53)$$

$$\forall\, pp = 1..NPP, t = 1..NT$$

The following constraint manages the number of active physical piles. If no material comes in or out of the pileline, the number of active physical piles must match the one in the previous time period. If material enters or leaves the pileline, the number of active physical piles must change accordingly.

$$\sum_{ppp=1}^{NPPP} y_{ppp,t} \cdot XFR_{pp} - xh_{pp,t-1} = XFR_{pp} \cdot \left(\sum_{pl=1}^{NPL} y_{i,pl,t} - \sum_{pl=1}^{NPL} y_{j,pl,t} \right), \qquad (54)$$

$$\forall\, pp = 1..NPP, t = 1..NT$$

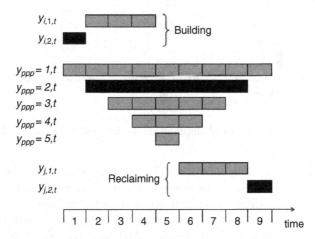

Fig. 14. Solution profile for a pileline containing product 1 (gray) and product 2 (black)

5.3 Solution Profile

A typical solution profile for the five-pile two-product pileline of Figure 13 can be seen in Figure 14.

Initially pile 1 contains product 1 ($y_{ppl,t} = y_{1,1} = 1$). As product 2 is introduced in the pileline ($y_{i,pl,t} = y_{i,2,1} = 1$) it is placed on logical pile 2 while pile 1 still contains product 1. Three piles of product 1 are then built in time periods 2, 3 and 4 and are placed in logical piles 3, 4 and 5, respectively. The reclaiming of product from time period 6 to time period 9 is done in the opposite sequence as in the building of the pileline: instead of one pile of product 2 and three piles of product 1 (as in the building phase), three piles of product 1 and one of product 2 are reclaimed. Note that the initial inventory of the pileline is taken into account (in this example, product 1 in logical pile 1) and that the pileline does not need to be completely empty at the last time period (one pile of product one remained in the pileline in time period 9).

6 Parcels

A parcel unit models the transportation or distribution of material from one point or location to another. There are three main types of shipping models: industrial, liner and tramp shipping. In industrial shipping the vessel is loaded at a single point or dock[10] and has a fixed route to the destination docks where the material or cargo is unloaded (Figure 15). The vessel must then return to

---

[10] Dock in this context is also synonymous with wharf, berth and jetty where a collection of docks can be considered as a harbor or port.

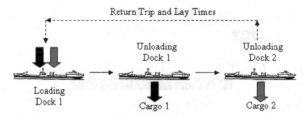

Fig. 15. Schematic of industrial shipping model using a parcel unit with multiple cargos

the original loading or source dock which is known as the return trip. In liner shipping the vessel also has a fixed route which has to be followed. However, this shipping model differs from industrial shipping in that the cargo may be loaded to and/or unloaded from the vessel at any dock. An example of liner shipping is a bus where its route is fixed and passengers may embark or disembark the bus at any bus stop. Tramp shipping does not assume a fixed route for the vessel and may be loaded or unloaded at any dock similar to the workings of a taxi cab where a dispatcher determines the next pick-up and drop-off locations. Given that the focus of this chapter is process industry specific, where liner and tramp shipping are rare, only the modeling of industrial shipping is highlighted.

Essentially, a parcel unit can be represented as a variable-size fixed-time batch process unit. In the case of parcels, hold time can be referred to as haul time given that during the hold, the parcel unit is traveling from one location point to another while holding the cargos until they are ready to be unloaded. A particularity of parcel units is that there is only one logic variable $y_{pl,t}$ for the set up of the logical parcel unit at each time period relating to one or more different cargos of varying sizes.

In terms of how the cargo inventories are managed, the parcel unit fills (lifts or loads) and draws (delivers or unloads) material subject to the same logic setup variable for a particular mode operation. These different mode operations for the parcel unit are required when, for the same set of cargos, different shipping routes can be used to lift, deliver and return to the source dock. Within the haul time of the logical parcel unit the different logical cargos can be filled and drawn independently using the same notion of start-to-fill (SFD) and start-to-draw (SDD) delays on inlet and outlet ports, respectively, that were presented in Section 2.2. The actual load balancing of the ship, barge or tanker to maintain the proper draught and tilt, usually requiring the management of ballast water, is considered as a near-term operating detail and is proxied through the mode operation logic by incorporating them into the specific SFD, SDD and the uptime[11] parameters.

---

[11] Uptime is the collection of all of the loading, unloading, hauling, lay and return trip times.

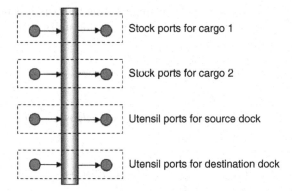

Fig. 16. Inlet and outlet ports on a parcel

Each physical parcel unit must have the same number of inlet and matching outlet ports with each matching set of ports representing one cargo or dock. Different mode operations on the parcel can also be used to model different cargo-size lower and upper limits. For example, this may be required when a different shipping route with the same source and destination docks requires more or less cargo given the available fuel capacity of the parcel unit. A vessel with two cargos with matching inlet and outlet stock ports is shown in Figure 16. This structure enables the representation of different SFD and SDD for each cargo. One inlet and one outlet utensil port are also required whenever there is the need to manage the docks at a source or destination point given that only one physical parcel unit can be docked at a time.

6.1 Logic Balances

The logic balances for parcel units are identical to those of fixed-time batch processes (equations (1)-(4), Section 2.1) and will therefore not be repeated. The main difference between a parcel and a batch process is that the former can be seen as a collection of several batch processes (corresponding to the cargos or docks) that share the same variable $y_{pl,t}$ for the setup of the logical parcel. The parcel in Figure 16 for instance could be interpreted as a combination of four batch processes grouped together with one mode operation logic setup variable.

6.2 Logistics Balances

The logistics balances for parcel units are a simplification of the ones for variable-size and fixed-time batch processes except that the inlet and outlet ports have to be matched to each one of the logical cargos cl. In the following equations the notation $i \Rightarrow cl$ indicates that inlet port i is associated with cargo cl. Since the cargo sizes do not change during the trip, it is not necessary to consider yields given that these are defined to be unity.

$$xf_{i,t} \geq \sum_{tt=\tau_{SFD,i}^{\min}}^{\tau_{SFD,i}^{\max}} XH_{cl,pl}^{\min} \cdot yu_{pl,t-tt}, \tag{55}$$

$\forall \, cl = 1..NCL, pl = 1..NPL$ and $t = 1..NT$, where $i \Rightarrow cl$ and $cl \in pl$

$$xf_{i,t} \leq \sum_{tt=\tau_{SFD,i}^{\min}}^{\tau_{SFD,i}^{\max}} XH_{cl,pl}^{\max} \cdot yu_{pl,t-tt}, \tag{56}$$

$\forall \, cl = 1..NCL, pl = 1..NPL$ and $t = 1..NT$, where $i \Rightarrow cl$ and $cl \in pl$

$$xf_{j,t} \geq \sum_{tt=\tau_{SDD,j}^{\min}}^{\tau_{SDD,j}^{\max}} XH_{cl,pl}^{\min} \cdot yd_{pl,t+tt}, \tag{57}$$

$\forall \, cl = 1..NCL, pl = 1..NPL$ and $t = 1..NT$, where $j \Rightarrow cl$ and $cl \in pl$

$$xf_{j,t} \leq \sum_{tt=\tau_{SDD,j}^{\min}}^{\tau_{SDD,j}^{\max}} XH_{cl,pl}^{\max} \cdot yd_{pl,t+tt}, \tag{58}$$

$\forall \, cl = 1..NCL, pl = 1..NPL$ and $t = 1..NT$, where $j \Rightarrow cl$ and $cl \in pl$

In order to handle the return trip time the hold time of the parcel unit should be equal to the round-trip time. The SDD determines when within the parcel unit's hold time can its cargos be unloaded. This also enables the configuration of diverse unloading times for the cargos on a single parcel unit by establishing different SDD for the cargos. Similar to Section 2.2, the following equations represent the overall quantity balances for each logical parcel's cargo. This is to ensure that when the cargo sizing is variable, the loading amount and the unloading amount are identical.

$$\sum_{tt=0}^{\tau_{pl}-1} xf_{j,t+tt} - \sum_{tt=0}^{\tau_{pl}-1} xf_{i,t+tt} \leq (1 - yu_{pl,t}) \cdot XH_{cl,pl}^{\max}, \tag{59}$$

$\forall \, cl = 1..NCL, pl = 1..NPL, t = 1..NT$ and $i, j \Rightarrow cl$ and $cl \in pl$

$$\sum_{tt=0}^{\tau_{pl}-1} xf_{j,t+tt} - \sum_{tt=0}^{\tau_{pl}-1} xf_{i,t+tt} \geq -(1 - yu_{pl,t}) \cdot XH_{cl,pl}^{\max}, \tag{60}$$

$\forall \, cl = 1..NCL, pl = 1..NPL, t = 1..NT$ and $i, j \Rightarrow cl$ and $cl \in pl$

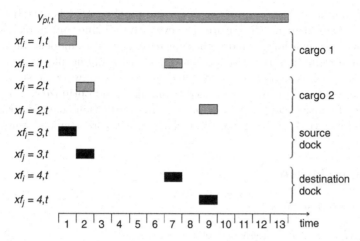

Fig. 17. Solution profile for a parcel unit with stock ports (gray) and utensil ports (black)

6.3 Solution Profile

The solution profile for a parcel unit is presented in Figure 17. In this case, the total trip time is represented by the setup of the logical parcel unit variable $y_{pl,t}$ and the two cargos are loaded onto the vessel at the source dock with the same destination. Multiple destinations can be considered by including additional inlet and outlet utensil ports for each extra destination dock when the docks are a limited resource. This is the case when cargo 1 is delivered to customer 1 (destination dock 1) and cargo 2 is delivered to customer 2 (destination dock 2) where the customers are at different locations. In Figure 17 the source dock is occupied for two time periods and occurs in time periods 1 and 2. The destination dock is occupied for three time periods (in this case from time periods 7 to 9). The first material to be loaded is cargo 1 in time period 1 and it is unloaded at the destination dock in time period 7. As for cargo 2 it is loaded in time period 2 and unloaded in time period 9. Since this is an industrial shipping model, the vessel has to return to the source dock to be loaded again if required. The return trip in this case takes four time periods and happens between time periods 10 and 13.

7 Summary

This chapter presents details on the modeling of multiproduct inventory logistics systems in the process industries suitable for simulation and optimization using MILP over some discretized decision-making horizon. The resulting MILP models consist of a combination of quantity balances, logic balances and logistics balances. Five inventory models were formulated with varying levels of complexity. Batch process, pools and pipeline models have been previously presented in the literature; however, the pileline and parcel models are original formulations. Finally, these inventory logistics models can be considered as production and distribution objects or building blocks which can be embedded, linked and interconnected to form a network or superstructure. These networks can then be used to simulate and optimize complex and integrated facilities comprising a supply chain, production chain, distribution chain and eventually a complete value chain or enterprise-wide system.

8 Nomenclature

Indices and Range-Sets

$i = 1..NI$ or $1..NI_{pl}$ – logical inlet ports

$j = 1..NJ$ or $1..NJ_{pl}$ – logical outlet ports

$pp = 1..NPP$ – physical units

$pl = 1..NPL$ – logical units

$ppp = 1..NPPP$ – physical pipe or pile units

$ppl = 1..NPPL$ – logical pipe or pile units

$cl = 1..NCL$ – logical cargos

$t = 1..NT$ – time periods

Parameters

XFR_{pp} – flow rate on physical pipeline or pileline pp

$XFY_{i,t}$ – fixed inverse yield on inlet port i

$XFY_{j,t}$ – fixed yield on outlet port j

$XVY_{i,t}^{\min}$ – minimum variable inverse yield on inlet port i

$XVY_{j,t}^{\min}$ – minimum variable yield on outlet port j

$XVY_{i,t}^{\max}$ – maximum variable inverse yield on inlet port i

$XVY_{j,t}^{\max}$ – maximum variable yield on outlet port j

$XVY_{pl,t}^{sum,\max}$ – maximum summation of inlet or outlet port variable yields on logical unit pl

$XH_{pl}^{\min}$ – minimum holdup or inventory of logical unit pl

$XH_{pl}^{\max}$ – maximum holdup or inventory of logical unit pl

XH_{pl}^{FTF} – fill-to-full lot size of logical pool unit pl

XH_{pl}^{DTE} – draw-to-empty lot size of logical pool unit pl

XH_{pl}^{SUWA} – start-up-when-above lot size of logical pool unit pl

XH_{pl}^{SUWB} – start-up-when-below lot size of logical pool unit pl

XH_{pl}^{SDWA} – shut-down-when-above lot size of logical pool unit pl

XH_{pl}^{SDWB} – shut-down-when-below lot size of logical pool unit pl

$XH_{cl,pl}^{\min}$ – minimum cargo size of logical cargo cl in logical unit pl

$XH_{cl,pl}^{\max}$ – maximum cargo size of logical cargo cl in logical unit pl

τ_{pl} – hold time for logical unit pl

$\tau_{FDD,pl}^{\min}$ – minimum fill-draw delay on logical unit pl

$\tau_{FDD,pl}^{\max}$ – maximum fill-draw delay on logical unit pl

$\tau_{DFD,pl}^{\min}$ – minimum draw-fill delay on logical unit pl

$\tau_{DFD,pl}^{\max}$ – maximum draw-fill delay on logical unit pl

$\tau_{SFD,i}^{\min}$ – minimum start-to-fill delay on inlet port i

$\tau_{SFD,i}^{\max}$ – maximum start-to-fill delay on inlet port i

$\tau_{SDD,j}^{\min}$ – minimum start-to-draw delay on outlet port j

$\tau_{SDD,j}^{\max}$ – maximum start-to-draw delay on outlet port j

Variables

$xf_{i,t}$ – flow entering the unit through port i at time period t

$xf_{j,t}$ – flow leaving the unit through port j at time period t

$xh_{pp,t}$ – holdup of physical unit pp at time period t

$xh_{pl,t}$ – holdup of logical unit pl at time period t

$y_{pl,t}$ – logic variable for the setup of logical unit pl at time period t

$y_{ppl,t}$ – logic variable for the setup of logical pipe or pile ppl at time period t

$yu_{pl,t}$ – logic variable for the start-up of logical unit pl at time period t

$yd_{pl,t}$ – logic variable for the shutdown of logical unit pl at time period t

$yw_{pl,pl,t}$ – logic variable for the switch-over-to-itself of logical unit pl at time period t

$y_{pl,t}^{FD}$ – logic variable for logical pool if in the suboperation of filling (0) or drawing (1)

$y_{j,i,t}$ – logic variable for the movement from outlet port j to inlet port i at time period t

$y_{i,pl,t}$ – logic variable for the movement to inlet port i in logical unit pl at time period t

$y_{i,ppl,t}$ – logic variable for the movement from inlet port i to logical pipe ppl at time period t

$y_{j,pl,t}$ – logic variable for the movement from outlet port j in logical unit pl at time period t

$y_{ppl,j,t}$ – logic variable for the movement from logical pipe ppl to outlet port j at time period t

References

1. Hopp W, Spearman M (2001) Factory physics. Mc-Graw Hill, New York, USA
2. Kelly JD (2005) The unit-operation-stock superstructure (UOSS) and the quantity-logic-quality paradigm (QLQP) for production scheduling in the process industries. In MISTA 2005 Conference Proceedings 327–333
3. Kondili E, Pantelides CC, Sargent RWH (1993) A general algorithm for short-term scheduling of batch operations - I MILP formulation. Comp Chem Eng 17:211–227
4. Kelly JD, Zyngier D (2007) A projectional model of data for complexity management of enterprise-wide optimization models. Submitted to Comp Chem Eng
5. Realff MJ, Shah N, Pantelides CC (1996) Simultaneous design, layout and scheduling of pipeless batch plants. Comp Chem Eng 20:869–883
6. Van den Akker JM, Hurkens CAJ, Savelsbergh MWP (2000) Time-indexed formulations for single-machine scheduling problems: Column generation. INFORMS J Comput 12:111–124
7. Lee K-H, Park HI, Lee I-B (2001) A novel nonuniform discrete time formulation for short-term scheduling of batch and continuous processes. Ind Eng Chem Res 40:4902–4911
8. Wolsey LA (1998) Integer programming. John Wiley and Sons, Ltd., New York, USA
9. Prasad P, Maravelias CT, Kelly JD (2006) Optimization of aluminum smelter casthouse operations. Ind Eng Chem Res 45:7603–7617
10. Jain V, Grossmann IE (1998) Cyclic scheduling of continuous parallel-process units with decaying performance. AIChE J 44:1623–1636
11. Georgiadis MC, Papageorgiou LG, Macchietto S (2000) Optimal cleaning policies in heat exchanger networks under rapid fouling. Ind Eng Chem Res 39:441–454
12. Savelsbergh MWP (1994) Preprocessing and probing techniques for mixed integer programming problems. ORSA J Comput 6:445–454
13. Relvas S, Matos HA, Barbosa-Povoa APFD, Fialho J, Pinheiro AS (2006) Pipeline scheduling and inventory management of a multiproduct distribution oil system. Ind Eng Chem Res 45:7841–7855
14. Rejowski R, Pinto JM (2003) Scheduling of a multiproduct pipeline system. Comp Chem Eng 27:1229–1246
15. Rejowski R, Pinto JM (2004) Efficient MILP formulations and valid cuts for multiproduct pipeline scheduling. Comp Chem Eng 28:1511–1528
16. Cafaro DC, Cerda J (2004) Optimal scheduling of multiproduct pipeline systems using a non-discrete MILP formulation. Comp Chem Eng 28:2053–2068

17. Hane CA, Ratliff HD (1995) Sequencing inputs to multi-commodity pipelines. Ann Oper Res 57:73–101
18. Kelly JD, Zyngier D (2007) An improved MILP modeling of sequence-dependent switchovers for discrete-time scheduling problems. Ind Eng Chem Res 46:4964–4973
19. Sahinidis NV, Grossmann IE (1991) MINLP model for cyclic multiproduct scheduling on continuous parallel lines. Comp Chem Eng 15:85–103
20. Wolsey LA (1997) MIP modelling of changeovers in production planning and scheduling problems. Eur J Oper Res 99:154–165
21. Williams HP (1999) Model building in mathematical programming. John Wiley and Sons, Ltd., 4th ed, West Sussex, England

Modeling and Managing Uncertainty in Process Planning and Scheduling

Marianthi Ierapetritou[1] and Zukui Li[2]

[1] Department of Chemical and Biochemical Engineering
   Rutgers University, Piscataway, NJ 08854
   `marianth@soemail.rutgers.edu`
[2] Department of Chemical and Biochemical Engineering
   Rutgers University, Piscataway, NJ 08854
   `zukui@eden.rutgers.edu`

Summary Uncertainty appears in all the different levels of the industry from the detailed process description to multisite manufacturing. The successful utilization of process models relies heavily on the ability to handle system variability. Thus modeling and managing uncertainty in process planning and scheduling has received a lot of attention in the open literature in recent years from chemical engineering and operations research communities. The purpose of this chapter is to review the main methodologies that have been developed to address the problem of uncertainty in production planning and scheduling as well as to identify the main challenges in this area. The uncertainties in process operations are first analyzed, and the different mathematical approaches that exist to describe process uncertainties are classified. Based on the different descriptions for the uncertainties, alternative planning and scheduling approaches and relevant optimization models are reviewed and discussed. Further research challenges in the field of planning and scheduling under uncertainty are identified and some new ideas are discussed.

1 Introduction

Modern industry faces major new challenges through increased global competition, greater regulatory pressures and uncertain prices of energy, raw materials and products. These competitive concerns increase the focus on integrated processes, information technology, and consideration of multiple decision criteria including profitability, flexibility, quality, and the environment. The success of any industrial sector will depend on how efficiently it generates value by dynamically optimizing deployment of its supply chain resources. Among the challenges for the dynamic optimization of the entire supply chain enterprises are the rigorous but tractable optimization of process operations and the efficient integration of different decision-making stages as well as the consideration of uncertainty and risk factors. Uncertainty

W. Chaovalitwongse et al. (eds.), *Optimization and Logistics Challenges in the Enterprise*, Springer Optimization and Its Applications 30, DOI 10.1007/978-0-387-88617-6_3, © Springer Science+Business Media, LLC 2009

appears in all the different levels of the industry from the detailed process description to multisite manufacturing. Therefore, the successful utilization of process models relies heavily on the ability to handle system variability. In this chapter, we presented an overview of the work along the directions of scheduling and planning optimization and uncertainty considerations within the decision-making process.

Both planning and scheduling problems deal with the allocation of available resources over time to perform a set of tasks required to manufacture one or more products. Production planning problems correspond to a higher level of process operation decision making since they consider longer time horizons and multiple orders that involve different operating conditions as well as unit changes, price and cost variability. Production planning determines the optimal allocation of resources within the production facility over a time horizon of a few weeks up to few months. On the other hand, scheduling provides the feasible production schedules to the plant. In an industrial process, each task requires certain amounts of specified resources for a specific time interval called the processing time. The resources include the use of equipment, the utilization of raw material or intermediates, the employment of operators, etc., and tasks involve the chemical or physical transformation of materials, transportation of products or intermediates, cleaning, and maintenance operations, etc. The purpose of scheduling is to optimally allocate limited resources to processing tasks over time. The scheduling objective can take many forms such as minimizing the time required to complete all the tasks (the make span), minimizing the number of orders completed after their committed due dates, maximizing customer satisfaction by completing orders in a timely fashion, maximizing plant throughput, maximizing profit or minimizing production costs. Scheduling decisions to be determined include the optimal sequence of tasks taking place in each unit, the amount of material being processed at each time in each unit and the processing time of each task in each unit.

Planning and scheduling can be distinguished based on various characteristics. First in terms of the considered time horizon, long-term planning problems deal with longer time horizons (e.g., months or years) and are focused on higher-level decisions such as timing and locations of additional facilities and levels of production. The area of medium-term scheduling involves medium time horizons (e.g., weeks or months) and aims to determine detailed production schedules, which can result in very large-scale problems. Moreover, short-term scheduling models address short time horizons (e.g., days or weeks) and are focused on determining detailed sequencing of various operational tasks. Second in terms of the decisions involved, short-term scheduling provides feasible production schedule considering the detailed operation conditions; while planning involves consideration of financial and business decisions over extended periods of time. Lastly considering uncertainty, short-term scheduling needs to consider the disturbing events such as rush orders, machine breakdown and attempts to absorb the impact, while the planning needs to foresee the possible changes in the future and the effects of the current decisions, thus achieving an optimal solution for the benefits of the entire planning time horizon. However, planning and scheduling decisions are always closely coupled in the practical industrial plant, which requires the integration of the decisions process for planning and scheduling.

A large number of publications are devoted to modeling and solution of the planning and scheduling problems. Early work has focused on the development of mathematical models based on discretization of a time horizon into a number of intervals of equal duration [68]. Sahinidis et al. [96] presented a multiperiod mixed integer linear programming (MILP) model for long-range planning. In order to reduce the computational expense, several strategies were investigated in their work, including branch and bound, the use of integer cuts, strong cutting planes, Benders decomposition and heuristics. Liu and Sahinidis [74] reformulated this MILP and developed a solution approach based on a constraints generation scheme and projection in conjunction with the strong cutting plane algorithm. Heever and Grossmann [52] proposed a disjunctive outer approximation algorithm for the solution of a multiperiod design and planning problem, which is an extension of the logic-based outer approximation algorithm for single period mixed integer nonlinear programming (MINLP) [112]. The main limitations of these time discretization methods are that (a) they require all the tasks to start and finish at the boundaries of time intervals, thus resulting in suboptimal solutions, and (b) they require a large number of binary variables due to time discretization that depends on the size of the largest common time-period duration in the system that usually results in large mathematical models difficult to be addressed compared with continuous time methods.

The solution methodologies of addressing planning problems can be distinguished into two main categories: the simultaneous and the hierarchical approaches [17]. The hierarchical approaches involve the problem decomposition into planning and scheduling-level problems which can be decoupled. Following the simultaneous approach the whole problem is considered and solved for the entire horizon. Papageorgiou and Pantelides [86] proposed a single-level formulation for a campaign planning problem. The algorithm determines the campaigns (i.e., duration and constituent products) as well as the production schedule simultaneously. Orcun et al. [82] developed a unified continuous time model of MINLP for planning problem. After the MINLP is reformulated as MILP by using linearization techniques, the problem is still addressed as "extremely difficult, almost impossible" to solve.

Due to the complexity and size of the problem these models are reported to be hard to solve without decomposition. Periodic scheduling is developed in the context of campaign-mode operation [68, 105]. Schilling and Pantelides [102] presented a mathematical programming formulation which is based on a continuous representation of time. A novel branch-and-bound algorithm that branches on both discrete and continuous variables was proposed. This work was extended to multipurpose plants, periodic scheduling problems. The proposed model resulted in the determination of both the optimal duration of the operating cycle and the detailed schedule in each cycle. The objective function was to minimize the average cost which corresponds to a nonlinear function. A relaxed form of the optimization problem was generated after replacing the definition of the objective function by a set of linear constraints. Castro et al. [18] modified their short-term scheduling formulation to fit periodic scheduling requirements for an industrial application.

There is a plethora of different approaches that appear in the literature to address the problem of scheduling formulation, and a recent review about classification of scheduling problems is given by Méndez et al. [79]. One major classification is based on the nature of the production facility to manufacture the required number of products utilizing a limited set of units. If every job consists of the same set of tasks to be performed in the same order and the units are accordingly arranged in production lines, the problem is classified as a multiproduct plant (also called flow-shop problem). If production orders have different routes (require different sequences of tasks) and some orders may even visit a given unit several times, the problem is known as multipurpose plant (also called job-shop problem).

A number of alternative ways of formulating the scheduling problem exist in the open literature. One distinguishing characteristic is the time representation, according to which the approaches are classified into two broad categories. Early attempts of formulating the scheduling problem were mainly concentrated on the discrete-time formulation, where the time horizon is divided into a number of intervals of equal duration [7, 30, 68]. Recently, a large number of publications have focused on developing efficient methods based on a continuous-time representation [38]. Following the continuous-time representation a number of alternatives have appeared in the literature targeting the reduction of computational complexity of the resulting model. Thus there are formulations using global time intervals that coincide with the discrete time representation, global time points that uses a common time grid for all resources, unit-specific time events that utilize different event points for different units in the production facility, as well as synchronous/asynchronous time slots that use a set of predefined time slots of unknown duration.

Most of the work in the area of planning and scheduling deals with the deterministic optimization model where all the parameters are considered known. Along with the studies in deterministic operations, consideration of uncertainties in the planning and scheduling problem has gotten more attention in recent years. In real plants uncertainty is a very important concern that is coupled with the scheduling process since many of the parameters that are associated with scheduling are not known exactly. Parameters like raw material availability, prices, machine reliability, and market requirements vary with respect to time and are often subject to unexpected deviations. Having ways to systematically consider uncertainty is as important as having the model itself. Methodologies for process scheduling under uncertainty aim at producing feasible, robust and optimal schedules. In essence, uncertainty consideration plays the role of validating the use of mathematical models and preserving plant feasibility and viability during operations.

In the context of a campaign-mode production where demands are relatively stable and uncertainty is minimum over the planning period, a periodic scheduling model is presented to address the simultaneous consideration of scheduling and planning problem. A continuous-time formulation is exploited based on a scheduling formulation of Ierapetritou and Floudas [58]. New constraints are developed to determine the scheduling decisions between cycles and incorporated into the continuous-time planning model. This model results in an efficient solution of large-scale planning problems where scheduling decisions are simultaneously determined [117].

For the case where demand is distributed within the time horizon and changes frequently, a hierarchical solution framework is presented. The planning and scheduling levels are considered within a recursive algorithm that converges to the final optimal schedule. In this framework, uncertainty is considered in the planning problem using scenario-based multistage optimization modeling. Although future time periods are considered in the planning model, only the decisions for the current time period are made and the required production is transferred to the scheduling problems. Short-term scheduling model is utilized to generate an optimal schedule that satisfies the production from planning results. In the case where a discrepancy appears such as in overoptimistic or underestimated planning results, an iterative procedure is employed to resolve the difference with necessary adjustments until the results become consistent [117].

Lagrangian relaxation and Lagrangian decomposition have been used to decompose the problem as other authors reported in their work [35, 49]. However, the performance of the Lagrangian approach for practical problems are not always satisfying due to its poor convergence. As a result, an improved Nelder-Mead-based algorithm is developed to update the Lagrangian multipliers which guarantees the bound generated is at least as good as that of the previous iteration. This approach provides a good alternative to the current prevalent subgradient method, and can be exploited when subgradient method fails to improve the Lagrangian objective function as illustrated in the case studies.

The scope of this chapter is to provide a review of the work that has been proposed in the literature to deal with the problem of scheduling, planning and the consideration of uncertainty in process operations decision making. Following these introductory remarks Section 2 is devoted to the uncertain parameter description. Section 3 discusses the alternative approaches that exist in the literature for scheduling under uncertainty, whereas Section 4 presents two alternative approaches to deal with the integration of planning and scheduling and the incorporation of uncertainty, followed by a discussion regarding the remaining challenges and future research directions that are presented in Sections 5 and 6, respectively.

2 Uncertainties in Process Operations

2.1 Uncertainty Description

To include the description of uncertain parameters within the optimization model of the scheduling problem, several methods have been used:

Bounded form. In many cases, there is not enough information in order to develop an accurate description of the probability distribution that characterizes the uncertain parameters, but only error bounds can be obtained. In this case interval mathematics can be used for uncertainty estimation, as this method does not require information about the type of uncertainty in the parameters. An uncertain parameter is described by an interval $\theta \in [\theta_{min}, \theta_{max}]$ or $|\tilde{\theta} - \theta| \leq \varepsilon |\theta|$, where $\tilde{\theta}$ is the "true" value, θ is the nominal value, and $\varepsilon > 0$ is a given uncertainty level. This is the typical and readily applicable method to describe the uncertain parameters. The bounds represent the ranges of all possible realizations of the uncertain parameters. The upper and lower bounds can be determined with the analysis of the historical data, customer's orders and market indicators.

Probability distribution function. This is a common approach for the treatment of uncertainties when information about the behavior of uncertainty is available since it requires the use of probabilistic models to describe the uncertain parameters. The uncertainty is modeled using either discrete probability distributions or the discretization of continuous probability distribution function. Uncertainties such as equipment breakdown and failure of process operations are generally described with discrete parameters.

An interesting outcome of a probabilistic analysis framework was described by Isukapalli [61] that presented an approach to allow the uncertainty propagation of input uncertainties to the output of any model using the ideas of stochastic response surface method (SRSM).

Fuzzy description. Fuzzy sets allow modeling of uncertainty in cases where historical (probabilistic) data are not readily available. The resulting scheduling models based on fuzzy sets have the advantage that they do not require the use of complicated integration schemes needed for the continuous probabilistic models and they do not need a large number of scenarios as the discrete probabilistic uncertainty representations [5].

In the classic set theory, the truth value of a statement can be given by the membership function $\mu_A(x)$ in the following way:

$$\mu_A(x) = \begin{cases} 1 \text{ iff } x \in A \\ 0 \text{ iff } x \notin A \end{cases}$$

Fuzzy theory allows for a continuous value of $\mu_A(x)$ between 0 and 1:

$$\mu_A(x) = \begin{cases} 1 \text{ iff } x \in A \\ 0 \text{ iff } x \notin A \\ p; \ 0 < p < 1 \text{ if } x \text{ paritically bilongs to } A \end{cases}$$

Instead of probability distributions, these quantities make use of membership functions based on possibility theory. A fuzzy set is a function that measures the degree of membership to a set. A high value of this membership function implies a high possibility, while a low value implies a poor possibility. For example, consider a linear constraint $ax \leq \theta$ in terms of decision vector x and assume that the random right-hand-side θ can take values in the range from θ to $\theta + \Delta\theta$, $\Delta\theta \geq 0$, then the linear membership function of this constraint is defined as:

$$\mu_A(x) = \begin{cases} 1 \text{ if } ax \leq \theta \\ 0 \text{ if } \theta + \Delta\theta \leq ax \\ 1 - \frac{ax-\theta}{\Delta\theta} \text{ if } \theta \leq ax \leq \theta + \Delta\theta \end{cases}$$

Following the alternative description methods for uncertainty, different scheduling models and optimization approaches have been developed that are reviewed in the next section.

3 Scheduling under Uncertainty

The two key elements in scheduling are the schedule generation and the scheduling revisions [19, 95]. Scheduling generation acts as a predictive mechanism that determines planned start and completion times of production tasks. Schedule revision is a reactive part, which monitors execution of schedule and deals with unexpected events. In the literature, scheduling approaches are divided using off-line/online scheduling [66] or predictive/reactive scheduling [108]. For off-line scheduling, all available jobs are scheduled all at once for the entire planning horizon, whereas online scheduling makes decisions at a time that are needed. Predictive scheduling creates the optimal schedule based on given requirements and constraints prior to the production process. Reactive scheduling is a process to modify the created schedule during the manufacturing process to adapt change in production environment, which is also refereed to as rescheduling.

Thus, reactive scheduling deals with uncertainties, such as disruptive events, rush order arrivals, order cancellations or machine breakdowns. For this type of uncertainty there is not enough information prior to realization of the uncertain parameters that will allow a protective action. Preventive scheduling, on the other hand, can deal with parameter uncertainties, such as processing times, demand of products or prices. For this type of uncertainty, historical data and forecasting techniques can be used to derive information about the behavior of uncertain parameters in future in the form of a range of parameters or stochastic distributions as described in the previous section.

Preventive scheduling, although the decisions may be modified as time passes, serves as the basis for planning support activities because commitments are made based on preventive schedule. For preventive scheduling we distinguish the following approaches: stochastic-based approaches, robust optimization methods, fuzzy programming methods, sensitivity analysis and parametric programming methods.

3.1 Reactive Scheduling

In the operations research community, most existing reactive scheduling methods are characterized by least commitment strategies such as real-time dispatching that create partial schedules based on local information. One extension of this dispatching approach is to allow the system to select a dispatching rule dynamically as the state of the shop changes. Another extension of these completely reactive approaches are those based on a number of independent, intelligent agents each trying to optimize its own objective function [3].

The approaches that exist in the process scheduling literature to address the problem of reactive scheduling deal with the following two types of disturbances: (a) machine breakdown or changes in machine operation that affects the processing times of the tasks in these units; and (b) order modification or cancellation that changes the product demands and due dates. The purpose of the proposed approaches is thus to update the current production schedule in order to provide an immediate response to the unexpected event. The original schedule is obtained in a deterministic manner and the reactive scheduling corrections are performed either at or right before the execution of scheduled operations. Reactive scheduling activity is by itself a short-term scheduling problem with some additional characteristics, mainly the possibility that all due dates not be fulfilled.

The reactive scheduling actions are based on various underlying strategies. It relies either on very simple techniques aimed at a quick schedule consistency restoration or it involves a full scheduling of the tasks that have to be executed after the unexpected event occurs. Such an approach will be referred to as rescheduling and it can use any deterministic performance measure, such as the make span of the new project. Cott and Macchietto [21] considered fluctuations of processing times and used a shifting algorithm to modify the starting times of processing steps of a batch by the maximum deviation between the expected and actual processing times of all related processing steps. The proposed approach is easy to implement but is limited in terms of the unexpected events it can address.

A number of the techniques presented in the literature solve the reactive scheduling problem through mathematical programming approaches relying mostly on MILP and the application of heuristic rules. Rodrigues et al. [91] considered uncertain processing times and proposed a reactive scheduling technique based on a modified batch-oriented MILP model according to the discrete-time state-task network (STN) formulation proposed by Kondili et al. [68]. A rolling horizon approach is utilized to determine operation starting times with look-ahead characteristics taking into account possible violations of future due dates. Honkomp et al. [55] proposed a reactive scheduling framework for processing time variations and equipment break-down by coupling a deterministic schedule optimizer with a simulator incorporating stochastic events. Vin and Ierapetritou [114] considered two types of disturbances in multiproduct batch plants: machine breakdown and rush order arrival. They applied a continuous-time scheduling formulation and reduced the computational effort required for the solution of the resulting MILP problems by fixing binary variables involved in the period before an unexpected event occurs. Méndez and Cerdá [77] used several different rescheduling operations to perform reactive scheduling in multiproduct, sequential batch plants. They considered start time shifting, local reordering, and unit reallocation of old batches as well as insertion of new batches. In Méndez and Cerdá's work [78], they extended their work to include limited discrete renewable resources where only start time shifting, local batch reordering, and resource reallocation of existing batches are allowed. Rescheduling was performed by first reassigning resources to tasks that still need to be processed and then reordering the sequence of processing tasks for each resource item.

Ruiz et al. [93] presented a fault diagnosis system (FDS) that interacts with a schedule optimizer for multipurpose batch plants to perform reactive scheduling in the event of processing time variability or unit unavailability. The proposed system consists of an artificial neural network structure supplemented with a knowledge-based expert system. The information needed to implement the FDS includes a historical database, a hazard and operability analysis, and a model of the plant. When a deviation from the predicted schedule occurs, the FDS activates the reactive scheduling tools to minimize the effect of this deviation on the remaining schedule.

There are several dispatching rules which are considered as heuristics in reactive scheduling. These rules use certain empirical criteria to prioritize all the batches that are waiting for processing on a unit. Kanakamedala et al. [65] considered the problem of reactive scheduling in batch processes where there are deviations in processing times and unit availabilities in multipurpose batch plants. They developed a least-impact heuristic search approach for schedule modification that allowed time shifting and unit replacement. Huercio et al. [57] and Sanmarti et al. [100] proposed reactive scheduling techniques to deal with variations in task processing times and equipment availability. They used heuristic equipment selection rules for modification of task starting times and reassignment of alternative units. Two rescheduling strategies were employed: shifting of task starting times and reassignment of tasks to alternative units. Their method generates a set of decision trees using alternative unit assignments, each based on a conflict in the real production schedule caused by a deviation in the real schedule from the nominal schedule. Branches of the trees are then pruned according to heuristic equipment selection rules.

Sanmartí et al. [99] presented a different approach for the scheduling of production and maintenance tasks in multipurpose batch plants in the face of equipment failure uncertainty. They computed a reliability index for each unit and for each scheduled task and formulated a nonconvex MINLP model to maximize the overall schedule reliability. Because of the significant difficulty in the rigorous solution of the resulting problem, a heuristic method was developed to find solutions that improve the robustness of an existing schedule. Roslöf et al. [92] developed an MILP-based heuristic algorithm that can be used to improve an existing schedule or to reschedule jobs in the case of changed operational parameters by iteratively releasing a set of jobs in an original schedule and optimally reallocating them.

3.2 Stochastic Scheduling

Stochastic scheduling is the most commonly used approach in the litera-ture for preventive scheduling, in which the original deterministic scheduling model is transformed into a stochastic model treating the uncertainties as stochastic variables. In this type of approach, either discrete probability dis-tributions or the discretization of continuous probability distribution functions is used. The expectation of a certain performance criterion, such as the expected make span, is optimized with respect to the scheduling decision variables. Stochastic programming models are divided into the following cat-egories: recourse model (two-stage or multistage stochastic programming) or chance constraint programming-based approach. In these stochastic schedul-ing formulations, either discrete probability distributions or the discretization of continuous probability distribution function of the uncertain parameters is needed.

In the two-stage stochastic programming, the first stage variables are those that have to be decided before the actual realization of the uncertain parameters; a recourse decision can then be made in the second stage that compensates for any bad effects that might have been experienced as a result of the first-stage decision. The optimal policy from such a model is a single first-stage policy and a collection of recourse decisions (a decision rule) defining which second-stage action should be taken in response to each random out-come. Using the same idea, two-stage stochastic programming is also extended to multi-stage approach. Ierapetritou and Pistikopoulos [60] addressed the scheduling of single-stage and multistage multiproduct continuous plants for a single production line. At each stage they considered uncertainty in product demands. They used Gaussian quadrature integration to evaluate the expected profit and formulated MILP models for the stochastic scheduling problem. Bonfill et al. [14] used a stochastic optimization approach to manage risk in the short-term scheduling of multiproduct batch plants with uncertain demand. The problem is modeled using a two-stage stochastic optimization approach accounting for the maximization of the expected profit. Management of risk is addressed by adding a control measure as a new objective to be considered, thus leading to multiobjective optimization formulations. Bonfill et al. [15] addressed the short-term scheduling problem in chemical batch processes with variable processing times to identify robust schedules able to face the major effects driving the operation of batch processes with uncertain times, i.e., idle and waiting times. The problem is modeled using a two-stage stochastic approach accounting for the minimization of a weighted combination of the expected make span and the expected waiting times. Balasubramanian and Grossmann [6] presented a multistage stochastic MILP model based on the one formulated by Goel and Grossmann [42], wherein certain decisions are made irrespective of the realization of the uncertain parameters and some decisions are made upon realization of the uncertainty. They proposed an approximation strategy that consists of solving a series of two-stage stochastic programs within a shrinking-horizon framework to overcome the large computational expense associated with the solution of the multistage stochastic program.

Following the probabilistic or chance-constraint-based approach, the focus is on the reliability of the system, i.e., the system's ability to meet feasibility in an uncertain environment. The reliability is expressed as a minimum requirement on the probability of satisfying constraints. Orçun et al. [83] proposed a mathematical programming model for optimal scheduling of the operations of a batch processing chemical plant. They considered uncertain processing times in batch processes and employed chance constraints to account for the risk of violation of timing constraints under certain conditions such as uniform distribution functions. Petkov and Maranas [88] addressed the multiperiod planning and scheduling of multiproduct plants under demand uncertainty. The proposed stochastic model is an extension of the deterministic model introduced by Birewar and Grossmann [13]. The stochastic elements of the model are expressed with equivalent deterministic forms of the chance constraints, eliminating the need for discretization or sampling techniques. The resulting equivalent deterministic optimization models are MINLP problems with convex continuous parts.

Other than the previous two methods in modeling scheduling under uncertainty using stochastic programming ideas, other methods involve simulation-based-approaches such as the approach proposed by Bassett et al. [8] that takes into account processing time fluctuations, equipment reliability/availability, process yields, demands, and manpower changes. They used Monte Carlo sampling to generate random instances of the uncertain parameters, determined a schedule for each instance, and generated a distribution of aggregated properties to infer operating policies. In the reactive scheduling framework of Honkomp et al. [56], a deterministic schedule optimizer and a simulator incorporating stochastic events were developed. Replicated simulations were used to determine the performance of fixed deterministic schedules in light of uncertainty and for the validation of reactive scheduling techniques. Subramanian et al. [107] studied the stochastic optimization problem inherent to the management of an R&D pipeline and developed a computing architecture, Sim-Opt, which combines combinatorial optimization and discrete event system simulation to assess the uncertainty and control the risk present in the R&D pipeline.

3.3 Robust Optimization Method

Robust optimization has been applied to several areas in research and practice, such as production planning [34], machine scheduling [23, 70] and logistics [33, 118].

Robust scheduling focuses on building the predictive scheduling to minimize the effects of disruptions on the performance measure. It differs from the predictive scheduling approaches in the sense that it tries to ensure that the predictive and realized schedules do not differ drastically, while maintaining a high level of schedule performance. For the issue of robust schedule generation, different methods have been proposed in the literature. Generally, the formulations can be classified into two groups: (a) scenario-based stochastic programming formulation; (b) robust counterpart optimization formulation. For the first type of method, either discrete probability distributions or the discretization of continuous probability distribution function of the uncertain parameters is needed, but only a bound form uncertainty description is needed for the latter one.

In the literature, most existing work on robust scheduling has followed the scenario-based formulation. Mulvey et al. [80] developed the scenario-based robust optimization to handle the trade-off associated with solution and model robustness. A solution to an optimization is considered to be solution robust if it remains close to the optimal for all scenarios and model robust if it remains feasible for most scenarios. Kouvelis et al. [69] made the first attempts to introduce the concept of robustness for scheduling problems. They suggest a robust schedule when processing times are uncertain and compute a robust schedule based on maximum absolute deviation between the robust solution and all the possible scenarios, but this requires knowledge of all possible scenarios. Moreover, the optimal solution of each scenario is supposed to be known a priori. Vin and Ierapetritou [115] addressed the problem of quantifying the schedule robustness under demand uncertainty, introduced several metrics to evaluate the robustness of a schedule and proposed a multiperiod programming model using extreme points of the demand range as scenarios to improve the schedule performance of batch plants under demand uncertainty. Using flexibility analysis, they observed that the schedules from the multiperiod programming approach were more robust than the deterministic schedules. Balasubramanian and Grossmann [4] proposed a multiperiod MILP model for scheduling multistage flow shop plants with uncertain processing times. They minimized expected make span and developed a special branch-and-bound algorithm with an aggregated probability model. The scenario-based approaches provide a straightforward way to implicitly incorporate uncertainty. However, they inevitably enlarge the size of the problem significantly as the number of scenarios increases exponentially with the number of uncertain parameters. This main drawback limits the application of these approaches to solve practical problems with a large number of uncertain parameters. Jia and Ierapetritou [64] proposed a multiobjective robust optimization model to deal with the problem of uncertainty in scheduling considering the expected performance (make span), model robustness and solution robustness. Normal boundary intersection (NBI) technique is utilized to solve the multiobjective model and successfully produce a Pareto optimal surface that captures the trade-off among different objectives in the face of uncertainty. The schedules obtained by solving this multiobjective optimization problem include robust assignments that can accommodate the demand uncertainty.

As an alternative to the scenario-based formulation, a relatively recent framework called robust counterpart optimization has been proposed which avoids the shortcomings of the scenario-based formulation. The underlying framework of robust counterpart scheduling formulation is based on solving robust counterpart optimization problem for the uncertain scheduling problem.

One of the earliest papers on robust counterpart optimization, by Soyster [106], considered simple perturbations in the data and aimed to find a reformulation of the original problem such that the resulting solution would be feasible under all possible perturbations. The pioneering work by Ben-Tal and Nemirovski [10], El-Ghaoui et al. [29], and Bertsimas and Sim [11] extended the framework of robust counterpart optimization and included sophisticated solution techniques with nontrivial uncertainty sets describing the data. The major advantages of robust optimization compared to scenario-based stochastic programming are that no assumptions are needed regarding the underlying probability distribution of the uncertain data and that it provides a way of incorporating different attitudes toward risk. However, on the application of process scheduling problem, only very few works have been done in a robust counterpart optimization for generating robust schedules. Line Janak et al. [73] extended work by Ben-Tal and Nemirovski [10] and proposed a robust optimization method to address the problem of scheduling with uncertain processing times, market demands, or prices.

To improve the schedule flexibility prior to its execution, it is important to measure the performance of a deterministic schedule under changing conditions due to uncertainty. Standard deviation (SD) is one of the most commonly used metrics for evaluating the robustness of a schedule. To evaluate the SD, the deterministic model with a fixed sequence of tasks $(wv_{i,j,n})$ is solved for different realizations of uncertain parameters that define the set of scenarios k that results in different make spans H_k. The SD is the defined as following if the scheduling objective is make span minimization:

$$SD_{avg} = \sqrt{\sum_k \frac{(H_k - H_{avg})^2}{(p_{tot} - 1)}}, \quad H_{avg} = \frac{\sum_k H_k}{p_{tot}}$$

where H_{avg} is the average make span over all the scenarios, and p_{tot} denotes the total number of scenarios. Another similar robustness metric is proposed by Vin and Ierapetritou [115], where the infeasible scenarios were taken into consideration. In case of infeasibility, the problem is solved to meet the maximum demand possible by incorporating slack variables in the demand constraints. Then the inventory of all raw materials and intermediates at the end of the schedule are used as initial conditions in a new problem with the same schedule to satisfy the unmet demand. The make span under infeasibility (H_{corr}) is determined as the sum of those two make spans. Then the robustness metric is defined as:

$$SD_{corr} = \sqrt{\sum_k \frac{(H_{act} - H_{avg})^2}{(p_{tot} - 1)}}$$

where $H_{act} = H_k$ if scenario k is feasible and $H_{act} = H_{corr}$ if scenario k is infeasible.

3.4 Fuzzy Programming Method

The approaches presented so far rely on the use of probabilistic models that describe the uncertain parameters in terms of probability distributions. However, sometimes such information is not available. For such cases an alternative approach is the use of fuzzy set theory and interval arithmetic to describe the imprecision and uncertainties in process parameters.

Fuzzy programming also addresses optimization problems under uncertainty. Although the uncertainties in process scheduling are generally described through probabilistic models, fuzzy set theory has been applied to scheduling optimization using heuristic search techniques during the past decades. The principal difference between the stochastic and fuzzy optimization approaches is in the way uncertainty is modeled. Here, fuzzy programming considers random parameters as fuzzy numbers and constraints are treated as fuzzy sets. Some constraint violation is allowed and the degree of satisfaction of a constraint is defined as the membership function of the constraint. Objective functions in fuzzy mathematical programming are treated as constraints with the lower and upper bounds of these constraints defining the decision makers' expectations.

Balasubramanian and Grossmann [5] applied a nonprobabilistic approach to the treatment of processing time uncertainty in two scheduling problems: (i) flow-shop plants and (ii) the new product development process. The examples considered show that very good estimates of the uncertain make span and income can be obtained by using fairly coarse discretizations and that these models can be solved with little computational effort. In these examples the improvement in the estimation of the completion time by using a denser discretization was not significant enough to warrant the order of magnitude increase in computation time required. Wang [116] developed a robust scheduling methodology based on fuzzy set theory for uncertain product development projects. The imprecise temporal parameters involved in the project were represented by fuzzy sets. The measure of schedule robustness was proposed to guide the genetic algorithm (GA) to determine the schedule with the best worst-case performance. The proposed GA approach can obtain the robust schedule with acceptable performance. Petrovic and Duenas [89] recently used a fuzzy programming method to deal with parallel machine scheduling/rescheduling in the presence of uncertain disruptions. A predictive reactive approach is defined as a two-step process, where the first step consists of adding idle time to the jobs' processing times in order to generate a schedule capable of absorbing the negative effects of uncertain material shortages. The second step is rescheduling where two questions are addressed, namely when to reschedule and which rescheduling method to apply.

3.5 Sensitivity Analysis and Parametric Programming

In addition to scheduling problems, most problems in the area of process design and operations are commonly formulated as MILP problems. An alternative way, compared to the approaches presented before, to incorporate uncertainty into these problems is using MILP sensitivity analysis and parametric programming methods. These methods are important as they can offer significant analytical results to problems related to uncertainty. What's more, only bound information of the uncertain parameters is needed in these types of methods.

Sensitivity Analysis

Sensitivity analysis (SA) is used to ascertain how a given model output depends upon the input parameters. This is an important method for checking the quality of a given model, as well as a powerful tool for checking the robustness and reliability of any solution. Due to the combinatorial nature of the scheduling problem, SA poses some unique issues.

Not much work has appeared in the literature in SA for the scheduling problem. A recent review in the literature by Hall and Posner [50] pointed out a number of issues associated with the application of SA in scheduling problems that involve: the applicability of SA to special classes of scheduling problems; the efficiency of SA approaches when simultaneous parameter changes occur; the selection of a schedule with minimum sensitivity; the computational complexity of answering SA questions for intractable scheduling problems; etc.

Samikoglu et al. [97] studied the effect of perturbations on the optimal solution and developed a series of perturbed solutions that span a specified bounded region of the parameter space. This work attempts to reveal parameters most sensitive to perturbations and those that have the greatest impact on the solution. The interaction between the resource constraints and the objective function are explored using single parameter variations. A branch-and-bound procedure enhanced with logical programming is suggested to reveal these interactions in a more general framework. Penz et al. [87] studied the performance of static scheduling policies in the presence of online

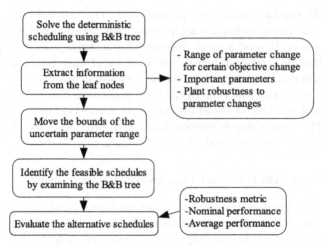

Fig. 1. Flowchart of the sensitivity analysis method [62]

disturbances using SA and showed that in the case of independent tasks, the sensitivity can be guaranteed not to exceed the square root of the magnitude of the perturbation. Guinand et al. [48] considered the problem of SA of statically computed schedules for the problem when the actual communication delays differ from the estimated ones.

Jia and Ierapetritou [62] handled uncertainty in short-term scheduling based on the idea of inference-based sensitivity analysis for MILP problems and utilization of a branch-and-bound solution methodology. The basic idea of the proposed method is to utilize the information obtained from the SA of the deterministic solution to determine (a) the importance of different parameters and constraints and (b) the range of parameters where the optimal solution remains unchanged. The main steps of the proposed approach are shown in Figure 1. More specifically, the proposed analysis consists of two parts. In the first part, important information about the effects of different parameters is extracted following the SA step, whereas in the second part, alternative schedules are determined and evaluated for different uncertainty ranges. The proposed method leads to the determination of the importance of different parameters and the constraints to the objective function and the generation and evaluation of a set of alternative schedules given the variability of the uncertain parameters. The main advantage of the proposed method is that no substantial complexity is added compared with the solution of the deterministic case since the only additional required information is the dual information at the leaf nodes of the branch-and-bound tree.

Parametric Programming

Many process design and operation problems are commonly formulated as MILP problems. Using parametric programming methods to address the uncertainty in these problems provides the best available information regarding the solution behavior. Parametric optimization also serves as an analytic tool in process synthesis under uncertainty mapping the uncertainties in the definition of the synthesis problem to optimal design alternatives. From this point of view, it is the exact mathematical solution of the uncertainty problem.

Thompson and Zawack [110] proposed a zero-one integer programming model for the job shop scheduling problem with minimum make span criterion presented. The algorithm consists of two parts: (a) a branch-and-bound parametric linear programming code for solving the job shop problem with fixed completion time; and (b) a problem expanding algorithm for finding the optimal completion time. Ryu et al. [94] addressed the problem of bilevel decision making under uncertainty in the context of enterprise-wide supply chain optimization with one level corresponding to a plant planning problem, while the other to a distribution network problem. The corresponding bilevel programming problem under uncertainty has been solved by parametric optimization techniques. Acevedo and Pistikopoulos [1] addressed linear process engineering problems under uncertainty using a branch-and-bound algorithm based on the solution of multiparametric linear programs at each node of the tree and the evaluation of the uncertain parameters space for which a node must be considered. It should be noticed that although solution degeneracy is common in scheduling problem, which will bring a certain difficulty for the study of parametric solution (e.g., in the process of basis variable determination for linear programming problem), it will not affect the result of parametric information which represents the relationship between the scheduling objective and uncertain parameters.

Recently, Jia and Ierapetritou [63] proposed a new method of uncertainty analysis on the right-hand side (RHS) for MILP problems. The proposed solution procedure starts with the branch-and-bound (B&B) tree of the MILP problem at the nominal value of the uncertain parameter and requires two iterative steps, linear programming (LP)/multiparametric linear programming (mpLP) sensitivity analysis and updating the B&B tree (Figure 2).

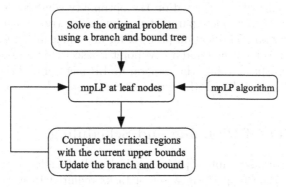

Fig. 2. Flowchart of solving mp-MILP problem [63]

Based on this framework for uncertainty analysis of scheduling, our current work [72] involves the extension of this approach in two fundamental directions. The first one addresses the issue of infeasibility by providing a description of the feasible region prior to the implementation of the parametric MILP algorithm. They have developed a preprocessing step that enables the consideration of a wide uncertainty range by exploiting the idea of projection into the constraint set. For an mpLP problem with RHS type uncertainty:$\min_{x}\{ z = cx|\ Ax \leq D\theta + b\}$, θ is the uncertain parameter vector. The feasible region is given by $\{\theta|Ax \leq D\theta + b\}$. To get the feasible region of uncertain parameters, the polytope formed by $\{(x, \theta)|Ax - D\theta \leq b\}$ is projected to the θ hyperplane using projection algorithm [2, 101]. A vertex enumeration/convex hull-based projection algorithm first enumerates the vertices of the polytope, then projects these vertices onto the subhyperplane, and finally the projection is gotten by calculating the convex hull from these projected vertices. All these computations can be done via the double description method [39]. Thus the complete description of the feasible region was determined before solving the parametric MILP. Following is an example of identifying the feasible region for an mpLP problem. The second extension is the consideration of uncertainty in the objective function coefficients and problem constraints. The problem of mpLP problems considering uncertainty in the right-hand side (RHS) and objective function coefficients have been well studied in the literature [16, 40]. The case of uncertainty at the objective function coefficients and RHS simultaneously was treated and solved as special multiparametric quadratic programming (mpQP) problems by Seron et al. [103], Bemporad et al. [9], Tondel et al. [111]. Very few papers, however, have appeared in the literature that address the mpLP problem where uncertainty appears at the RHS, left-hand side–constraint coefficients (LHS) and objective function coefficients at the same time. Because most of existing mpMILP methods are based on solving mpLP subproblems, existing mpMILP approaches have mainly focused on the special case of RHS and objective uncertainty. A general algorithm for mpLP problems was developed to address the problem with the simultaneous uncertain parameters in RHS, LHS and objective. The algorithm is based on the optimality condition of a standard LP problem and the complete parametric solution is retrieved through an iterating critical region identification process: for any point in the feasible region a critical region is identified and a new point is selected in an uncovered area of the uncertain space. Using this general mpLP method and the established framework [63], general mpMILP problem can be solved.

4 Integration of Planning and Scheduling

Both planning and scheduling problems deal with the allocation of available resources over time to perform a set of tasks required to manufacture one or more products. A production planning problem corresponds to a higher level of process operation decision making since it considers a longer time horizon and multiple orders that involve different operating conditions as well as unit changes, price and cost variability. On the other hand, process scheduling addresses the optimal assignment of tasks to units over the allotted time horizon in the operations of multiproduct and multipurpose plants that manufacture a variety of products through several sequences of operations that operate in batch, semicontinuous, and continuous mode. Planning and scheduling decisions are always closely coupled in a practical industrial plant, which requires the integration of the decisions process for planning and scheduling. Typically, planning models are linear and simplified representations that are used to predict production targets over several months. Also, at this level effects of changeovers are only roughly approximated as compared to the scheduling models. This, however, results in optimistic estimates of production which may not be realized at the scheduling level and may yield infeasible schedules. In order to avoid infeasible schedules and guarantee consistency between production planning and scheduling levels, planning and scheduling should be effectively integrated.

In this section two methods are presented for the integration of planning and scheduling decisions. The first approach is based on simultaneous consideration and uses the concept of periodic scheduling. For the scheduling decisions, we use the continuous-time representation model of Ierapetritou and Floudas [58] that requires fewer variables and constraints compared to discrete time formulations.

4.1 Periodic Scheduling Approach

The planning problem considered in this work is defined as follows. Given data are: the production recipe (i.e., the processing times for each task at the suitable units and the amounts of the materials required for the production of each product); the available units and their capacity limits; the available storage capacity for each of the materials; the time horizon under consideration; and the market requirements of products. The objective is to determine the optimal operational plan to meet a specified economic criterion such as maximal profit or minimal cost while satisfying all the production requirements. It should be noted, however, that the product demands are considered at the end of time horizon and all of the above constraints are fixed within the time horizon.

The idea of periodic scheduling is frequently utilized for the solution of the planning problem described above. The optimal solution of a planning problem implies that the schedule does not exhibit any periodicity [85]. However, one has to balance against the computational complexity of solving nonperiodic schedules for a long time horizon. The presented periodic scheduling approach resides on the following assumption. For the case that the time horizon is long compared with the duration of individual tasks, a proper time period exists. The period is much smaller than the whole time horizon, within which some

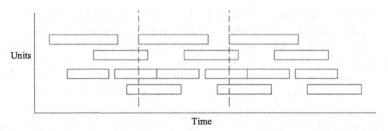

Fig. 3. Periodic schedule

maximum capacities or crucial criteria have been reached so that the periodic execution of such a schedule will achieve results very close to the optimal one by solving the original problem without any periodicity assumption. Thus the size of the problem is reduced to a much smaller one that can be efficiently solved. Besides its computation efficiency the proposed operation plan is more convenient and easier to implement since it assumes repetition of the same schedule. In this approach, the variables include the cycle time length as well as the detailed schedule of this period, which are defined as unit period and unit schedule, respectively. Unlike the short-term scheduling where all intermediates other than those provided initially have to be produced before the beginning of the tasks, unit schedule can start with certain amounts of intermediates as long as storage capacity constraints are not violated. The initial amounts of intermediates are equal to the amounts stored at the end of unit period, so as to preserve the material balance across the boundaries as shown in Figure 3.

It should be noticed that in periodic scheduling, each processing unit may have an individual cycle as long as the cycle time is equal to the duration of the unit period, so as illustrated in Figure 4a, all the units do not necessarily share the same starting and ending time points. This concept can be found in Shah and Pantelides [104] in their discrete time representations for periodic scheduling problem as "wraparound." Schilling and Pantelides [102] incorporated the same concept into their continuous time formulation based on the resource-task network (RTN) representation [85]. In this work, the same concept is used together with the continuous time representation using the idea of event points.

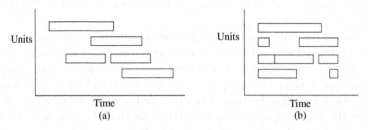

Fig. 4. Unit schedule

Figure 4a illustrates the unit schedule that corresponds to the periodic schedule of Figure 3. When a larger time period has to be scheduled using the unit schedule, overlapping is allowed in order to achieve better resource utilization. In this way the equivalent unit schedule is determined as shown in Figure 4b. Note that by using this idea better schedules are determined since tasks are allowed to cross the unit schedule boundaries. If time decomposition was applied even using the optimal cycle time length, the resulting schedule would be inferior since only small batches are allowed.

Mathematical Model

In order to represent the features of periodic scheduling for planning, the following constraints are introduced that enforce the continuity in plant operation between cycles.

- Material Balances between Cycles

$$STIN(s) = ST(s, n), \quad \forall s \in S, n = N \tag{1}$$

These constraints (1) represent the key feature of periodic scheduling. The intermediates stored at the last event point of the previous cycle should equal the amount of material needed to start the next cycle in order to maintain smooth operation without any accumulation or shortage in between. Raw material and product are calculated based on the consumed or produced amounts in the first cycle.

- Demand Constraints

$$\sum_{n \in N} d(s, n) \geq r_s H, \quad \forall s \in S \tag{2}$$

where r_s represents the average requirement. These constraints (2) express the requirement of meeting demand specifically for all products. Note that the requirements for the time horizon of planning problem are assumed to be evenly distributed to each cycle.

- Cycle Timing Constraints
  Cycle timing constraints (3) and (4) express the timing relationship of the last task in the previous cycle and the first task in the current cycle so as to maintain continuity of operation between cycles.
  Cycle Timing Constraints: Task in the same unit

$$T^s(i', j, n0) \geq T^f(i, j, n) - H, \quad \forall j \in J, \forall i, i' \in I_j, n = N \tag{3}$$

where n_0 stands for the first event point in the current cycle. $T^f(i,j,n)$-H corresponds to the time of the last event point in the previous cycle. These constraints represent that task i' performing at the beginning of the cycle has to start after the end of task i at the previous cycle. Since only one task can take place in the same unit at each event point n, constraints (3) also express the correct recipe sequence for the same unit.

Cycle Timing Constraints: Task in the different units

$$T^s(i', j', n0) \geq T^f(i, j, n) - H, \qquad \forall j, j' \in J, \forall i, i' \in I_j, i = i', n = N \quad (4)$$

These constraints represent the requirement of the first task in a new cycle to start after the completion of the tasks in different units in a previous cycle based on the recipe requirements.

- Time Horizon Constraints

$$T^s(i, j, n) \leq 2H, \quad \forall i \in I, j \in J_i, n \in N \tag{5}$$
$$T^f(i, j, n) \leq 2H, \quad \forall i \in I, j \in J_i, n \in N \tag{6}$$

Since the starting points of a cycle are not necessarily synchronized for all units, some units may start performing tasks later than others. The maximum idle time, however, will not be greater than a cycle period. Therefore the time horizon constraints represent the requirement of each task i to start and finish before two cycle lengths $2H$.

Cycle Length Constraints

$$\sum_{n \in N} \sum_{i \in I_j} \left(T^f(i, j, n) - T^s(i, j, n) \right) \leq H, \quad \forall i \in I_j, n \in N \tag{7}$$

The cycle length constraints (7) state that the duration of all tasks performed in the same unit must be less than the cycle length H, which ensures that cycle of each unit cannot be longer than the cycle length.

Objective: Maximization of average profit

$$\frac{\sum_s \sum_n price_s \times d(s, n)}{H} \tag{8}$$

The objective function for the planning problem is to maximize the production in terms of profit due to product sales. Assuming periodic scheduling this objective is transformed to maximize the average profit. The average profit is considered to express the dependence of the profit over the whole time horizon on both the production during each cycle and the cycle time. Note that the objective function involves fractional terms $d(s, n)/H$, thus giving rise to a MINLP problem. Alternative objectives can be also incorporated to express different scheduling targets such as cost minimization.

Decomposition Approach

In order to consider the whole planning problem the time horizon is divided into three periods: the initial period when the necessary amounts of intermediates are produced to start the periodic schedule, the main period when periodic scheduling is applied and the final period to wrap up all the intermediates. The initial and final periods are bounded by a time range and solved independently. The sum of time lengths of all three periods equals that of the whole time horizon. Given the optimal cycle length resulting from solving the periodic scheduling problem, the problem for the initial period is solved first with the objective function of minimum make span so as to ensure the existence of feasible solution in order to provide those intermediates for periodic scheduling. Then the same problem is solved with the objective of maximizing the profit with the time horizon as obtained from the first solution. The problem for the final period can be solved in parallel once the time horizon for the cycle length and the initial period are determined. The intermediates considered for the final period are obtained from the unit schedule and the time horizon is the time left for the planning problem. Both initial and final problems are using the same set of constraints presented in the previous section except that for the final period the time horizon is fixed. Figure 5 schematically shows the overall approach.

4.2 Hierarchical Approach for Planning and Scheduling

As discussed in the previous section, the simultaneous approach emphasizes that planning and scheduling decisions need to be considered in a single model in order to achieve the optimality. A periodic scheduling formulation as well as the solution approach is presented. However, cycle operation is more appropriate for plants operating under more stable demand conditions [85],

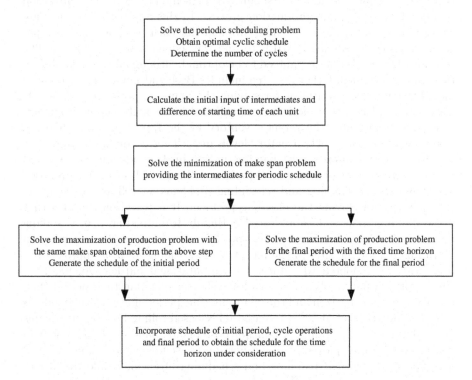

Fig. 5. Flowchart of periodic scheduling decomposition approach

thus limiting the applicability of the periodic scheduling. Uncertainty is also a concern when longer planning horizons are considered. While for a smaller time horizon demands and prices can be considered deterministic, this is not the case for a time horizon of a month or larger. Additional disturbances may also upset the production schedule, as for example rush order arrival and machine breakdown.

The hierarchical approaches involve the problem decomposition into planning and scheduling level problems which can be decoupled. McDonald and Karimi [76] developed production planning and scheduling models for application-using single-stage processor. The planning model divides the time horizon into a number of time periods with demands due at the end of each time period and compares different timescale cases of planning periods and individual production events. Two continuous-time formulations are presented for the short-term scheduling problem where discrete time events can be accommodated. Papageorgiou and Pantelides [86] presented a hierarchical approach attempting to exploit better the inherent flexibility with respect to intermediate storage policies and multiusage of the equipment. A three-step procedure was proposed. First, a feasible solution to the campaign planning problem subject to restrictive assumptions is obtained. Second, the production rate in each campaign is improved by removing these assumptions. Finally, the timing of the campaigns is revised to take advantage of the improved production rates. Harjunkoski and Grossmann [51] presented a bilevel decomposition strategy for a steel plant production process. In this approach, products are grouped into blocks and scheduling problems for each block are solved separately followed by solving a MILP to find the sequence of these blocks. The solution obtained is not guaranteed to be optimal, although near optimal solutions are determined. Other research includes Bose and Pekny [17], who used model predictive control ideas for solving the planning problem. A forecasting and an optimization model are established in a simulation environment. The former calculates the target inventory in the future periods while the latter tries to achieve such inventory levels in each corresponding period. The advantage of this approach is that fluctuation in demands and prices could be incorporated into planning level problem. Zhu and Majozi [119] proposed an integration of planning and scheduling problems as well as a decomposition strategy for solving the planning problem. It is based on the idea that if the raw materials can be allocated optimally to individual plants, solving individual models for each plant can produce the same results as solving an overall model for the site. In the recent work, a rolling horizon approach has been widely considered to reduce the computational burden [26, 98, 113]. This approach only makes decisions for a shorter planning time period than the planning time horizon and this planning time period moves as the model is solved. Dimitriadis et al. [26] presented RTN-based rolling horizon algorithms for medium-term scheduling of multipurpose plants. Sand et al. [98] and Van den Heever [113] employed a rolling horizon as well as a Lagrangian type of decomposition for their planning and scheduling problems, while the former considered uncertainty on the planning level. Most of the existing approaches, however, are limited due to an overly simplified planning level problem, the lack of uncertainty and task sequence feasibility consideration.

A general hierarchical framework for the solution of planning and scheduling problems is briefly presented in this section that allows the consideration of uncertainty using a multistage planning model where a parameter denoted as a sequence factor is introduced to simplify the computational complexity and to account for recipe time constraints.

Solution Approach

The overall decision process is based on the idea of a rolling horizon strategy. The planning time horizon is decomposed into three stages with various durations based on the orders and market uncertainty. The first stage with the smallest duration is denoted as the "current" period where operating parameters are considered deterministic. The second stage with larger duration is subject to small variability of demands and prices, and the final stage with the largest duration has a higher level of fluctuations regarding demands and prices. Uncertainty is modeled using the ideas of multistage programming [24], where each planning period corresponds to a different stage. Uncertainty is expressed by incorporating a number of scenarios at each stage. More scenarios are considered towards the last stage in order to represent the increasing level of uncertainty. Each scenario is associated with a weight representing the probability of the scenario realization. Moreover, it is assumed that each unit will process a certain number of batches at a full capacity and a single batch at flexible size at every stage, which is a valid assumption for realistic case studies. Using this basic idea the following planning model is developed.

Stochastic Multiperiod Planning-Level Formulation

- Capacity Constraints

$$V_{ij}^{\min} wv^k(i,j,q^k) \leq B^k(i,j,q^k) \leq V_{ij}^{\max} wv^k(i,j,q^k),$$
$$\forall i \in I, j \in J_i, q^k \in Q^k \qquad (9)$$

These constraints (9) enforce the requirement for minimum batch size, (V_{ij}^{min}), in order for the batch to be executed and to put a limit in the maximum batch size due to unit capacities when task i is performed in unit j at period k under scenario q^k.

- Material Balances

$$Input^k(s,q^k) = Input^{k-1}(s,q^k) - d^k(s,q^k) + \sum_{i \in I_s} \rho_{si}^p \sum_{j \in J_i} B^k(i,j,q^k)$$

$$- \sum_{i \in I_s} \rho_{si}^c \sum_{j \in J_i} B^k(i,j,q^k), \forall s \in S, q^k \in Q^k, k > 1 \quad (10)$$

$$Input^1(s) = Init_s, \qquad s \in S \qquad (11)$$

The material balances (10) and (11) state that the amount of material of state s at the end of period k is equal to that at period $k-1$ adjusted by any amounts produced or consumed between the period k and the amount delivered to the market in period k. The objective of the planning-level problem is to determine the amount of materials after the first period, $Input^2$, for which the scheduling-level problem will generate an optimal production schedule in the next step.

- Demand Constraints

$$d^k(s, q^k) \geq r^k_{s,q^k} - sk^k(s, q^k), \qquad \forall s \in S, q^k \in Q^k \tag{12}$$

These constraints (12) express the requirement to produce the maximum amount of state s towards satisfying the required demand, (r^k_{s,q^k}). The amount that cannot be produced (sk^k_{s,q^k}) is thus denoted as a back order amount and is considered with an associated cost in the objective function.

- Duration Constraints

$$Tp^k(i, j, q^k) = \alpha_{ij} wv^k(i, j, q^k), +\beta_{ij} B^k(i, j, q^k), \quad \forall i \in I, j \in J_i, q^k \in Q^k \tag{13}$$

These constraints (13) express the processing time of task i in unit j given the amount of material being processed, α_{ij}, β_{ij}. Note that these constraints are only enforced for one batch at each stage under each scenario q^k.

- Time Horizon Constraints

$$\sum_{i \in I_j} n^k(i, j, q^k) \left(\alpha_{ij} + \beta_{ij} V_{ij}^{\max} \right) + Tp^k(i, j, q^k) \leq \mu^k \times H^k,$$

$$\forall i \in I, j \in J_i, q^k \in Q^k \tag{14}$$

Time horizon constraints (14) require that all the tasks performed in unit j have to be completed within the time horizon of each stage H^k; μ^k is the sequence factor, which is used to indicate the effects of the sequence constraints on make span. The use of the sequence factor is considered to reduce the infeasibilities at the scheduling level where the detailed sequence constraints representing the production recipe are considered.

- Objective: Minimization of Cost

The objective function (15) consists of minimizing the overall cost during the whole time horizon including raw material cost, back order cost and operating cost as follows:

$$Cost = \sum_{k=1}^{3} Raw\ Material\ Cost^k + Inventory\ Cost^k$$

$$+ Backorder\ Cost^k + Operating\ Cost^k \tag{15}$$

In particular raw material cost is given by the following equation:

$Raw\ Material\ Cost =$

$$\sum_{k=1}^{3}\sum_{s\in S}\cos t_s^k \sum_{q^1\in Q^1} w_{q^k}^k \sum_{i\in I_s}\rho_{si}^c \sum_{j\in J_i}\left(n^k(i,j,q^k)\times V_{ij}^{\max} + B^k(i,j,q^k)\right)$$

(16)

where $w^1=1$ for the first stage where no scenarios are considered.
The inventory cost (17) expresses the cost of storing the product at the end of the second and third periods. Since in the planning level it is hard to calculate the exact inventory time due to aggregation of the orders, we use an approximation of half-length of the consecutive periods to represent the inventory time.

$Inventory\ Cost =$

$$\sum_{s\in S} hin_s^1 \times Input^2(s) * tiv^1 + \sum_{s\in S}\sum_{q^1\in Q^1} w_{q^1}^1 \times hin_s^2 \times Input^3(s,q^1) * tiv^2$$

(17)

where tiv^1 and tiv^2 is the time interval from period 1 to 2 and period 2 to 3, respectively, equal to half of $H^1 + H^2$ and $H^2 + H^3$.
Back order cost (18) is considered to penalize the partial order satisfaction.

$$Back\ orderCost = \sum_{k=1}^{3}\sum_{s\in S}\sum_{q^k\in Q^k} w_{q^k}^k \times b\cos t_s^k \times sk^k(s,q^k)$$

(18)

Operating cost (19) considers the cost of equipment utilization which is related to a fixed cost f^k and a varying cost v^k.

$$Operating\ Cost = \sum_{k=1}^{3}\sum_{i\in I}\sum_{j\in J_i}\sum_{q^k\in Q^k} w_{q^k}^k \times \left(n^k(i,j,q^k)\times(f_{ijq^k}^k + v_{ijq^k}^k V_{ij}^{\max})\right.$$
$$\left. + \left(f_{ijq^k}^k wv^k(i,j,q^k) + v_{ijq^k}^k B^k(i,j,q^k)\right)\right)$$

(19)

Sequence Factor

The sequence factor μ^k represents the effect of sequence constraints in the planning problem. Since the scheduling problem is not simultaneously solved with planning problem, the sequence factor is introduced such that the planning results are close to the scheduling solution. In this section, a general procedure is presented for estimating the sequence factor. However, realizing that a gap always exists between the planning problem solution involving the sequence factor and the short-term scheduling problem, an iterative procedure is developed within the planning and scheduling framework, which dynamically adjusts the sequence factor such that it always represents the best estimation.

Thus the following procedure is developed that provides a reliable estimate. First the planning and scheduling problems are solved for a test case where the planning problem is solved using only one stage and one scenario. The ratio of the objective functions from the solutions of the planning problem and the corresponding scheduling problem is used as an approximation of the sequence factor. If the scheduling problem cannot be efficiently solved, a smaller time horizon is considered which will provide a lower bound on the sequence factor since sequence constraints are increasingly more important as the time horizon decreases. Moreover, the LP relaxation can be used to estimate the sequence factor for two reasons: (1) the solution can be achieved very efficiently; and (2) the ratio of the objective functions usually is larger than that from solving the original problems since the LP relaxation of the planning problem produces a tighter relaxation. Based on these approximations, a good estimate of the sequence factor can be obtained.

It should be noticed, however, that the target is not to obtain the exact value of μ^k since its value is updated within the overall proposed framework as explained in the next section.

Scheduling Model

The scheduling problem is solved after the solution of the planning model to ensure a feasible production schedule for the current period given the production requirement determined by the planning problem. Since planning takes into consideration the future time periods, the production required by scheduling could exceed the orders imposed by the market. Assuming that parameters in the first stage are deterministic, the scheduling problem is solved using a continuous time formulation of Ierapetritou and Floudas [58]. In order to incorporate the planning considerations, demand constraints as well as the objective function are modified as follows:

- Demand Constraints

$$\sum_{n \in N, n \leq n'} d(s, n) \geq r_{s,n'} - sk(s), \quad \forall s \in S, n, n' \in N \tag{20}$$

The solution of the scheduling problem is required to determine a feasible production schedule satisfying the planning production results. Thus, all the orders are required to be satisfied by their due dates. According to these constraints, the production of material s by event point n should satisfy the individual order of product s by the due date that corresponds to event point n' [59]. Slack variables sk are considered to represent the amount of back orders that are penalized in the objective function.

- Required Production Constraints

  In order to consider the requirements imposed by the planning problem, additional production may be required at the current period. This requirement is considered separately as follows:

$$\sum_{n \in N, n \leq n'} d(s, n) \geq rp_s - slack(s) \quad \forall s \in S, n, n' \in N \tag{21}$$

where rp_s is the production requirement of material s obtained from the solution of planning model.

- Objective: Minimization of cost

  The objective of the scheduling problem is to minimize the slack variables of required production, *slack(s)* as well as all the production cost similarly to the planning model.

$$Min \sum_{s \in S} \cos t_s \sum_{i \in I_s} \rho_{si}^c \sum_{j \in J_i} \sum_n B(i, j, n)$$

$$+ \sum_{s \in S} hin_s \times \sum_n d(s, n) * tiv^1 + \sum_{s \in S} b \cos t_s \times sk(s)$$

$$+ \sum_{i \in I} \sum_{j \in J_i} \sum_n (f_{ij} \times wv(i, n) + v_{ij} B(i, j, n))$$

$$+ \sum_s plt_s \times slack(s) \qquad (22)$$

The first term represents the cost of raw materials, whereas the second and third terms consider inventory cost and back order cost, respectively. Since the time length of storing the inventory in the next period is not known, tiv^1 is used in the scheduling problem as well. Operating cost is involved in the fourth term. The last term enforces the consideration of the additional production, where plt_s is a penalty considered to indicate which products or intermediate materials are desired to be produced first.

Solution Framework

In this section, the overall hierarchical framework is presented that addresses the solution of planning and scheduling problem. An iterative procedure ensures consistent results between planning and scheduling stages.

The proposed framework is based on a rolling time horizon approach, which considers several periods of the entire planning horizon at a time. The optimal production schedule of these periods is determined and a new planning problem is formulated and solved following the same procedure. The advantages of a rolling horizon approach are the following:

It is adaptive to the dynamic production environment. Long-term forecast usually suffers large uncertainty. Therefore when planning over a large time horizon, one may not be able to predict the fluctuating demands or price well and foresee all the disturbances such as machine breakdown or rush orders. Consequently the optimized results will not represent the optimal schedule if these care.

At each decision-making point, the future periods are grouped into three stages. Product demands are aggregated at each stage and a scenario tree is generated to represent uncertainty. The formulated planning model thus takes into consideration a number of periods, although only the decisions for the current period (stage 1) are implemented. The planning model is solved as a MILP problem, or alternatively utilizing a heuristic by solving the LP relaxation of the original problem and fixing the number of full-size batches. The solution of the planning model could result in the following two cases: (1) the production for the current period could not satisfy the aggregated demand in this period; and (2) the production meets or exceeds the aggregated demand.

In the first case, we need to further identify if the shortage is due to capacity limitation or inaccurate parameters in the planning model. Therefore, the demands are disaggregated and the short-term scheduling problem is solved in order to obtain a feasible production schedule. If the scheduling problem generates an optimal schedule that satisfies all the orders, this means that the infeasibility of a planning problem is due to an underestimated sequence factor. In this case, the sequence factor is increased and the problem is resolved until the convergence between planning and scheduling models is achieved. The sequence factor is updated according to the following equation:

$$\mu^l = Min\left(\mu^{l-1} \times \frac{P_{sche}}{P_{plan}}, 1\right) \tag{23}$$

where μ^k is the value of the sequence factor at iteration l, P_{sche}, P_{plan} is the production determined by scheduling and planning models, respectively; and they can be derived from the solution but not necessarily to be the objective to be optimized. Note that the sequence factor should not exceed the value of one since this would mean unrealistic production capacity. If the short-term scheduling solution cannot satisfy all the orders, this means that the problem is infeasible, and thus back orders are allowed. The optimal solution is the best production schedule for the current period since the objective function minimizes the amount of back orders. Assuming we want to satisfy these back orders as early as possible, the amount of back order is added into the market demands for the next period. In both cases, the inventory level at the end of a current period is updated and the model is reformulated utilizing the rolling horizon approach so that the schedule of the following period can be determined.

In the second case where the planning solution satisfies all the orders, it might result in additional production beyond the demands. In order to achieve the additional production as well as satisfy the orders, we incorporate the modified demand constraints into the scheduling model. The scheduling problem can lead to one of the following cases:

1. The schedule meets the orders and the additional production requirements. In this case, the solution represents the optimal production schedule

for the current period and the inventory at the end of current period is updated.

2. The optimal production from the scheduling problem cannot satisfy the orders although the planning model suggests the opposite. This means that the sequence factor underestimates the effects of sequence constraints in production capabilities of the plant. Consequently in this case, the result of the scheduling problem is accepted as optimal allowing back orders adjusting the demand for the next period. Since the results for this period reveals an underestimate of the sequence factor, to avoid such a case in the future the sequence factor is updated. The inventory is updated as well.

3. The optimal production satisfies all the orders but not the additional production determined by the planning model. In this case, the sequence factor is adjusted and the planning model is resolved for the same time period to determine more realistic production targets.

At each time point, this iterative procedure continues until convergence is achieved between the planning and scheduling problems. Demand and inventory are updated and the same procedure is followed for the next time period in the rolling horizon approach.

Lagrangian Decomposition Approaches

The complexities of realistic problems require the use of large-scale models that often prevent the convergence to the optimal solution. Lagrangian relaxation and Lagrangian decomposition are promising decomposition techniques that reduce the problem size and achieve a solution in affordable computational time. As an extended research to the previous scheduling work, this section discusses in depth the Lagrangian technique and an alternative Lagrangian multiplier updating method is presented.

Lagrangian relaxation was original developed by Held and Karp [53] and successfully applied to many optimization problems such as production scheduling [35], planning [49, 43] and lot-sizing problems [25, 109]. The idea of Lagrangian relaxation is based on the characteristic that mathematical problem formulations involve "hard" constraints, the existence of which increases the complexity of problem. As shown in Figure 6, assuming that the first set of constraints are the complicating "hard" constraints, Lagrangian relaxation proceeds by relaxing these constraints and penalizing the constraint violation in the objective function, thus reducing the computational complexity

$$(P) \quad \max \quad z = cx \qquad (LR_u) = \max \quad cx + u(b - Ax)$$
$$\text{s.t.} \quad Ax \leq b \qquad \Longrightarrow \qquad \text{s.t.} \quad Cx \leq d \qquad \Longrightarrow \quad (LR) \quad \min_{u} \ (LR_u)$$
$$Cx \leq d \qquad \qquad x \in X, b \geq 0$$
$$x \in X$$

Fig. 6. Lagrangian relaxation

of the solution since the problem with the remaining constraints is easier to solve.

Since the set of Lagrangian multipliers are chosen to be nonnegative ($u \geq 0$), for every optimal solution x of the original optimization problem (P) we have: $V(LR_u) \geq V(LR) \geq V(P)$, where the operator $V(.)$ denotes the optimal value. Therefore the resulting objective function from Lagrangian relaxation is an upper bound of the original optimization problem. Let operator $Co(.)$ denote the convex hull and P^* represent the following problem:

$$\begin{aligned} (P^*) \quad & \max \quad z = cx \\ s.t. \quad & Ax \leq b \\ & x \in Co\{\, Cx \leq d,\ x \in X\} \end{aligned}$$

It can be further shown that (P^*) and (LR) are duals [41]. Thus: $V(LP) \geq V(LR) = V(P^*) \geq V(P)$, where ($LP$) is the LP relaxation of problem (P). If the multipliers of the complicating constraints from the optimal LP relaxation solution are used to solve $V(LR_u)$, the solutions corresponds to an upper bound of the original problem at least as tight as the bound from (LP), i.e., $V(LP) \geq V(LR_u)$. Geoffrion [41] proved the following integrality property: the optimal value of (LR_u) is not changed by dropping the integrality condition on the x variables, i.e., $Co\{Cx \leq d,\ x \in X\} = \{Cx \leq d,\ x \in X\}$. Only when this property holds, the following equalities hold: $V(LP) = V(LR) = V(P^*) \geq V(P)$. However, for practical problems, this integrality property usually does not hold. Thus this allows the application of Lagrangian relaxation to provide a tighter bound. A drawback of the Lagrangian relaxation is that the problem loses its original structure since the complicating constraints are removed from the constraints set and embedded into the objective function. As a method to avoid this, Lagrangian decomposition is presented by Guignard and Kim [46]. It can be regarded as an extension of the Lagrangian relaxation since instead of relaxing constraints it is based on the idea of relaxing a set of variables responsible for connecting important model components. Considering the problem (P) shown in Figure 7 the set of variables x that connect the set of constraints

$$\underset{I:P}{\left\{\begin{aligned} \max \quad & z = cx \\ s.t. \quad & Ax \leq b \\ & Cx \leq d \\ & x \in X \end{aligned}\right\}} \Leftrightarrow \underset{II:P'}{\left\{\begin{aligned} & \max \quad cx \\ & s.t.\ Ax \leq b \\ & \quad Cy \leq d \\ & \quad x = y \\ & x \in X, y \in X \end{aligned}\right\}} \Rightarrow \underset{III:LD_u}{\left\{\begin{aligned} & LD_u = \max \quad cx + u(y - x) \\ & s.t.\ Ax \leq b \\ & \quad Cy \leq d \\ & \quad x \in X, y \in X \end{aligned}\right\}}$$

$$\Leftrightarrow \left\{ \underset{LD_u^1}{\left[\begin{aligned} & \max\ (c-u)x \\ & s.t.\ Ax \leq b \\ & \quad x \in X \end{aligned}\right]} + \underset{LD_u^2}{\left[\begin{aligned} & \max\ uy \\ & s.t.\ Cy \leq d \\ & \quad y \in X \end{aligned}\right]} \right\} \Rightarrow (LD) \quad \min_u (LD_u)$$

Fig. 7. Lagrangian decomposition

are duplicated by introducing an identical set of variables y and an equality constraint $x = y$ as shown in the problem (P'). Lagrangian relaxation is then applied to the new set of equality constraints, resulting in problem (LD_u), which can be further decomposed into two subproblems (LD_u^1) and (LD_u^2) as shown in Figure 7.

Therefore $V(LD) \leq V(LR)$, which means that the upper bound generated from Lagrangian decomposition is always at least as tight as that from Lagrangian relaxation. Similarly, if an integrality property holds in one of the subproblems, for instance, $Co\{Cx \leq d, x \in X\} = \{Cx \leq d, x \in X\}$, then (LD) is equivalent to (LR) and $V(LD) = V(LR)$. If an integrality property holds in both subproblems, i.e., $Co\{Ax \leq b, x \in X\} = \{Ax \leq b, x \in X\}$ and $Co\{Cx \leq d, x \in X\} = \{Cx \leq d, x \in X\}$, then (LD) is equivalent to (LP) and $V(LD) = V(LP)$. Otherwise (LD) is strictly tighter than (LR), which is usually the practical case. Adding surrogate constraints, which means both subproblems have overlapped constraints, can further tighten the bound of Lagrangian decomposition. However, this increases the problem complexity, and thus the solution computational time. Figure 8 provides a geometric interpretation of the Lagrangian decomposition [46].

In general, the function of Lagrangian decomposition is nondifferentiable at the optimal point since it can have multiple optimal solutions. Thus the traditional gradient methods could not work for updating the Lagrangian multipliers in order to minimize the duality gap. Subgradient technique is typically employed to update the Lagrangian multiplier [31, 75, 90]. The subgradient method utilizes the distance between the objective value at the current iteration Z^k and the estimated optimum Z^* to calculate a step size t_k which is used to update the Lagrangian multipliers as follows:

$$u^{k+1} = u^k + t_k(y^k - x^k) \qquad (24)$$

$$t_k = \frac{\lambda(Z^k - Z^*)}{\|y^k - x^k\|^2} \qquad 0 \leq \lambda \leq 2 \qquad (25)$$

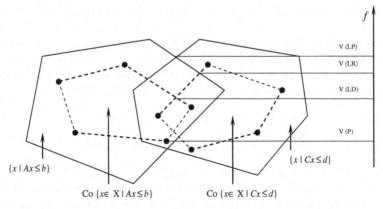

Fig. 8. Geometric interpretation of Lagrangian decomposition

where superscript k corresponds to the iteration number and λ is a scaling factor of the step size to control the convergence, normally considered to be between 0 and 2.

A number of problems, however, arise with the use of the subgradient method. Since Lagrangian decomposition is based on duality theory, theoretically the method converges the dual variables to the same value (more generally, $u(x-y)=0$), which results in the optimal solution of the original optimization problem. However, in practice using subgradient method is reported to have unpredictable convergence [45].

For some problems, the subgradient method generates monotonic improvement in Lagrangian objective function and the convergence is quick and reliable. Other problems result in an erratic multiplier sequence and the Lagrangian function value keeps deteriorating. According to the complementary slackness condition, subgradient method updates the Lagrangian multiplier u until the convergence of the dual variables. The gap between the decomposition variables x and y may exist even when the (LD) converges with the optimal objective value of the original optimization problem. We will illustrate this case in the following section through solution of an example problem. Other drawbacks of the subgradient method are (a) the lack of convergence criterion, (b) the lack of estimate of the optimal objective $Z*$ and (c) dependency to the heuristic choice of the step size sequence [22]. Due to these problems the subgradient method is not stable when applied to large-scale problems. Another issue related to Lagrangian decomposition is that the dualized variable set must be appropriately chosen so that the resulting subproblems are easy to solve and the solution converges fast, which is case-dependent [49, 84].

The bundle method is an extension of the subgradient method presented by Lemarechal [71] and Zowe [120]. This method considers improving Lagrangian function as well as staying close to the approximated optimal solution. At each iteration the method either updates the Lagrangian multipliers or improves the function approximation. A trade-off exists between the size of the region within which the bundle method allows to move and the bound improvement.

Realizing these deficiencies a number of techniques have been developed to update the Lagrangian multipliers. The constraint generation method [67] considers a family of k existing solutions (x^k, y^k) and generates a new Lagrangian multiplier u by solving a restricted LP master problem MP^k:

$$\begin{aligned} \min_{u,\eta} \quad & \eta \\ s.t. \quad & \eta \geq cx^k + u(y^k - x^k), \quad k = 1, ..., K \end{aligned}$$

The resulting u is then used in the LD_u and a new cut of (x^{k+1}, y^{k+1}) is obtained from solving LD_u, which is added in the master problem [44]. The process terminates when $V(MP^k)=V(LD_u)$. However, there is no guarantee that the new Lagrangian multiplier will generate an improved solution; thus the problem of cycling needs to be resolved. This method also depends on the cuts of (x^k, y^k) in the master problem.

Multiplier adjustment method, also referred to as dual ascent/descent algorithms, was presented by Bilde and Krarup [12] and successful application reported by Erlenkotter [32], Fisher and Hochbaum [36], Fisher and Kedia [37], Guignard and Rosenwein [47]. This method generates a sequence of u^k by using the following relationship:

$$u^{k+1} = u^k + t^k d^k \tag{26}$$

where t^k is a positive scalar and d^k is a descent direction; d^k is usually determined from a finite set of directions by evaluating the directional derivative of (LD_u). Typically the direction of the steepest descent is chosen and the step size t^k is chosen to minimize $V(LD_u{}_{k+td}{}_k)$. Although this method is reported to work better than the subgradient method for some problems [32], the set of directions to choose from may involve specific knowledge of the problem such that it is minimized but still contains direction to descent. A good review of the methods solving for Lagrangian multipliers is given by Guignard [45].

These considerations initiate the efforts towards an improved method for updating the Lagrangian multipliers based on direct search in Lagrangian multiplier space.

Update the Lagrange Multipliers

The proposed approach uses a direct search method to update the Lagrangian multipliers in order to improve the performance of the Lagrangian decomposition. The main idea is that given a fairly good estimation of Lagrangian multipliers, only the promising directions need to be searched. Thus the computational complexity decreases and the objective of the Lagrangian decomposition is improved at each iteration.

The Nelder-Mead method [81] is used to determine the promising search directions since it is proven to be a very efficient direct search algorithm. For a function of n variables, the algorithm maintains a set of $n+1$ points forming the vertices of a simplex or polytope in n-dimensional space. The result of each iteration is either (1) a single new vertex which replaces the one having the worst function value in the set of vertices for the next iteration, or (2) if a shrink is performed, a set of $n+1$ new points form the simplex at the next iteration. Four scalar parameters must be specified to define a complete Nelder-Mead method: coefficients of reflection, expansion, contraction, and shrinking. As with every direct search method the Nelder-Mead method has the advantage of not requiring derivative computations, but they tend to be efficient in relatively low dimensions.

In order to be able to efficiently use the Nelder-Mead method we need to determine a good initial set of Lagrangian multipliers in order to reduce the search space. Moreover, the promising new search directions should be easily determined. These two questions are addressed as follows. As a common practice an effective way to generate an initial set of Lagrangian multipliers for integer linear problems is to use the dual values of LP relaxation. This is because the Lagrangian decomposition problem is equivalent to LP relaxation if we drop the integrality constraints; thus most of these values are already near the optimum and we only need to examine the reduced Lagrangian multiplier space for the next update. The second question is most critical to the proposed algorithm. A promising search direction is defined as the direction resulting in the improvement of the objective function. In order to adjust the Lagrangian multipliers independently, the axes in the Lagrangian multiplier space are used as the possible directions. Thus we have n orthogonal directions to search where n is the number of the Lagrangian multipliers and at each iteration we should choose among the directions that lead to the objective function improvement. In practice the directions examined are much less than the actual number of Lagrangian multipliers because of their good initial values.

The steps of the proposed approach are shown in Figure 9. In particular first the LP relaxation of the original optimization problem is solved and the

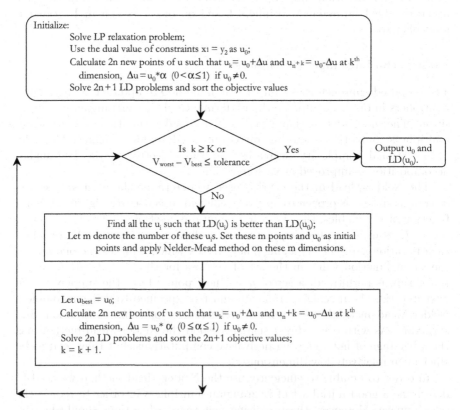

Fig. 9. Lagrangian decomposition using the Nelder-Mead algorithm

Lagrangian multipliers are initialized as the dual values of the corresponding dual equality constraints. Assuming that there are n pairs of dual variables, the Lagrangian multiplier space has n dimensions. To generate the initial points for the Nelder-Mead algorithm, we fix the n-1 variables at their original values and change only one dimension by a value of $\pm\Delta u$, which is α times the original value $\Delta u = \alpha^* u_0$. In this way two new points are generated. We apply the same procedure for all dimensions and generate $2n$ points. Each point corresponds to a set of Lagrangian multipliers. The next step is to solve $2n + 1$ Lagrangian decomposition problems associated with these points and sort them based on their objective values of Lagrangian decomposition problem. Although this might be a time-consuming step, a potential advantage is that such a multidirection search can be computed in parallel since these problems are completely independent of each other. Those with improved objective values are selected to form the reduced space where the Nelder-Mead algorithm is applied. The final point is guaranteed to be at least as good as the previous best point, and thus it is used as the new starting point to generate the next $2n$ neighboring points. The iterations continue until the difference in the objective function values is within tolerance or the number of iterations reaches a limit.

Since each Nelder-Mead iteration returns a new point with equal or better objective function than the previous point, convergence is guaranteed. However, the algorithm's efficiency depends on the value α. A large value of α gives more emphasis in the most promising directions in Lagrangian multiplier space and results in larger changes; while small values of α concentrates on small areas and attempts to find all promising directions, although it may result in slower convergence. An adaptive strategy is thus proposed starting with a value of α at the initial iterations in order to improve the values of the Lagrangian multiplier faster and reducing α when no new improving directions can be found.

As an alternative to the Nelder-Mead method, a more recent and improved direct search method is derivative-free optimization (DFO) [20], which is characterized by global convergence to a local optimum and is computationally cheaper than the Nelder-Mead simplex.

5 Challenges

In the past three decades, the research area of plant planning and scheduling has received great attention from both academia and industry. Managing the uncertainty in planning and scheduling gets more focus in these years as an advanced topic.

Regarding the consideration of planning and scheduling decisions simultaneously, the solution procedures reviewed and outlined in the previous sections are only economically justified as the optimum, but they may not be the best choice when evaluated from all perspectives of a plant. Ideally we would like to model every constraint and solve for a solution balancing all considerations. Unfortunately this is not the case for realistic industry problems due to the following reasons. First, modeling all the considerations is elaborate work. Due to the management policies and other ad hoc situations, the constraints are unique for every plant, which makes the problem case-dependent. Thus it is hard to utilize a general short-term scheduling model for all production problems. Second, it is not easy to set priorities for all these considerations. For example, some are hard constraints while the others can be relaxed under certain circumstances. This will be translated into the mathematical model using a large number of binary variables, which dramatically increases the complexity of the problem. Third and most importantly, current optimization techniques are still far from being able to handle large-scale MILP and MINLP problems. Comprehensive models do not necessarily result in good solutions in an affordable time using the available commercial solvers. All these limitations bring us a question: how can we make the scheduling model take part in business decisions for realistic industry problems? The consideration of uncertainty as an integrated factor for realistic decision-making problems brings an additional complexity to an already complex problem.

Existing research in the area of scheduling under uncertainty falls into the category of a predictive-reactive mode, where a predictive schedule is generated and then the schedule is modified during execution in response to unexpected disruptions. Since most scheduling formulations belong to the class of NP-complete problems even when simplifications in comparison to practical problems are introduced, it is often argued that only the use of problem-specific heuristics can lead to efficient solution procedures. Such heuristic algorithms incorporate specific knowledge which often leads to good solutions that can be obtained in an acceptable amount of time. However, the use of heuristics has the disadvantage that they cannot usually guarantee convergence and good quality solutions.

Current capabilities of optimization methods to reactive scheduling problems are still very restrictive and mostly focused on sequential batch processes. More general, efficient and systematic rescheduling tools are required for recovering feasibility and efficiency with short reaction time and minimum additional cost. The main effort should be oriented towards avoiding a time-expensive full-scale rescheduling, allowing during the rescheduling process only limited changes to the scheduling decisions already made at the beginning of the time horizon.

Research on sensitivity analysis and parametric programming has emerged in the area of scheduling. However, most of the existing methods are dealing with linear models that represent a simplification of the realistic processes. Thus in order to be able to address nonlinear models, effective sensitivity analysis and parametric programming for MINLP problems are needed. Towards this direction most of the progress done recently is due to the work of Pistikopoulos and co-workers [27, 28, 54] and Ierapetritou and co-workers [63, 72].

Another challenging issue is to be able to handle a large number of uncertain parameters. Since uncertainties appear in different levels of the process and in different forms as described in the previous sections, a systematic consideration of all uncertain parameters is of vital importance. However, most of the existing work addresses just part of the problem due to increasing problem complexity.

6 Summary

This chapter presented a review of the planning and scheduling techniques under uncertainties. A detailed description of process uncertainty was given as the basis to model the planning and scheduling process. According to the different treatment of uncertainty, scheduling methods were divided into two groups: reactive scheduling and preventive scheduling. Reactive scheduling, dynamic scheduling, rescheduling, and online scheduling deal with the problem of modifying the original scheduling policy or generating scheduling policy on time when uncertainty occurs. Preventive scheduling or predictive scheduling generates robust scheduling policy before the uncertainty occurs. Recent progress of applying sensitivity analysis and parametric programming is also reported in this chapter. Integration of the scheduling level within the planning problem is a way to managing the uncertainties. The integration of planning and scheduling usually leads to intractable models in terms of the required computational time. Thus one has to trade-off optimality with computational efficiency. Two alternative approaches—periodic scheduling and hierarchical approach—are presented to deal with the integration of planning and scheduling and the incorporation of uncertainty. Future work in the field of planning and scheduling under uncertainty requires extended work in the direction of more effective and general methods for dealing with the uncertainties in the process industry.

Nomenclature

Indices

i	task
j	unit
n	event point representing the beginning of a task
s	state
q^k	scenario at period k

Sets

I	tasks
I_j	tasks that can be performed in unit j
I_s	tasks that process state s and either produce or consume
J	units

J_i units that are suitable for performing task i
N event points within the time horizon
S states
IS subset of all involved intermediate states s
Q^k scenarios at period k

Parameters

$V_{ij}^{\min}$	minimum amount of material processed by task i required to start operating unit j
$V_{ij}^{\max}$	minimum amount of material processed by task i in unit j
$ST_s^{\max}$	available maximum storage capacity for state s
r_s	market requirement for state s at event point n
r_{s,q^k}^k	market requirement for state s at the end of period k under scenario q^k
ρ_{si}^p, ρ_{si}^c	proportion of state s produced, consumed from task i, respectively
α_{ij}	constant term of processing time of task i in unit j
β_{ij}	variable term of processing time of task i in unit j
U	upper bound of cycle time length
$price_s$	price of state s
P_s^{res}	required production of state s
H_k	time horizon of period k
f_{i,j,q^k}^k	fixed cost of task i at unit j at period k
v_{i,j,q^k}^k	variable cost of task i at unit j at period k
$\cos t_s^k$	cost of state s at period k
$b\cos t_s^k$	back order cost of state s at period k
hin_s^k	inventory cost of state s at period k
$Init_s$	initial input of state s at the first period
wk_q^k	weight coefficient of scenario q^k at period k
μ^k	sequence factor for period k

Variables

H	cycle time length
$wv(i,n)$	binary variables that assign the beginning of task i at event point n
$yv(j,n)$	binary variables that assign the utilization of unit j at event point n
$B(i,j,n)$	amount of material undertaking task i in unit j at event point n
$d(s,n)$	amount of state s being delivered to the market at event point n
$ST(s,n)$	amount of state s at event point n
$STIN(s)$	amount of state s imputed initially
$T^s(i,j,n)$	time that task i starts in unit j at event point n
$T^f(i,j,n)$	time that task i finishes in unit j while it starts at event point n
$P(s,n)$	production of state s at event point n
$slack(s)$	slack variable of required production for state s

$sk(s)$ slack variable orders for state s

$wv^k(i,j,q^k)$ binary variables that assign task i in unit j at period k

$b^k(i,j,q^k)$ amount of material undertaking task i in unit j at period k

$d^k(s,q^k)$ amount of state s being delivered to the market at period k

$Tp^k(i,j,q^k)$ processing time of task i in unit j at period k

$Input^k(s,q^k)$ input of state s at period k

$sk^k(s,q^k)$ back order of state s at period k

$n^k(i,j,q^k)$ number of full batches of task i in unit j at period k

Acknowledgments

The authors gratefully acknowledge financial support from the National Science Foundation under Grants CTS 0625515 and 0224745.

References

1. J. Acevedo and E.N. Pistikopoulos. A multiparametric programming approach for linear process engineering problems under uncertainty. *Industrial and Engineering Chemistry Research*, 36:717–728, 1997.
2. D. Avis and K. Fukuda. A pivoting algorithm for convex hulls and vertex enumeration of arrangements and polyhedra. *Discrete Computational Geometry*, 8:295–313, 1992.
3. H. Aytug, M.A. Lawley, K. McKay, S. Mohan, and R. Uzsoy. Executing production schedules in the face of uncertainties: A review and some future directions. *European Journal of Operational Research*, 161(1):86–110, 2005.
4. J. Balasubramanian and I.E. Grossmann. A novel branch and bound algorithm for scheduling flowshop plants with uncertain processing times. *Computers and Chemical Engineering*, 26:41–57, 2002.
5. J. Balasubramanian and I.E. Grossmann. Scheduling optimization under uncertainty - an alternative approach. *Computers and Chemical Engineering*, 27(4):469–490, 2003.
6. J. Balasubramanian and I.E. Grossmann. Approximation to multistage stochastic optimization in multiperiod batch plant scheduling under demand uncertainty. *Industrial and Engineering Chemistry Research*, 43(14):3695–3713, 2004.
7. M.H. Bassett, J.F. Pekny, and G.V. Reklaitis. Decomposition techniques for the solution of large-scale scheduling problems. *AIChE Journal*, 42:3373–3387, 1996.
8. M.H. Bassett, J.F. Pekny, and G.V. Reklaitis. Using detailed scheduling to obtain realistic operating policies for a batch processing facility. *Industrial and Engineering Chemistry Research*, 36:1717–1726, 1997.

9. A. Bemporad, M. Morari, V. Dua, and E.N. Pistikopoulos.The explicit linear quadratic regulator for constrained systems. *Automatica*, 38(1):3–20, 2002.

10. A. Ben-Tal and A. Nemirovski. Robust solutions to uncertain programs. *Operations Research Letters*, 25:1–13, 1999.

11. D. Bertsimas and M. Sim. Robust discrete optimization and network flows. *Mathematic Progress*, 98:49–71, 2003.

12. O. Bilde and J. Krarup. Sharp lower bounds and efficient algorithms for the simple plant location problem. *Annals of Discrete Mathematics*, 1: 79–97, 1977.

13. D.B. Birewar and I.E. Grossmann. Simultaneous production planning and scheduling in multiproduct batch plants.*Industrial and Engineering Chemistry Research*, 29(4):570–580, 1990.

14. A. Bonfill, M. Bagajewicz, A. Espuna, and L. Puigjaner.Risk management in the scheduling of batch plants under uncertain market demand.*Industrial and Engineering Chemistry Research*, 43(3):741–750, 2004.

15. A. Bonfill, A. Espuna, and L. Puigjaner. Addressing robustness in scheduling batch processes with uncertain operation times. *Industrial and Engineering Chemistry Research*, 44(5):1524–1534, 2005.

16. E. Borrelli, A. Bemporad, and M. Morari. A geometric algorithm for multi-parametric linear programming. *Journal of Optimization Theory and Applications*, 118:515–540, 2003.

17. S. Bose and J. Pekny. A model predictive framework for planning and scheduling problems: a case study of consumer goods supply chain. *Computers and Chemical Engineering*, 24:329–335, 2000.

18. P. Castro, A. Barbosa-Povoa, and H. Matos. Optimal periodic scheduling of batch plants using rtn-based discrete and continuous-time formulations: A case study approach.*Industrial and Engineering Chemistry Research*, 42:3346–3360, 2003.

19. L. Churh and R. Uzsoy. Analysis of periodic and event-driven rescheduling policies in dynamic shops. *International Computer Integrated Manufacturing*, 5:153–163, 1992.

20. A.R. Conn, K. Scheinberg, and P.L. Toint On the convergence of derivative-free methods for unconstrained optimization. In *Invited Presentation at the Powellfest*, Report 96/10, Cambridge, 1996.

21. B.J. Cott and S. Macchietto.Minimizing the effects of batch process variability using online schedule modification. *Computers and Chemical Engineering*, 13:105–113, 1989.

22. H. Crowder. *Computational Improvements for Subgradient Optimization*, volume XIX of *Symposia Mathematica*.Academic Press, London, 1976.

23. R.L. Daniels and P. Kouvelis. Robust scheduling to hedge against process-ing time uncertainty in single-stage production.*Management Science*, 41:363–376, 1995.

24. G.B. Dantzig. Linear programming under uncertainty. *Management Science*, 1:197–206, 1955.

25. M. Diaby, H.C. Bahl, M.H. Karwan, and S. Zionts. A lagrangean relaxation approach for very-large-scale capacitated lot-sizing. *Management Science*, 9:1329, 1992.

26. A.D. Dimitriadis, N. Shah, and C.C. Pantelides. Rtn-based rolling horizon algorithms for medium term scheduling of multipurpose plants. *Computers and Chemical Engineering*, 21:S1061–S1066, 1997.Suppl. S.

27. V. Dua and E.N. Pistikopoulos. An outer-approximation algorithm for the solution of multiparametric minlp problems. *Computers and Chemical Engineering*, 22:S955 – S958, 1998.

28. V. Dua and E.N. Pistikopoulos. Algorithms for the solution of multiparametric mixed-integer nonlinear optimization problems. *Industrial and Engineering Chemistry Research*, 38:3976 – 3987, 1999.

29. L. El-Ghaoui, F. Oustry, and H. Lebret. Robust solutions to uncertain semidefinite programs. *SIAM Journal of Optimization*, 9:33–52, 1998.

30. A. Elkamel, M. Zentner, J.F. Pekny, and G.V. Reklaitis. A decomposition heuristic for scheduling the general batch chemical plant. *Engineering Optimization*, 28(4):299–330, 1997.

31. L. Equi, G. Giorgio, S. Marziale, and A. Weintraub. A combined transportation an scheduling problem. *European Journal of Operational Research*, 97: 94, 1997.

32. D. Erlenkotter. A dual-based procedure for uncapacitated facility location. *Operational Research*, 26(1):992–1009, 1978.

33. L. Escudero, F.J. Quintana, and J. Salmeron. A modeling and an algorithmic framework for oil supply, transformation and distribution optimization under uncertainty. *European Journal of Operational Research*, 114:638–656, 1999.

34. L.F. Escudero, P.V. Kamesan, A.J. King, and R.J-B. Wets. Production planning via scenario modeling. *Annals of Operations Research*, 43:311–335, 1993.

35. M.L. Fisher. Optimal solution of scheduling problems using lagrange multipliers: Part i. *Operational Research*, 21:1114–1127, 1973.

36. M.L. Fisher and D.S. Hochbaum. Database location in a computer network. *Journ. ACM*, 27(718-735), 1980.

37. M.L. Fisher and P. Kedia. Optimal solution of set covering/partitioning problems using dual heuristics. *Management Science*, 39(67-88), 1990.

38. C.A. Floudas and X. Lin. Continuous-time versus discrete-time approaches for scheduling of chemical processes: A review. *Computers and Chemical Engineering*, 28:2109–2129, 2004.

39. K. Fukuda and A. Prodon. Double description method revisited. In M. Deza, R. Euler, and I. manoussakis, editors, *Combinatorics and Computer Science*, volume 1120 of *Lecture Notes in Computer Science*, pages 91–111. Springer-Verlag, 1996.

40. T. Gal. *Postoptimal Analyses, Parametric Programming and Related Topics*. deGruyter, 2 edition, 1995.

41. A.M. Geoffrion. Lagrangean relaxation and its uses in integer programming. *Mathematic Programmatic Studies*, 2:82, 1974.

42. V. Goel and I.E. Grossmann. A stochastic programming approach to planning of offshore gas field developments under uncertainty in reserves. *Computers and Chemical Engineering*, 28:1409–1429, 2004.

43. S.C. Graves. Using lagrangean techniques to solve hierarchical production planning problems. *Management Science*, 28:260, 1982.

44. M. Guignard. Lagrangean relaxation: A short course. *Belgian Journal of Operational Research*, 35(3-4):95, 1995.

45. M. Guignard. Lagrangean relaxation. *Top*, 11(2):151–228, 2003.

46. M. Guignard and S. Kim. Lagrangean decomposition: A model yielding stronger lagrangean bounds. *Mathematic Programing*, 39:215, 1987.

47. M. Guignard and M.B. Rosenwein. An application of lagrangean decomposition to the resource-constrained minimum weighted arborescence problem. *Networks*, 20:345–359, 1990.

48. F. Guinand, A. Moukrim, and E. Sanlaville. Sensitivity analysis of tree scheduling on two machines with communication delays. *Parallel Computing*, 30(1):103–120, 2004.

49. A. Gupta and C.D. Maranas. A hierarchical lagrangean relaxation procedure for solving midterm planning problem. *Industrial and Engineering Chemistry Research*, 38:1937, 1999.

50. N.G. Hall and M.E. Posner. Sensitivity analysis for scheduling problems. *Journal of Scheduling*, 7(1):49–83, 2004.

51. I. Harjunkoski and I.E. Grossmann. A decomposition approach for the scheduling of a steel plant production. *Computers and Chemical Engineering*, 25(11-12):1647–1660, 2001.

52. S.A.V.D. Heever and I.E. Grossmann. Disjunctive multiperiod optimization methods for design and planning of chemical process systems. *Computers and Chemical Engineering*, 23:1075–1095, 1999.

53. M. Held and R.M. Karp. The traveling salesman problem and minimum spanning trees. *Operational Research*, 18:1138–1162, 1970.

54. T.S. Hene, V. Dua, and E.N. Pistikopoulos. A hybrid parametric/stochastic programming approach for mixed- integer nonlinear problems under uncertainty. *Industrial and Engineering Chemistry Research*, 41:67–77, 2002.

55. S.J. Honkomp, L. Mockus, and G.V. Reklaitis. Robust scheduling with processing time uncertainty. *Computers and Chemical Engineering*, 21(1):S1055–S1060, 1997.

56. S.J. Honkomp, L. Mockus, and G.V. Reklaitis. A framework for schedule evaluation with processing uncertainty. *Computers and Chemical Engineering*, 23:595–609, 1999.

57. A. Huercio, A. Espuna, and L. Puigjaner. Incorporating on-line scheduling strategies in integrated batch production control. *Computers and Chemical Engineering*, 19:S609–S615, 1995.

58. M.G. Ierapetritou and C.A. Floudas. Effective continuous-time formulation for short-term scheduling. 1. multipurpose batch processes. *Industrial and Engineering Chemistry Research*, 37(11):4341–4359, 1998.

59. M.G. Ierapetritou, T.S. Hene, and C.A. Floudas. Effective continuous-time formulation for short-term scheduling. 3. multiple intermediate due dates. *Industrial and Engineering Chemistry Research*, 38(9):3446–3461, 1999.

60. M.G. Ierapetritou and E.N. Pistikopoulos. Batch plant design and operations under uncertainty. *Industrial and Engineering Chemistry Research*, 35(3): 772–787, 1996.

61. S.S. Isukapalli. *Uncertainty Analysis of Transport-Transformation Models*. PhD thesis, Rutgers University, 1999.

62. Z. Jia and M.G. Ierapetritou. Short-term scheduling under uncertainty using MILP sensitivity analysis. *Industrial and Engineering Chemistry Research*, 43:3782, 2004.

63. Z. Jia and M.G. Ierapetritou. Uncertainty analysis on the right-hand-side for MILP problems. *AIChE Journal*, 52:2486, 2006.

64. Z. Jia and M.G. Ierapetritou. Generate pareto optimal solutions of scheduling problems using normal boundary intersection technique. *Computers and Chemical Engineering*, 31(4):268–280, 2007.

65. K.B. Kanakamedala, G.V. Reklaitis, and V. Venkatasubramanian. Reactive schedule modification in multipurpose batch chemical plants. *Industrial and Engineering Chemistry Research*, 30:77–90, 1994.

66. S. Karabuk and I. Sabuncuoglu. Rescheduling frequency in an fms with uncertain processing times and unreliable machines. *Research Report*, pages IEOR–9704, 1997.

67. J.E. Kelley. The cutting-plane method for solving convex programs. *Journal SIAM*, 8:703–712, 1960.

68. E. Kondili, C.C. Pantelides, and R.W.H. Sargent. A general algorithm for short-term scheduling of batch operations. i. MILP formulation. *Computers and Chemical Engineering*, 17:211–227, 1993.

69. P. Kouvelis, R.L. Daniels, and G. Vairaktarakis. Robust scheduling of a two-machine flow shop with uncertain processing times. *IIE Trans*, 32:421–432, 2000.

70. M. Laguna. Applying robust optimization to capacity expansion of one location in telecommunications with demand uncertainty. *Management Science*, 44:S101–S110, 1998.

71. C. Lemarechal. An algorithm for minimizing convex functions. In *Proceedings IFIP'74 Congress*, pages 552–556. North Holland, 1974.

72. Z. Li and Marianthi G. Ierapetritou. Process scheduling under uncertainty using multiparametric programming. *AIChE Journal*, 53:3183–3203, 2007.

73. X. Lin, S.L. Janak, and C.A. Floudas. A new robust optimization approach for scheduling under uncertainty: I. bounded uncertainty. *Computers and Chemical Engineering*, 28:1069–1085, 2004.

74. M.L. Liu and N.V. Sahinidis. Long range planning in the process industries: A projection approach. *Computers and Operations Research*, 23(3): 237–253, 1996.

75. A. Marin and B. Pelegrin. The return plant location problem: Modeling and resolution. *European Journal of Operational Research*, 104:375, 1998.

76. C.M. McDonald and I.A. Karimi. Planning and scheduling of parallel semi-continuous processes. 1. production planning. *Industrial and Engineering Chemistry Research*, 36:2691–2700, 1997.

77. C.A. Mendez and J. Cerda. Dynamic scheduling in multiproduct batch plants. *Computers and Chemical Engineering*, 27:1247–1259, 2003.

78. C.A. Mendez and J. Cerda. An MILP framework for batch reactive scheduling with limited discrete resources. *Computers and Chemical Engineering*, 28:1059–1068, 2004.

79. C.A. Mendez, J. Cerda, I.E. Grossmann, I. Harjunkoski, and M. Fahl. State-of-the-art review of optimization methods for short-term scheduling of batch processes. *Computers and Chemical Engineering*, 30(6-7):913–946, 2006.

80. J. Mulvey, R. Vanderbei, and S. Zenios. Robust optimization of large scale systems. *Operations Research*, 43:264–281, 1995.

81. J.A. Nelder and R. Mead. A simplex method for function minimization. *Computer Journal*, 308–313, 1965.

82. S. Orcun, I.K. Altinel, and O. Hortacsu. General continuous-time models for production planning and scheduling of batch processing plants: Mixed integer linear program formulations and computational issues. *Computers and Chemical Engineering*, 25:371–389, 2001.

83. S. Orcun, K. Altinel, and O. Hortacsu. Scheduling of batch processes with operational uncertainties. *Computers and Chemical Engineering*, 20:S1191–S1196, 1996.

84. S.O. Orero and M.R. Irving. A combination of the genetic algorithm and lagrangian relaxation decomposition techniques for the generation unit commitment problem. *Electric Power Systemic Research*, 43:149–156, 1997.

85. C.C. Pantelides. Unified frameworks for the optimal process planning and scheduling. *Proceedings on the Second Conference on Foundations of Computer Aided Operations*, pages 253–274, 1994.

86. L.G. Papageorgiou and C.C. Pantelides. A hierarchical approach for campaign planning of multipurpose batch plants. *Computers and Chemical Engineering*, 17:S27–S32, 2000.

87. B. Penz, C. Rapine, and D. Trystram. Sensitivity analysis of scheduling algorithms. *European Journal of Operational Research*, 134(3):606–615, 2001.

88. S.B. Petkov and C.D. Maranas. Multiperiod planning and scheduling of multiproduct batch plants under demand uncertainty. *Industrial and Engineering Chemistry Research*, 36(11):4864–4881, 1997.

89. D. Petrovic and A. Duenas. A fuzzy logic based production scheduling/rescheduling in the presence of uncertain disruptions. *Fuzzy Sets and Systems*, 2006.

90. K. Rana and R.G. Vickson. Routing container ships using lagrangean relaxation and decomposition. *Transactions of Science*, 25(3), 1991.

91. M.T.M. Rodrigues, L. Gimeno, C.A.S. Passos, and M.D. Campos. Reactive scheduling approach for multipurpose chemical batch plants. *Computers and Chemical Engineering*, 20:S1215–S1220, 1996.

92. J. Roslof, I. Harjunkoski, J. Bjorkqvist, S. Karlsson, and T. Westerlund. An milp-based reordering algorithm for complex industrial scheduling and rescheduling. *Computers and Chemical Engineering*, 25:821–828, 2001.

93. D. Ruiz, J. Canton, J.M. Nougues, A. Espuna, and L. Puigjaner. On-line fault diagnosis system support for reactive scheduling in multipurpose batch chemical plants. *Computers and Chemical Engineering*, 25:829–837, 2001.

94. J.H. Ryu, V. Dua, and E.N. Pistikopoulos. A bilevel programming framework for enterprise-wide process networks under uncertainty. *Computers and Chemical Engineering*, 28(6-7):1121–1129, 2004.

95. I. Sabuncuoglu and M. Bayiz. Analysis of reactive scheduling problems in a job shop environment. *Research Report*, pages IEOR–9826, 1998.

96. N.V. Sahinidis, I.E. Grossmann, R.E. Fornari, and M. Chathrathi. Optimization model for long range planning in the chemical industry. *Computers and Chemical Engineering*, 13(9):1049–1063, 1989.

97. O. Samikoglu, S.J. Honkomp, J.F. Pekny, and G.V. Reklaitis. Sensitivity analysis for project planning and scheduling under uncertain completions. *Computers and Chemical Engineering*, 22:S871–S874, 1998. Suppl. S.

98. G. Sand, S. Engell, A. Markert, R. Schultz, and C. Schultz. Approximation of an ideal online scheduler for a multiproduct batch plant. *Computers and Chemical Engineering*, 24:361–367, 2000.

99. E. Sanmarti, A. Espuna, and L. Puigjaner. Batch production and preventive maintenance scheduling under equipment failure uncertainty. *Computers and Chemical Engineering*, 21(10):1157–1168, 1997.

100. E. Sanmarti, A. Huercio, A. Espuna, and L. Puigjaner. A combined scheduling/reactive scheduling strategy to minimize the effect of process operations uncertainty in batch plants. *Computers and Chemical Engineering*, 20:1263–1268, 1996.

101. V. Saraswat and P.V. Hentenryck. Principles and practice of constraint programming. pages 245–268. The MIT Press, 1995.

102. G. Schilling and C.C. Pantelides. Optimal periodic scheduling of multipurpose plants. *Computers and Chemical Engineering*, 23(4-5):635–655, 1999.

103. M. Seron, J.A. De Dona, and G.C. Goodwin. Global analytical model predictive control with input constraints. In *Proceedings of IEEE Conference Decision and Control*, Sydney, 2000.

104. N. Shah and C.C. Pantelides. A general algorithm for short term scheduling of batch operations – II. Computational issues. *Computers and Chemical Engineering*, 17:229, 1993.

105. N. Shah, C.C. Pantelides, and R.W.H. Sargent. Optimal periodic scheduling of multipurpose batch plants. *Annals of Operations Research*, 42: 193–228, 1993.

106. A.L. Soyster. Convex programming with set-inclusive constraints and applications to inexact linear programming. *Operations Research*, 21:1154–1157, 1973.

107. D. Subramanian, J.F. Pekny, and G.V. Reklaitis. A simulation-optimization framework for addressing combinatorial and stochastic aspects of an r&d pipeline management problem. *Computers and Chemical Engineering*, 24: 1005–1011, 2000.

108. V. Suresh and D. Chaudhuri. Dynamic scheduling-a survey of research. *International Journal of Production Economics*, 32(1):53–63, 1993.

109. J.M. Thizy and L.N.V. Wassenhove. Lagrangean relaxation for the multi-item capacitated lot-sizing problem: A heuristic implementation. *IIE Trans*, 17: 308, 1985.

110. G.L. Thompson and D.J. Zawack. A problem expanding parametric programming method for solving the job shop scheduling problem. *Annals of Operations Research*, 4(1):327 – 342, 1985.

111. P. Tondel, T.A. Johansen, and A. Bemporad. An algorithm for multi-parametric quadratic programming and explicit MPC solutions. In *IEEE Conference on Decision and Control*, pages 1199–1204, 2001.

112. M. Turkay and I.E. Grossmann. Logic-based MINLP algorithms for the optimization of process systems with discontinuous investment costs-multiple size regions. *Industrial and Engineering Chemistry Research*, 35:2611–2623, 1996.

113. S.A.V.E. Heever and I.E. Grossmann. A strategy for the integration of production planning and reactive scheduling in the optimization of a hydrogen supply network. *Computers and Chemical Engineering*, 27(12):1813–1839, 2003.

114. J.P. Vin and M.G. Ierapetritou. A new approach for efficient rescheduling of multiproduct batch plants. *Industrial and Engineering Chemistry Research*, 39:4228–4238, 2000.

115. J.P. Vin and M.G. Ierapetritou. Robust short-term scheduling of multiproduct batch plants under demand uncertainty. *Industrial and Engineering Chemistry Research*, 40:4543–4554, 2001.

116. J. Wang. A fuzzy robust scheduling approach for product development projects. *European Journal of Operational Research*, 152(1):180–194, 2004.

117. D. Wu and M.G. Ierapetritou. Hierarchical approach for production planning andscheduling under uncertainty using continuous-time formulation. In *Foundations of Computer-Aided Process Design 2004 Conference Proceedings*, Princeton University, 2004.
118. C. Yu and H. Li. A robust optimization model for stochastic logistic problems. *International Journal of Production Economics*, 64:385–397, 2000.
119. X.X. Zhu and T. Majozi. Novel continuous-time milp formulation for multipurpose batch plants. 2. integrated planning and scheduling. *Industrial and Engineering Chemistry Research*, 40:5621–5634, 2001.
120. J. Zowe. *Nondifferentiable Optimization.* Comput. Math. Progrm. in Computer and Systems Science, NATO ASI Series. Springer-Verlag, 1985.

A Relative Robust Optimization Approach for Full Factorial Scenario Design of Data Uncertainty and Ambiguity

Tiravat Assavapokee[1], Matthew J. Realff[2], and Jane C. Ammons[3]

[1] Department of Industrial Engineering
University of Houston, Houston, TX 77201
tiravat.assavapokee@mail.uh.edu
[2] Department of Chemical and Biomolecular Engineering
Georgia Institute of Technology, Atlanta, GA 30332
matthew.realff@chbe.gatech.edu
[3] Department of Industrial and Systems Engineering
Georgia Institute of Technology, Atlanta, GA 30332
jane.ammons@isye.gatech.edu

Summary This chapter presents a relative robust optimization algorithm for two-stage decision making under uncertainty (ambiguity) where the structure of the first-stage problem is a mixed integer linear programming model and the structure of the second-stage problem is a linear programming model. In the structure of the considered problem, each uncertain parameter can take its value from a finite set of real numbers with unknown probability distribution independently of other parameters' settings. This structure of parametric uncertainty is referred to in this chapter as the full-factorial scenario design of data uncertainty. The algorithm is shown to be efficient for solving large-scale relative robust optimization problems under this structure of the parametric uncertainty. The algorithm coordinates three computational stages to efficiently solve the overall optimization problem. Bi-level programming formulations are the main components in two of these three computational stages. The main contributions of this chapter are the theoretical development of the robust optimization algorithm and its applications in robust strategic decision making under uncertainty (e.g., supply chain network infrastructure design problems).

1 Introduction

In this work, we address the problem of two-stage decision making under uncertainty, where the uncertainty appears in the values of key parameters of a mixed integer linear programming (MILP) formulation presented as

W. Chaovalitwongse et al. (eds.), *Optimization and Logistics Challenges in the Enterprise*, Springer Optimization and Its Applications 30,
DOI 10.1007/978-0-387-88617-6_4, © Springer Science+Business Media, LLC 2009

$$\min \ \vec{\mathbf{c}}^T \vec{\mathbf{x}} + \vec{\mathbf{q}}^T \vec{\mathbf{y}}$$
$$s.t. \ \mathbf{W}\vec{\mathbf{y}} \geq \vec{\mathbf{h}} + \mathbf{T}\vec{\mathbf{x}}$$
$$\mathbf{V}\vec{\mathbf{y}} = \vec{\mathbf{g}} + \mathbf{S}\vec{\mathbf{x}}$$
$$\vec{\mathbf{x}} \in \{0,1\}^{|\vec{\mathbf{x}}|}, \vec{\mathbf{y}} \geq \vec{\mathbf{0}}$$

In this chapter, bolded lower case letters with vector cap such as $\vec{\mathbf{x}}$ represent vectors and the notation x_i represents the ith element of the vector $\vec{\mathbf{x}}$. The corresponding bolded upper case letters such as $\mathbf{W}$ denote matrices and the notation W_{ij} represents the $(i,j)^{th}$ element of the matrix $\mathbf{W}$.

In the considered model, let the vector $\vec{\mathbf{x}}$ represent the first-stage decision setting made before the realization of uncertainty and the vector $\vec{\mathbf{y}}$ represent the second-stage decision setting made after the realization of uncertainty. Each uncertain parameter in the model can independently take its value from a finite set of real values with unknown probability distribution. In other words, let vector $\boldsymbol{\xi} = [\vec{\mathbf{c}}, \vec{\mathbf{q}}, \vec{\mathbf{h}}, \vec{\mathbf{g}}, \mathbf{T}, \mathbf{S}, \mathbf{W}, \mathbf{V}]$ denote the parameters defining the objective function and the constraints of the optimization problem. For each element p of the vector $\boldsymbol{\xi}$, p can take any value from the finite set of real values: $\{p_{(1)}, p_{(2)}, \ldots, p_{(\bar{p})}\}$ such that $p_{(1)} < p_{(2)} < \ldots < p_{(\bar{p})}$ with unknown probability distribution where the notation $\bar{p}$ represents the number of possible values for the parameter p. This type of parameter setting can happen in decision-making problems when each uncertain parameter is classified into many possible levels (e.g., low, medium, and high) based on expert's opinion with unknown joint probability distribution. For simplicity of notations, we will equivalently use the notations p^U and p^L as $p_{(\bar{p})}$ and $p_{(1)}$, respectively, for any parameter p in this chapter.

Because of the lack of complete knowledge about the joint probability distribution of uncertain parameters in the model, decision makers cannot search for the first-stage decision setting with the best long-run average performance. Instead, decision makers can search for the first-stage decision setting that performs reasonably well (reasonable objective function value) across all possible input scenarios without attempting to assign assumed probability distributions to any ambiguous parameters. This resulting first-stage decision setting is referred to as the robust decision setting. In this work, we develop an optimization algorithm for assisting decision makers who search for the robust first-stage decision setting under the relative robustness definition (min-max relative regret robust solution) defined in [21].

Traditionally, a relative robust solution can be obtained by solving a scenario-based extensive form model of the problem which is also a MILP model (explained in detail in Section 3). The size of this extensive form model grows substantially with the number of possible scenarios used to represent uncertainty as does the required computation time to find optimal solutions. Under the full factorial scenario design, the number of possible scenarios grows exponentially with the number of uncertain parameters and the number of possible values for each uncertain parameter (e.g., a problem with 20 uncertain parameters each with 5 possible values has over 95,367 billion scenarios). Solving the extensive form model directly obviously is not the efficient way for solving this type of robust decision problems even with the use of Benders' decomposition technique [11]. For example, if the problem contains 95,367 billion scenarios and a subproblem of Benders' decomposition can be solved within 0.0001 second, the algorithm would require approximately 302 years of computation time per iteration to generate a set of required cuts. For this reason, neither the extensive form model nor the direct application of the Benders' decomposition algorithm with one subproblem per scenario are efficient tools for solving robust decision problems under this type of parametric uncertainty. In addition, we also demonstrate a counterexample in Section 4 illustrating the inefficiency of the robust solution obtained by considering only scenarios generated by the combinations of end points of each uncertain parameter. This example illustrates that the end point robust solution may converge to a nonoptimal robust solution and may not be an efficient solution to the considered problem.

Because of the failure of the extensive form model, the Benders' decomposition algorithm and the end point robust solution for solving a large-scale problem of this type, the three-stage optimization algorithm is proposed in this chapter for solving this type of relative robust optimization problems. The algorithm can be used to determine the robust setting of the first-stage decisions when the only information available to the decision maker at the time is a finite set of possible real values for each uncertain parameter in the model with unknown probability distribution. The algorithm is designed explicitly to efficiently handle a combinatorial-sized set of possible scenarios. The proposed algorithm sequentially solves and updates a relaxation problem until both feasibility and optimality conditions are satisfied. The feasibility and optimality verification steps involve the use of bi-level programming formulations, which coordinate the Stackelberg game [28] between the decision environment and decision makers which is explained in detail in Section 3. Several preprocessing procedures and problem transformation steps are also presented to improve the computational tractability of the algorithm. The algorithm is proven to generate an optimal relative robust solution (if one exists) in a finite number of iterations.

In the following section, we summarize related literature in robust optimization and supporting topics. In Section 3, we detail the robust optimization approach and solution algorithm. In Section 4, we illustrate two small examples and an application of the proposed algorithm in solving the relative robust facility location problem under an incredibly large number of possible scenarios. The purpose of the first example is to demonstrate the mechanism of the proposed algorithm to readers on a small optimization problem. The second example presents the counterexample illustrating the inefficiency of the end point robust solution. All results illustrate good computational performance of the proposed algorithm on an example of practical size.

2 Background

Two-stage relative robust optimization addresses optimization problems where some of the model parameters are uncertain or ambiguous at the time of making the first-stage decisions. In many cases, decision makers have to make long-term decisions (e.g., capacity decisions and/or location decisions) long before the realization occurs for uncertain parameter values. Many of these long-term decisions are typically represented by binary and/or discrete variables. After the first-stage decisions have been made and the decision maker obtains the realization of uncertain parameters, the second-stage decisions will then be made under the fixed settings of the first-stage decisions. One criterion for the first-stage decisions can be to minimize the maximum relative regret between the optimal objective function value under perfect information and the resulting objective function value under the robust decisions over all possible realizations of the uncertain parameters (scenarios) in the model. The work of Kouvelis and Yu [21] summarizes the state-of-the-art in min-max regret and relative regret optimization up to 1997 and provides a comprehensive discussion of the motivation for the min-max regret and relative regret approaches and various aspects of applying them in practice. Ben-Tal and Nemirovski [12], [13], [14], [15] address robust solutions (min-max/max-min objective) by allowing the uncertainty sets for the data to be ellipsoids, and propose efficient algorithms to solve convex optimization problems under data uncertainty. Mausser and Laguna [22],[23],[24] present the mixed integer linear programming formulation to finding the maximum regret scenario for a given candidate robust solution of one-stage linear programming problems under interval data uncertainty for objective function coefficients. They also develop an iterative algorithm for finding the robust solution of one-stage linear programming problems under relative robust criterion and a heuristic algorithm under the min-max absolute regret criterion with the similar type of uncertainty. Averbakh [3], [4] shows that polynomial solvability is preserved for a specific discrete optimization problem (selecting p elements of minimum total weight out of a set of m elements with uncertainty in weights of the elements) when each weight can vary within an interval under the min-max regret robustness definition. Bertsimas and Sim [16], [17] propose an approach to address data uncertainty for discrete optimization and network flow problems that allows the degree of conservatism of the solution (min-max/max-min objective) to be controlled. They show that the robust counterpart of an NP-hard α-approximable 0-1 discrete optimization problem remains α-approximable. They also propose an algorithm for robust network flows that solves the robust counterpart by solving a polynomial number of nominal minimum cost flow problems in a modified network. Assavapokee et al. [1], [2] present a scenario relaxation algorithm for solving scenario-based min-max regret and min-max relative regret robust optimization problems for the mixed integer linear programming formulations. They also present a min-max regret robust optimization algorithm for two-stage decision making under interval data uncertainty for the mixed integer linear programming formulation. Theoretical concepts on Bi-level and Multi-level programming can be found in [7], [8], [9], [10], [19], [20], and [25]. The work by Bajalinov [6] thoroughly summarizes theoretical concepts on fractional programming up to 2003.

In the following section, we present the theoretical methodology of the relative robust optimization algorithm when the structure of parametric uncertainty allows any combinations of a discrete set of values: a "full-factorial uncertainty representation." Further in Section 3.1, we illustrate some example applications of the proposed algorithm on solving relative robust optimization problems under the full-factorial uncertainty representation with an extremely large number of scenarios.

3 Methodology

This section begins by reviewing key concepts of scenario-based relative robust optimization (i.e., types of decisions and its extensive form formulation). The methodology of the proposed algorithm is then summarized and explained in detail, and all of its three stages are specified. The section concludes with the proof that the algorithm always generates the relative robust optimal solution (if one exists) in a finite number of iterations. We address the problem where the basic components of the model's uncertainty are represented by a finite set of all possible scenarios of model parameters, referred to as the scenario set $\bar{\Omega}$. The problem contains two types of decision variables. The first-stage variables model binary choice decisions, which have to be made before the realization of uncertainty. The second-stage decisions are continuous recourse decisions, which can be made after the realization of uncertainty. Let vector $\overrightarrow{\mathbf{x}}_\omega$ denote binary choice decision variables and let vector $\overrightarrow{\mathbf{y}}_\omega$ denote continuous recourse decision variables and let $\overrightarrow{\mathbf{c}}_\omega, \overrightarrow{\mathbf{q}}_\omega, \overrightarrow{\mathbf{h}}_\omega, \overrightarrow{\mathbf{g}}_\omega, \mathbf{T}_\omega, \mathbf{S}_\omega, \mathbf{W}_\omega, \mathbf{V}_\omega$ denote model parameters setting for each scenario $\omega \in \bar{\Omega}$. If the realization of a model parameter is known to be scenario ω a priori, the optimal choice for the decision variables $\overrightarrow{\mathbf{x}}_\omega$ and $\overrightarrow{\mathbf{y}}_\omega$ can be obtained by solving the following model (1).

$$O_\omega^* = \min \overrightarrow{\mathbf{c}}_\omega^T \overrightarrow{\mathbf{x}}_\omega + \overrightarrow{\mathbf{q}}_\omega^T \overrightarrow{\mathbf{y}}_\omega$$
$$s.t. \ \mathbf{W}_\omega \overrightarrow{\mathbf{y}}_\omega - \mathbf{T}_\omega \overrightarrow{\mathbf{x}}_\omega \geq \overrightarrow{\mathbf{h}}_\omega \tag{1}$$
$$\mathbf{V}_\omega \overrightarrow{\mathbf{y}}_\omega - \mathbf{S}_\omega \overrightarrow{\mathbf{x}}_\omega = \overrightarrow{\mathbf{g}}_\omega$$
$$\overrightarrow{\mathbf{x}}_\omega \in \{0,1\}^{|\overrightarrow{\mathbf{x}}_\omega|}, \overrightarrow{\mathbf{y}}_\omega \geq \overrightarrow{\mathbf{0}}$$

When parameters' uncertainty (ambiguity) exists, the search for the relative robust solution comprises finding binary choice decisions, $\overrightarrow{\mathbf{x}}$, such that the function $\max_{\omega \in \bar{\Omega}}((Z_\omega^*(\overrightarrow{\mathbf{x}}) - O_\omega^*)/O_\omega^*)$ is minimized where

$$Z_\omega^*(\overrightarrow{\mathbf{x}}) = \min_{\overrightarrow{\mathbf{y}}_\omega \geq \overrightarrow{\mathbf{0}}} \{\overrightarrow{\mathbf{q}}_\omega^T \overrightarrow{\mathbf{y}}_\omega \mid \mathbf{W}_\omega \overrightarrow{\mathbf{y}}_\omega - \mathbf{T}_\omega \overrightarrow{\mathbf{x}} \geq \overrightarrow{\mathbf{h}}_\omega,$$

$$\mathbf{V}_\omega \overrightarrow{\mathbf{y}}_\omega - \mathbf{S}_\omega \overrightarrow{\mathbf{x}} = \overrightarrow{\mathbf{g}}_\omega\} + \overrightarrow{\mathbf{c}}_\omega^T \overrightarrow{\mathbf{x}} \qquad \forall \omega \in \bar{\Omega}.$$

In the case when the scenario set $\bar{\Omega}$ is a finite set, the optimal choice of decision variables $\overrightarrow{\mathbf{x}}$ (relative robust solution) can be obtained by solving the following model (2).

$$\min \delta$$
$$\begin{aligned} s.t. \quad & O_\omega^* \delta \geq \overrightarrow{\mathbf{c}}_\omega^T \overrightarrow{\mathbf{x}} + \overrightarrow{\mathbf{q}}_\omega^T \overrightarrow{\mathbf{y}}_\omega - O_\omega^* && \forall \omega \in \bar{\Omega} \\ & \mathbf{W}_\omega \overrightarrow{\mathbf{y}}_\omega - \mathbf{T}_\omega \overrightarrow{\mathbf{x}} \geq \overrightarrow{\mathbf{h}}_\omega && \forall \omega \in \bar{\Omega} \\ & \mathbf{V}_\omega \overrightarrow{\mathbf{y}}_\omega - \mathbf{S}_\omega \overrightarrow{\mathbf{x}} = \overrightarrow{\mathbf{g}}_\omega && \forall \omega \in \bar{\Omega} \\ & \overrightarrow{\mathbf{x}} \in \{0,1\}^{|\overrightarrow{\mathbf{x}}|}, \overrightarrow{\mathbf{y}}_\omega \geq \overrightarrow{\mathbf{0}} && \forall \omega \in \bar{\Omega} \end{aligned} \qquad (2)$$

This model (2) is referred to as the extensive form model of the problem. If an optimal solution for the model (2) exists, the resulting binary solution is the optimal setting of decision variables $\overrightarrow{\mathbf{x}}$. In this work, we consider the class of decision problems such that the O_ω^* value is nonnegative $\forall \omega \in \bar{\Omega}$, which is quite common for minimum cost decision problems under relative robust criterion.

Unfortunately, the size of the extensive form model can become unmanageably large and so does the required computation time to find the optimal setting of $\overrightarrow{\mathbf{x}}$. Because of the failure of the extensive form model and the Benders' decomposition algorithm for solving a large-scale problem of this type, a new algorithm, which can effectively overcome these limitations, is proposed in the following subsection. A key insight of this algorithm is to recognize that, even for a large set of scenarios, it is possible to identify a smaller set of important scenarios that actually need to be considered as part of the iteration scheme to solve the overall problem with the use of bi-level programming.

Let us define some additional notations which will be extensively used in this section and the following sections of the chapter. The uncertain parameters in the model (1) can be classified into eight major types. These uncertain parameters can be combined into a random vector $\boldsymbol{\xi} = [\overrightarrow{\mathbf{c}}, \overrightarrow{\mathbf{q}}, \overrightarrow{\mathbf{h}}, \overrightarrow{\mathbf{g}}, \mathbf{T}, \mathbf{S}, \mathbf{W}, \mathbf{V}]$. Because, in most cases, the values of parameters of type $\overrightarrow{\mathbf{c}}$ are known with certainty when making the first-stage decisions, the proposed algorithm handles uncertainty for seven other types of parameters. In this work, we assume that each component of $\boldsymbol{\xi}$ (except parameters of type $\overrightarrow{\mathbf{c}}$) can independently take its values from a finite set of real numbers with unknown probability distribution. This means that for any element p of the vector $\boldsymbol{\xi}$, p can take any value from the set of $\{p_{(1)}, p_{(2)}, \ldots, p_{(\bar{p})}\}$ such that $p_{(1)} < p_{(2)} < \ldots < p_{(\bar{p})}$ where $\bar{p}$ denotes the number of possible values for the parameter p. The scenario set $\bar{\Omega}$ is generated by all possible values of the parameter vector $\boldsymbol{\xi}$. Let us define $\boldsymbol{\xi}(\omega)$ as the specific setting of the parameter vector $\boldsymbol{\xi}$ under scenario $\omega \in \bar{\Omega}$ and $\Xi = \{\boldsymbol{\xi}(\omega) \mid \omega \in \bar{\Omega}\}$ as the support of the random vector $\boldsymbol{\xi}$. As described below, we propose a three-stage optimization algorithm for solving the relative robust optimization problem under scenario set $\bar{\Omega}$ that utilizes the creative idea based on the following inequality where $\Omega \subseteq \bar{\Omega}$.

$$\begin{aligned}
\triangle^U &= \max_{\omega \in \bar{\Omega}}\{(Z_\omega^*(\overrightarrow{\mathbf{x}}) - O_\omega^*)/O_\omega^*\} \\
&\geq \min_{\overrightarrow{\mathbf{x}}} \max_{\omega \in \bar{\Omega}}\{(Z_\omega^*(\overrightarrow{\mathbf{x}}) - O_\omega^*)/O_\omega^*\} \\
&\geq \min_{\overrightarrow{\mathbf{x}}} \max_{\omega \in \Omega}\{(Z_\omega^*(\overrightarrow{\mathbf{x}}) - O_\omega^*)/O_\omega^*\} = \triangle^L
\end{aligned}$$

In the considered problem, we would like to solve the middle problem, $\min_{\overrightarrow{\mathbf{x}}} \max_{\omega \in \bar{\Omega}}\{(Z_\omega^*(\overrightarrow{\mathbf{x}}) - O_\omega^*)/O_\omega^*\}$, which is intractable because $\mid \bar{\Omega} \mid$ is extremely large. Instead, we successively solve the left and right problems for $\triangle^U$ and $\triangle^L$. The left problem is solved by utilizing a reformulation as a tractable bi-level programming model. The right problem is solved by utilizing the extensive form model based on the fact that $\mid \Omega \mid$ is relatively small compared to $\mid \bar{\Omega} \mid$. The proposed algorithm proceeds as follows.

Three-Stage Algorithm

Step 0: (Initialization) Choose a subset $\Omega \subseteq \bar{\Omega}$ and set $\triangle^U = \infty$ and $\triangle^L = 0$. Let $\overrightarrow{\mathbf{x}}_{opt}$ denote the incumbent solution. Determine the value of $\epsilon \geq 0$, which is prespecified tolerance and proceed to Step 1.

Step 1: Solve the model (1) to obtain $O_\omega^* \ \forall \omega \in \Omega$. If the model (1) is infeasible for any scenario in the scenario set Ω, the algorithm is terminated; the problem is ill-posed. Otherwise the optimal objective function value to the model (1) for scenario ω is designated as O_ω^*. Proceed to Step 2.

Step 2: (Updating Lower Bound) Solve the smaller version of the model (2) by considering only the scenario set Ω instead of $\bar{\Omega}$. This smaller version of the model (2) is referred to as the relaxed model (2) in this chapter. If the relaxed model (2) is infeasible, the algorithm is terminated with the confirmation that no robust solution exists for the problem. Otherwise, set $\overrightarrow{\mathbf{x}}_\Omega = \overrightarrow{\mathbf{x}}^*$ which is an optimal solution from the relaxed model (2) and set $\triangle^L = \max_{\omega \in \Omega}\{(Z_\omega^*(\overrightarrow{\mathbf{x}}^*) - O_\omega^*)/O_\omega^*\}$ which is the optimal objective function value from the relaxed model (2). If $\{\triangle^U - \triangle^L\} \leq \epsilon$, the $\overrightarrow{\mathbf{x}}_{opt}$ is the globally ϵ-optimal robust solution and the algorithm is terminated. Otherwise the algorithm proceeds to Step 3.

Step 3: (Feasibility Check) Solve the bi-level-1 model described in detail in Section 3.2 by using the $\overrightarrow{\mathbf{x}}_\Omega$ information from Step 2. If the optimal objective function value of the bi-level-1 model is nonnegative (feasible case), proceed to Step 4. Otherwise (infeasible case), $\Omega \leftarrow \Omega \cup \{\omega_1^*\}$ where ω_1^* is the infeasible scenario for $\overrightarrow{\mathbf{x}}_\Omega$ generated by the bi-level-1 model in this iteration and return to Step 1.

First Stage

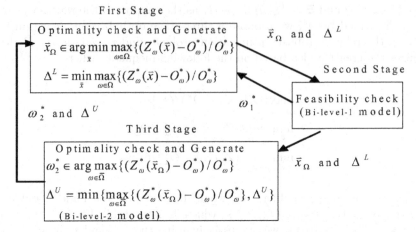

Fig. 1. Schematic structure of the algorithm

Step 4: (Updating Upper Bound) Solve the bi-level-2 model specified in detail in Section 3.3 by using the $\overrightarrow{x}_\Omega$ information from Step 2. Let $\omega_2^* \in$ arg $\max_{\omega \in \bar{\Omega}}\{(Z_\omega^*(\overrightarrow{x}_\Omega) - O_\omega^*)/O_\omega^*\}$ and $\triangle^{U*} = \max_{\omega \in \bar{\Omega}}\{(Z_\omega^*(\overrightarrow{x}_\Omega) - O_\omega^*)/O_\omega^*\}$ represent the results generated by the bi-level-2 model, respectively, in this iteration. If $\triangle^{U*} < \triangle^U$, then set $\overrightarrow{x}_{opt} = \overrightarrow{x}_\Omega$ and set $\triangle^U = \triangle^{U*}$. Otherwise $\Omega \leftarrow \Omega \cup \{\omega_2^*\}$ and return to Step 1.

We define the algorithm Steps 1 and 2 as the first stage of the algorithm and the algorithm Step 3 and Step 4 as the second stage and the third stage of the algorithm, respectively. Figure 1 illustrates a schematic structure of this algorithm. Each of these three stages of the proposed algorithm is detailed in the following subsections.

3.1 The First-Stage Algorithm

The purposes of the first stage algorithm are (a) to find the candidate robust decision from a considered finite subset of scenarios, Ω, (b) to find the lower bound on the min-max relative regret value, and (c) to determine if the algorithm has discovered a global optimal (or an ϵ-optimal) relative robust solution for the problem. The first stage algorithm consists of two main optimization models: the model (1) and relaxed model (2). The model (1) is used to calculate the value of O_ω^* $\forall \omega \in \Omega$. Note that this algorithm does not require the calculation of O_ω^* for all scenario $\omega \in \bar{\Omega}$. The algorithm only calculates the value of O_ω^* for scenarios in the small subset Ω of $\bar{\Omega}$. In each iteration, the set of scenarios in Ω is enlarged to include one additional scenario generated either by the second stage or the third stage of the algorithm. If the model (1) is infeasible for any generated scenario, the algorithm is terminated with the conclusion that there exists no robust solution to the problem. Otherwise the values of O_ω^* for all scenarios in the set Ω are used as the required parameters in the relaxed model (2). The relaxed model (2), which is the smaller version of the extensive form model by considering only the scenario set Ω instead of $\bar{\Omega}$, is then solved. If the relaxed model (2) is infeasible, the algorithm is terminated with the conclusion that there exists no robust solution to the problem. Otherwise, its results are the candidate robust decision, $\overrightarrow{\mathbf{x}}_\Omega$, such that

$$\max_{\omega \in \Omega}\{(Z_\omega^*(\overrightarrow{\mathbf{x}}_\Omega) - O_\omega^*)/O_\omega^*\} = \min_{\overrightarrow{\mathbf{x}}} \max_{\omega \in \Omega}\{(Z_\omega^*(\overrightarrow{\mathbf{x}}) - O_\omega^*)/O_\omega^*\}$$

and the lower bound value on the min-max relative regret value, Δ^L, which is the optimal objective function value of the relaxed model (2). The optimality condition is then checked. The optimality condition will be satisfied when $\{\Delta^U - \Delta^L\} \leq \epsilon$, where $\epsilon \geq 0$ is prespecified tolerance. If the optimality condition is satisfied, the algorithm is terminated with the solution $\overrightarrow{\mathbf{x}}_{opt}$ which is the ϵ-optimal robust solution. Otherwise the solution $\overrightarrow{\mathbf{x}}_\Omega$ and the value of Δ^L are forwarded to the second stage of the algorithm.

3.2 The Second-Stage Algorithm

The main purpose of the second-stage algorithm is to identify a scenario $\omega_1^* \in \bar{\Omega}$ which admits no feasible solution to $Z_\omega^*(\overrightarrow{\mathbf{x}}_\Omega)$ for $\omega = \omega_1^*$. To achieve this goal, the algorithm solves a bi-level programming problem referred to as the bi-level-1 model by following two main steps. In the first step, the algorithm starts by preprocessing model parameters. At this point, some model parameters' values in the original bi-level-1 model are predetermined at their optimal setting by following some simple preprocessing rules. In the second step, the bi-level-1 model is transformed from its original form into a single-level mixed integer linear programming structure. Next we describe the key concepts of each algorithm step and the structure of the bi-level-1 model.

One can find a model parameters' setting or a scenario $\omega_1^* \in \bar{\Omega}$ which admits no feasible solution to $Z_\omega^*(\overrightarrow{\mathbf{x}}_\Omega)$ for $\omega = \omega_1^*$ by solving the following bi-level programming problem referred to as the bi-level-1 model. The following model (3) demonstrates the general structure of the bi-level-1 model.

$$\min_\xi \quad \delta$$
$$\text{s.t.} \quad \xi \in \Xi$$
$$\max_{\overrightarrow{\mathbf{y}}, \overrightarrow{\mathbf{s}}, \overrightarrow{\mathbf{s}}_1, \overrightarrow{\mathbf{s}}_2, \delta} \quad \delta$$

$$\text{s.t.} - \mathbf{W}\overrightarrow{\mathbf{y}} + \overrightarrow{\mathbf{s}} = -\overrightarrow{\mathbf{h}} - \mathbf{T}\overrightarrow{\mathbf{x}}_\Omega \tag{3}$$
$$\mathbf{V}\overrightarrow{\mathbf{y}} + \overrightarrow{\mathbf{s}}_1 = \overrightarrow{\mathbf{g}} + \mathbf{S}\overrightarrow{\mathbf{x}}_\Omega$$
$$-\mathbf{V}\overrightarrow{\mathbf{y}} + \overrightarrow{\mathbf{s}}_2 = -\overrightarrow{\mathbf{g}} - \mathbf{S}\overrightarrow{\mathbf{x}}_\Omega$$
$$\delta\overrightarrow{\mathbf{1}} \leq \overrightarrow{\mathbf{s}}, \delta\overrightarrow{\mathbf{1}} \leq \overrightarrow{\mathbf{s}}_1, \delta\overrightarrow{\mathbf{1}} \leq \overrightarrow{\mathbf{s}}_2, \overrightarrow{\mathbf{y}} \geq \overrightarrow{\mathbf{0}}$$

In the bi-level-1 model, the leader's objective is to make the problem infeasible by controlling the parameters' settings. The follower's objective is to make the problem feasible by controlling the continuous decision variables, under the fixed parameters setting from the leader problem, when the setting of binary decision variables is fixed at $\overrightarrow{x}_\Omega$. In the model (3), δ represents a scalar decision variable and $\overrightarrow{0}$ and $\overrightarrow{1}$ represent the vector with all elements equal to zero and one, respectively. The current form of the model (3) has a nonlinear bi-level structure with a set of constraints restricting the possible values of the decision vectors $\boldsymbol{\xi} = [\overrightarrow{c}, \overrightarrow{q}, \overrightarrow{h}, \overrightarrow{g}, \mathbf{T}, \mathbf{S}, \mathbf{W}, \mathbf{V}]$. Because the structure of the follower problem of the model (3) is a linear program and it affects the leader's decisions only through its objective function, we can simply replace this follower problem with explicit representations of its optimality conditions. These explicit representations include the follower's primal constraints, the follower's dual constraints, and the follower's strong duality constraint.

Furthermore, from the special structure of the model (3), all elements in decision matrixes $\mathbf{T}, \mathbf{W}$, and vector $\overrightarrow{h}$ can be predetermined to either one of their bounds even before solving the model (3). For each element of the decision matrix $\mathbf{W}$ in the model (3), the optimal setting of this decision variable is the lower bound of its possible values. The correctness of these simple rules is obvious based on the fact that $\overrightarrow{y} \geq \overrightarrow{0}$. Similarly, for each element of the decision vector $\overrightarrow{h}$ and matrix $\mathbf{T}$, the optimal setting of this decision variable in the model (3) is the upper bound of its possible values.

Lemma 1. *The model (3) has at least one optimal solution* $\mathbf{T}^*, \overrightarrow{h}^*, \mathbf{W}^*, \mathbf{S}^*$, $\overrightarrow{g}^*$, *and* $\mathbf{V}^*$ *in which each element of these vectors takes on a value from one of its bounds.*

Proof. Because the optimal setting of each element of $\mathbf{T}, \overrightarrow{h}$, and $\mathbf{W}$ already takes its value from one of its bounds, we only need to prove this Lemma for each element of $\mathbf{S}, \overrightarrow{g}$, and $\mathbf{V}$. Each of these variables S_{il}, g_i, and V_{ij} appears in only two constraints in the model (3): $\sum_j V_{ij} y_j + s_{1i} = g_i + \sum_l S_{il} x_{\Omega l}$ and $-\sum_j V_{ij} y_j + s_{2i} = -g_i - \sum_l S_{il} x_{\Omega l}$. It is also easy to see that $s_{1i} = -s_{2i}$ and $\min\{s_{1i}, s_{2i}\} = -\mid s_{1i} - s_{2i} \mid /2$. This fact implies that the optimal setting of $\overrightarrow{y}$ which maximizes $\min\{s_{1i}, s_{2i}\}$ will also minimize $\mid s_{1i} - s_{2i} \mid /2$ and vise versa under the fixed setting of ξ. Because $\mid s_{1i} - s_{2i} \mid /2 = \mid g_i + \sum_l S_{il} x_{\Omega l} - \sum_j V_{ij} y_j \mid$, the optimal setting of S_{il}, g_i, and V_{ij} will maximize $\min_{\overrightarrow{y} \in \chi(\overrightarrow{x}_\Omega)} \mid g_i + \sum_l S_{il} x_{\Omega l} - \sum_j V_{ij} y_j \mid$ where $\chi(\overrightarrow{x}_\Omega) = \{\overrightarrow{y} \geq \overrightarrow{0} \mid \mathbf{W}\overrightarrow{y} \geq \overrightarrow{h} + \mathbf{T}\overrightarrow{x}_\Omega, \mathbf{V}\overrightarrow{y} = \overrightarrow{g} + \mathbf{S}\overrightarrow{x}_\Omega\}$. In this form, it is easy to see that the optimal setting of variables S_{il}, g_i, and V_{ij} will take on one of their bounds.

Let us define the notations L and E to represent sets of row indices associating with less-than-or-equal-to and equality constraints in the model (1), respectively. Let us also define the notations $w_{1i} \forall i \in L, w_{2i}^+$ and $w_{2i}^- \forall i \in E$ to represent dual variables of the follower problem in the model (3). Even though there are six sets of follower's constraints in the model (3), only three sets of dual variables are required. Because of the structure of dual constraints of the follower problem in the model (3), dual variables associated with the first three sets of the follower's constraints are exactly the same as those associated with the last three sets. After replacing the follower problem with explicit representations of its optimality conditions, we encounter a number of nonlinear terms in the model including: $V_{ij}y_j, V_{ij}w_{2i}^+, V_{ij}w_{2i}^-, g_iw_{2i}^+, g_iw_{2i}^-, S_{il}w_{2i}^+,$ and $S_{il}w_{2i}^-$. By utilizing the result from Lemma 1, we can replace these nonlinear terms with the new set of variables $VY_{ij}, VW_{2ij}^+, VW_{2ij}^-, GW_{2i}^+, GW_{2i}^-, SW_{2il}^+,$ and SW_{2il}^- with the use of binary variables. We introduce binary variables $biS_{il}, big_i,$ and biV_{ij} which take the value of zero or one if variables $S_{il}, g_i,$ and V_{ij}, respectively, take the lower or the upper bound values. The following three sets of constraints (4), (5), and (6) will be used to relate these new variables with nonlinear terms in the model. In these constraints, the notations $y_j^U, w_{2i}^{+U},$ and w_{2i}^{-U} represent the upper bound values of the variables $y_j, w_{2i}^+,$ and w_{2i}^-, respectively. Terlaky ([27]) describes some techniques on constructing these bounds of the primal and dual variables.

$$S_{il} = S_{il}^L + (S_{il}^U - S_{il}^L)biS_{il} \qquad \forall i \in E, \forall l$$
$$S_{il}^L w_{2i}^+ \leq SW_{2il}^+ \leq S_{il}^U w_{2i}^+ \qquad \forall i \in E, \forall l$$
$$S_{il}^L w_{2i}^- \leq SW_{2il}^- \leq S_{il}^U w_{2i}^- \qquad \forall i \in E, \forall l$$

$$SW_{2il}^+ \geq S_{il}^U w_{2i}^+ - (\mid S_{il}^U \mid + \mid S_{il}^L \mid)(w_{2i}^{+U})(1 - biS_{il}) \qquad \forall i \in E, \forall l \qquad (4)$$
$$SW_{2il}^+ \leq S_{il}^L w_{2i}^+ + (\mid S_{il}^U \mid + \mid S_{il}^L \mid)(w_{2i}^{+U})(biS_{il}) \qquad \forall i \in E, \forall l$$
$$SW_{2il}^- \geq S_{il}^U w_{2i}^- - (\mid S_{il}^U \mid + \mid S_{il}^L \mid)(w_{2i}^{-U})(1 - biS_{il}) \qquad \forall i \in E, \forall l$$
$$SW_{2il}^- \leq S_{il}^L w_{2i}^- + (\mid S_{il}^U \mid + \mid S_{il}^L \mid)(w_{2i}^{-U})(biS_{il}) \qquad \forall i \in E, \forall l$$
$$biS_{il} \in \{0,1\} \qquad \forall i \in E, \forall l$$

$$g_i = g_i^L + (g_i^U - g_i^L)big_i \qquad \forall i \in E$$
$$g_i^L w_{2i}^+ \leq GW_{2i}^+ \leq g_i^U w_{2i}^+ \qquad \forall i \in E$$
$$g_i^L w_{2i}^- \leq GW_{2i}^- \leq g_i^U w_{2i}^- \qquad \forall i \in E$$

$$GW_{2i}^+ \geq g_i^U w_{2i}^+ - (\mid g_i^U \mid + \mid g_i^L \mid)(w_{2i}^{+U})(1 - big_i) \qquad \forall i \in E \qquad (5)$$
$$GW_{2i}^+ \leq g_i^L w_{2i}^+ + (\mid g_i^U \mid + \mid g_i^L \mid)(w_{2i}^{+U})(big_i) \qquad \forall i \in E$$
$$GW_{2i}^- \geq g_i^U w_{2i}^- - (\mid g_i^U \mid + \mid g_i^L \mid)(w_{2i}^{-U})(1 - big_i) \qquad \forall i \in E$$
$$GW_{2i}^- \leq g_i^L w_{2i}^- + (\mid g_i^U \mid + \mid g_i^L \mid)(w_{2i}^{-U})(big_i) \qquad \forall i \in E$$
$$big_i \in \{0,1\} \qquad \forall i \in E$$

$$V_{ij} = V_{ij}^L + (V_{ij}^U - V_{ij}^L)biV_{ij} \qquad \forall i \in E, \forall j$$

$$V_{ij}^L y_j \le VY_{ij} \le V_{ij}^U y_j \qquad \forall i \in E, \forall j$$

$$VY_{ij} \ge V_{ij}^U y_j - (\mid V_{ij}^U \mid + \mid V_{ij}^L \mid)(y_j^U)(1 - biV_{ij}) \qquad \forall i \in E, \forall j$$

$$VY_{ij} \le V_{ij}^L y_j + (\mid V_{ij}^U \mid + \mid V_{ij}^L \mid)(y_j^U)(biV_{ij}) \qquad \forall i \in E, \forall j$$

$$V_{ij}^L w_{2i}^+ \le VW_{2ij}^+ \le V_{ij}^U w_{2i}^+ \qquad \forall i \in E, \forall j$$

$$V_{ij}^L w_{2i}^- \le VW_{2ij}^- \le V_{ij}^U w_{2i}^- \qquad \forall i \in E, \forall j \tag{6}$$

$$VW_{2ij}^+ \ge V_{ij}^U w_{2i}^+ - (\mid V_{ij}^U \mid + \mid V_{ij}^L \mid)(w_{2i}^{+U})(1 - biV_{ij}) \qquad \forall i \in E, \forall j$$

$$VW_{2ij}^+ \le V_{ij}^L w_{2i}^+ + (\mid V_{ij}^U \mid + \mid V_{ij}^L \mid)(w_{2i}^{+U})(biV_{ij}) \qquad \forall i \in E, \forall j$$

$$VW_{2ij}^- \ge V_{ij}^U w_{2i}^- - (\mid V_{ij}^U \mid + \mid V_{ij}^L \mid)(w_{2i}^{-U})(1 - biV_{ij}) \qquad \forall i \in E, \forall j$$

$$VW_{2ij}^- \le V_{ij}^L w_{2i}^- + (\mid V_{ij}^U \mid + \mid V_{ij}^L \mid)(w_{2i}^{-U})(biV_{ij}) \qquad \forall i \in E, \forall j$$

$$biV_{ij} \in \{0,1\} \qquad \forall i \in E, \forall j$$

After applying preprocessing rules, the follower's problem transformation, and the result from Lemma 1, the model (3) can be transformed from a bi-level nonlinear structure to a single-level mixed integer linear programming structure presented in the model (7). The table in the model (7) is used to identify some additional constraints and conditions for adding these constraints to the model (7). These results greatly simplify the solution methodology of the bi-level-1 model. After solving the model (7), if the optimal setting of the decision variable δ is negative, the algorithm will add scenario ω_1^*, which is generated by the optimal setting of $\overrightarrow{\mathbf{h}}, \overrightarrow{\mathbf{g}}, \mathbf{T}, \mathbf{S}, \mathbf{W}$, and $\mathbf{V}$ from the model (7) and any feasible combination of $\overrightarrow{\mathbf{c}}$ and $\overrightarrow{\mathbf{q}}$, to the scenario set Ω and return to the first-stage algorithm. Otherwise the algorithm will forward the solution $\overrightarrow{\mathbf{x}}_\Omega$ and the value of $\triangle^L$ to the third-stage algorithm.

$$\min \quad \delta$$

$$\text{s.t.} \quad -\sum_j W_{ij}^L y_j + s_i = -h_i^U - \sum_l T_{il}^U x_{\Omega l} \qquad \forall i \in L$$

$$\sum_j VY_{ij} + s_{1i} = g_i + \sum_l S_{il} x_{\Omega l} \qquad \forall i \in E$$

$$-\sum_j VY_{ij} + s_{2i} = -g_i - \sum_l S_{il} x_{\Omega l} \qquad \forall i \in E$$

$$\delta \le s_i \quad \forall i \in L, \qquad \delta \le s_{1i} \quad \forall i \in E$$

$$\delta \le s_{2i} \quad \forall i \in E$$

$$-\sum_{i \in L} W_{ij}^L w_{1i} + \sum_{i \in E} (VW_{2ij}^+ - VW_{2ij}^-) \ge 0 \qquad \forall j$$

$$\sum_{i \in L} w_{1i} + \sum_{i \in E} (w_{2i}^+ + w_{2i}^-) = 1 \tag{7}$$

$$\delta = \sum_{i \in L} (-h_{1i}^U - \sum_l T_{il}^U x_{\Omega l}) w_{1i} +$$

$$\sum_{i \in E} (GW_{2i}^+ - GW_{2i}^- + \sum_l (SW_{2il}^+ - SW_{2il}^-) x_{\Omega l})$$

$$w_{1i} \ge 0 \quad \forall i \in L, \qquad w_{2i}^+ \ge 0 \quad \forall i \in E$$

$$w_{2i}^- \ge 0 \quad \forall i \in E, \qquad y_j \ge 0 \quad \forall j$$

Condition for Constraints	Constraint Reference	Constraint Index Set
Always	(4)	$\forall i \in E, \forall l$
Always	(5)	$\forall i \in E$
Always	(6)	$\forall i \in E, \forall j$

3.3 The Third-Stage Algorithm

The main purpose of the third stage is to identify a scenario with the largest relative regret value for the current candidate robust decision, $\overrightarrow{\mathbf{x}}_\Omega$, overall possible scenarios in $\bar{\Omega}$. In other words, in this stage, we are searching for the scenario $\omega_2^* \in \arg\max_{\omega \in \bar{\Omega}}\{(Z_\omega^*(\overrightarrow{\mathbf{x}}_\Omega) - O_\omega^*)/O_\omega^*\}$. The mathematical model utilized by this stage is also a bi-level programming referred to as the bi-level-2 model with one leader problem and two follower problems. The leader problem is tasked with finding the setting of vector $\xi = [\overrightarrow{\mathbf{c}}, \overrightarrow{\mathbf{q}}, \overrightarrow{\mathbf{h}}, \overrightarrow{\mathbf{g}}, \mathbf{T}, \mathbf{S}, \mathbf{W}, \mathbf{V}]$ that results in the maximum relative regret value possible for the current candidate robust solution, $\overrightarrow{\mathbf{x}}_\Omega$. The first follower problem is tasked to respond with the setting of decision vector to correctly calculate the value of $Z_\omega^*(\overrightarrow{\mathbf{x}}_\Omega)$ under the fixed setting of vector ξ established by the leader problem. The second follower problem on the other hand is tasked with finding the setting of decision vectors $(\overrightarrow{\mathbf{x}}_1, \overrightarrow{\mathbf{y}}_1)$ that result in the correct calculation of O_ω^* under the fixed setting of vector ξ established by the leader problem. The general structure of the bi-level-2 model is represented in the following model (8) where θ is a very small constant positive real value. The term θ is added to the denominator in the objective function of the model (8) in order to prevent division by zero.

$$\max_\xi \left\{ \frac{\overrightarrow{\mathbf{q}}^T\overrightarrow{\mathbf{y}}_2 + \overrightarrow{\mathbf{c}}^T\overrightarrow{\mathbf{x}}_\Omega - \overrightarrow{\mathbf{q}}^T\overrightarrow{\mathbf{y}}_1 - \overrightarrow{\mathbf{c}}^T\overrightarrow{\mathbf{x}}_1}{\overrightarrow{\mathbf{q}}^T\overrightarrow{\mathbf{y}}_1 + \overrightarrow{\mathbf{c}}^T\overrightarrow{\mathbf{x}}_1 + \Theta} \right\}$$

$$\text{s.t.} \quad \xi \in \Xi$$

$$\min_{\overrightarrow{\mathbf{y}}_2} \overrightarrow{\mathbf{q}}^T\overrightarrow{\mathbf{y}}_2 \qquad\qquad \min_{\overrightarrow{\mathbf{x}}_1, \overrightarrow{\mathbf{y}}_1} \overrightarrow{\mathbf{q}}^T\overrightarrow{\mathbf{y}}_1 + \overrightarrow{\mathbf{c}}^T\overrightarrow{\mathbf{x}}_1 \qquad (8)$$

$$\text{s.t.} \quad \mathbf{W}\overrightarrow{\mathbf{y}}_2 \geq \overrightarrow{\mathbf{h}} + \mathbf{T}\overrightarrow{\mathbf{x}}_\Omega \quad \text{and} \quad \text{s.t.} \quad \mathbf{W}\overrightarrow{\mathbf{y}}_1 \geq \overrightarrow{\mathbf{h}} + \mathbf{T}\overrightarrow{\mathbf{x}}_1$$

$$\mathbf{V}\overrightarrow{\mathbf{y}}_2 = \overrightarrow{\mathbf{g}} + \mathbf{S}\overrightarrow{\mathbf{x}}_\Omega \qquad\qquad \mathbf{V}\overrightarrow{\mathbf{y}}_1 = \overrightarrow{\mathbf{g}} + \mathbf{S}\overrightarrow{\mathbf{x}}_1$$

$$\overrightarrow{\mathbf{y}}_2 \geq \overrightarrow{\mathbf{0}} \qquad\qquad\qquad \overrightarrow{\mathbf{y}}_1 \geq \overrightarrow{\mathbf{0}}, \overrightarrow{\mathbf{x}}_1 \in \{0,1\}^{|\overrightarrow{\mathbf{x}}_1|}$$

The solution methodology for solving the model (8) can be structured into two main steps. These two main steps include (1) parameter preprocessing step and (2) model transformation step. Each of these steps is described in detail in the following subsections.

3.3.1 Parameter Preprocessing Step

From the structure of the model (8), some elements of vector ξ can be predetermined to attain their optimal setting at one of their bounds. In many cases, simple rules exist to identify the optimal values of these elements of vector ξ when the information on $\vec{\mathbf{x}}_\Omega$ is given even before solving the model (8). The following section describes these preprocessing rules for uncertain parameters in matrix $\mathbf{T}$.

Preprocessing Step for Matrix T

The elements of vector ξ of type $\mathbf{T}$ represent the coefficients of the binary decision variables located in greater-than-or-equal-to constraints of the model (1). Each element T_{il} of the parameter of this type is presented in the functional constraint of the model (8) in the following form: $\sum_j W_{ij} y_{1j} \geq h_i + T_{il} x_{1l} + \sum_{k \neq l} T_{ik} x_{1k}$ and $\sum_j W_{ij} y_{2j} \geq h_i + T_{il} x_{\Omega l} + \sum_{k \neq l} T_{ik} x_{\Omega k}$. For any given $\vec{\mathbf{x}}_\Omega$ information, the value of T_{il} can be predetermined at $T_{il}^* = T_{il}^L$ if $x_{\Omega l} = 0$. In the case when $x_{\Omega l} = 1$, the optimal setting of T_{il} satisfies the following set of constraints illustrated in (9) where the new variable TX_{1il} replaces the nonlinear term $T_{il} x_{1l}$ in the model (8). The insight of this set of constraints (9) is that if the value of x_{1l} is set to be zero by the model, the optimal setting of T_{il} is T_{il}^U and $TX_{1il} = 0$. Otherwise the optimal setting of T_{il} cannot be predetermined and $TX_{1il} = T_{il}$.

$$TX_{1il} - T_{il} + T_{il}^L(1 - x_{1l}) \leq 0$$

$$-TX_{1il} + T_{il} - T_{il}^U(1 - x_{1l}) \leq 0 \tag{9}$$

$$T_{il}^L x_{1l} \leq TX_{1il} \leq T_{il}^U x_{1l}$$

$$T_{il}^U + x_{1l}(-T_{il}^U + T_{il}^L) \leq T_{il} \leq T_{il}^U$$

3.3.2 Problem Transformation Step

In order to solve the model (8) efficiently, the structure of the current formulation needs to be transformed. The first transformation is performed by using the assumption that the O_ω^* value is nonnegative $\forall \omega \in \bar{\Omega}$. By using this assumption, it is relatively easy to see that the setting of $\vec{\mathbf{x}}_1$ and $\vec{\mathbf{y}}_1$ vectors that minimize the function $\vec{\mathbf{q}}^T \vec{\mathbf{y}}_1 + \vec{\mathbf{c}}^T \vec{\mathbf{x}}_1$ will also maximize the function $\frac{\vec{\mathbf{q}}^T \vec{\mathbf{y}}_2 + \vec{\mathbf{c}}^T \vec{\mathbf{x}}_\Omega - \vec{\mathbf{q}}^T \vec{\mathbf{y}}_1 - \vec{\mathbf{c}}^T \vec{\mathbf{x}}_1}{\vec{\mathbf{q}}^T \vec{\mathbf{y}}_1 + \vec{\mathbf{c}}^T \vec{\mathbf{x}}_1 + \Theta}$ and vice versa. By using this observation, the model (8) can now be transformed into the following model (10) which is a bi-level programming model with one leader problem and one follower problem.

$$\max_{\xi, \vec{x}_1, \vec{y}_1} \left\{ \frac{\vec{q}^T \vec{y}_2 + \vec{c}^T \vec{x}_\Omega - \vec{q}^T \vec{y}_1 - \vec{c}^T \vec{x}_1}{\vec{q}^T \vec{y}_1 + \vec{c}^T \vec{x}_1 + \Theta} \right\}$$

s.t. $\quad \xi \in \Xi$

$$\mathbf{W}\vec{y}_1 \geq \vec{h} + \mathbf{T}\vec{x}_1$$
$$\mathbf{V}\vec{y}_1 = \vec{g} + \mathbf{S}\vec{x}_1$$
$$\vec{y}_1 \geq \vec{0}, \vec{x}_1 \in \{0,1\}^{|\vec{x}_1|}$$

$$\min_{\vec{y}_2} \vec{q}^T \vec{y}_2 \tag{10}$$

s.t. $\quad \mathbf{W}\vec{y}_2 \geq \vec{h} + \mathbf{T}\vec{x}_\Omega$
$$\mathbf{V}\vec{y}_2 = \vec{g} + \mathbf{S}\vec{x}_\Omega$$
$$\vec{y}_2 \geq \vec{0}$$

This model (10) is the bi-level programming model with the mixed integer fractional programming structure for its leader's problem and the linear programming structure for its follower's problem. Further transformations can still be applied to transform the model (10) structure into a more efficient formulation. By applying standard transformation techniques used in fractional programming and the fact that $\vec{q}^T \vec{y}_1 + \vec{c}^T \vec{x}_1 + \theta > 0$, the model (10) can further be transformed into the following model (11). This model (11) has a bi-level mixed integer structure with a linear programming structure for its follower's problem. In order to solve the model (11), the following two main tasks have to be accomplished. First, the efficient modeling technique is required to model the constraint $\xi \in \Xi$. Second, the effective transformation method is required to transform the current formulation of model (11) into a computationally efficient formulation. The following subsections summarize modeling techniques and methodologies for performing these two main tasks.

$$\max_{\xi, \vec{tx}_1, \vec{ty}_1, t_0} \vec{q}^T \vec{ty}_2 + (\vec{c}^T \vec{x}_\Omega)t_0 - \vec{q}^T \vec{ty}_1 - \vec{c}^T \vec{tx}_1$$
s.t. $\quad \xi \in \Xi$

$$\mathbf{W}\vec{ty}_1 \geq \vec{h}\, t_0 + \mathbf{T}\vec{tx}_1$$
$$\mathbf{V}\vec{ty}_1 = \vec{g}\, t_0 + \mathbf{S}\vec{tx}_1$$
$$\vec{ty}_1 \geq \vec{0}, \quad \vec{tx}_1 \in \{0, t_0\}^{|\vec{tx}_1|}, \quad t_0 \geq 0$$
$$\vec{q}^T \vec{ty}_1 + \vec{c}^T \vec{tx}_1 + \Theta t_0 = 1$$

$$\min_{\vec{ty}_2} \vec{q}^T \vec{ty}_2 \tag{11}$$

s.t. $\quad \mathbf{W}\vec{ty}_2 \geq \left(\vec{h} + \mathbf{T}\vec{x}_\Omega \right)t_0$
$$\mathbf{V}\vec{ty}_2 = \left(\vec{g} + \mathbf{S}\vec{x}_\Omega \right)t_0$$
$$\vec{ty}_2 \geq \vec{0}$$

Modeling Technique for the Constraint $\xi \in \Xi$

Consider a variable p which can only take its value from $\bar{p}$ distinct real values $p_{(1)}, p_{(2)}, \ldots, p_{(\bar{p})}$. We can model this variable p in the mathematical programming model as: $p = \sum_{i=1}^{\bar{p}} p_{(i)} bi_i$, $\sum_{i=1}^{\bar{p}} bi_i = 1, bi_i \geq 0 \quad \forall i = 1, \ldots, \bar{p}$ and $\{bi_1, bi_2, \ldots, bi_{\bar{p}}\}$ is SOS1. A special ordered set of type one (SOS1) is defined to be a set of variables for which not more than one member from the set may be nonzero. When these nonnegative variables $bi_i \quad \forall i = 1, \ldots, \bar{p}$, are defined as SOS1, there are only $\bar{p}$ branches required in the searching tree for these variables.

Final Transformation Steps for the Bi-level-2 Model

Because the structure of the follower's problem in the model (11) is a linear program, the final transformation steps start by replacing the follower's problem in the model (11) with explicit representations of its optimality conditions. These explicit representations include the follower's primal constraints, the follower's dual constraints, and the follower's strong duality constraint. The model (12) illustrates the formulation of the model (11) after this transformation where the decision variable $\overrightarrow{\mathbf{tw}_1}$ and $\overrightarrow{\mathbf{tw}_2}$ represent dual variables associated with follower's constraints and t_0^U is the upper bound value of the decision variable t_0. The model (12) is a single-level mixed integer nonlinear optimization problem. By applying results from parameter preprocessing steps and modeling technique previously discussed, the final transformation steps are completed and are summarized below.

$$\max \overrightarrow{\mathbf{q}}^T \overrightarrow{\mathbf{ty}_2} + (\overrightarrow{\mathbf{c}}^T \overrightarrow{\mathbf{x}}_\Omega) t_0 - \overrightarrow{\mathbf{q}}^T \overrightarrow{\mathbf{ty}_1} - \overrightarrow{\mathbf{c}}^T \overrightarrow{\mathbf{tx}_1}$$

$$\text{s.t.} \quad \xi \in \Xi$$

$$\mathbf{W} \overrightarrow{\mathbf{ty}_1} \geq \overrightarrow{\mathbf{h}} t_0 + \mathbf{T} \overrightarrow{\mathbf{tx}_1}$$

$$\mathbf{V} \overrightarrow{\mathbf{ty}_1} = \overrightarrow{\mathbf{g}} t_0 + \mathbf{S} \overrightarrow{\mathbf{tx}_1}$$

$$\overrightarrow{\mathbf{q}}^T \overrightarrow{\mathbf{ty}_1} + \overrightarrow{\mathbf{c}}^T \overrightarrow{\mathbf{tx}_1} + \Theta t_0 = 1 \tag{12}$$

$$\mathbf{W} \overrightarrow{\mathbf{ty}_2} \geq \left(\overrightarrow{\mathbf{h}} + \mathbf{T} \overrightarrow{\mathbf{x}}_\Omega \right) t_0$$

$$\mathbf{V} \overrightarrow{\mathbf{ty}_2} = \left(\overrightarrow{\mathbf{g}} + \mathbf{S} \overrightarrow{\mathbf{x}}_\Omega \right) t_0$$

$$\mathbf{W}^T \overrightarrow{\mathbf{tw}_1} + \mathbf{V}^T \overrightarrow{\mathbf{tw}_2} \leq \overrightarrow{\mathbf{q}} t_0$$

$$\left(\overrightarrow{\mathbf{h}} + \mathbf{T} \overrightarrow{\mathbf{x}}_\Omega \right)^T \overrightarrow{\mathbf{tw}_1} + \left(\overrightarrow{\mathbf{g}} + \mathbf{S} \overrightarrow{\mathbf{x}}_\Omega \right)^T \overrightarrow{\mathbf{tw}_2} = \overrightarrow{\mathbf{q}}^T \overrightarrow{\mathbf{ty}_2}$$

$$\overrightarrow{\mathbf{0}} \leq \overrightarrow{\mathbf{tx}_1} \leq t_0 \overrightarrow{\mathbf{1}}, \quad \overrightarrow{\mathbf{tx}_1} \leq t_0^U \overrightarrow{\mathbf{x}}_1, \quad \overrightarrow{\mathbf{tx}_1} \geq t_0 \overrightarrow{\mathbf{1}} - t_0^U (\overrightarrow{\mathbf{1}} - \overrightarrow{\mathbf{x}}_1)$$

$$\overrightarrow{\mathbf{ty}_1} \geq \overrightarrow{\mathbf{0}}, \quad \overrightarrow{\mathbf{ty}_2} \geq \overrightarrow{\mathbf{0}}, \quad t_0 \geq 0, \quad \overrightarrow{\mathbf{tw}_1} \geq \overrightarrow{\mathbf{0}}, \overrightarrow{\mathbf{x}}_1 \in \{0, 1\}^{|\overrightarrow{\mathbf{x}}_1|}$$

Final Transformation Steps

Parameter T_{il}: By applying previous results, if the parameter T_{il} can be preprocessed, then fix its value at the appropriate value of T_{il}^*. Otherwise, first add a decision variable TtX_{1il} and a set of constraints illustrated in (13) (modified from (9)) to replace the nonlinear term $T_{il}tx_{1l}$ in the model (12), then add a set of variables and constraints illustrated in (14) to replace the constraint $\xi \in \Xi$ for parameter T_{il} in the model (12), and finally add a variable TtW_{1il} and Tt_{il} and a set of variables and constraints illustrated in (15) to replace the nonlinear terms $T_{il}tw_{1i}$ and $T_{il}t_0$ in the model (12), where tw_{1i}^U and tw_{1i}^L represent the upper bound and the lower bound of the variable tw_{1i}, respectively.

$$TtX_{1il} - Tt_{il} + T_{il}^L(t_0 - tx_{1l}) \leq 0$$
$$-TtX_{1il} + Tt_{il} - T_{il}^U(t_0 - tx_{1l}) \leq 0$$

$$T_{il}^L tx_{1l} \leq TtX_{1il} \leq T_{il}^U tx_{1l} \tag{13}$$

$$T_{il}^U t_0 + tx_{1l}(-T_{il}^U + T_{il}^L) \leq Tt_{il} \leq T_{il}^U t_0$$
$$T_{il}^U + x_{1l}(-T_{il}^U + T_{il}^L) \leq T_{il} \leq T_{il}^U$$

$$T_{il} = \sum_{s=1}^{\bar{T}_{il}} T_{il(s)}biT_{il(s)}, \quad \sum_{s=1}^{\bar{T}_{il}} biT_{il(s)} = 1 \tag{14}$$

$$biT_{il(s)} \geq 0 \quad \forall s \in \{1, 2, \dots, \bar{T}_{il}\} \text{ and } \cup_{\forall s}\{biT_{il(s)}\} \text{ is SOS1}$$

$$tw_{1i}^L biT_{il(s)} \leq ZTtW_{1il(s)} \leq tw_{1i}^U biT_{il(s)} \quad \forall s \in \{1, 2, \dots, \bar{T}_{il}\}$$
$$ZTtW_{1il(s)} \leq tw_{1i} - tw_{1i}^L(1 - biT_{il(s)}) \quad \forall s \in \{1, 2, \dots, \bar{T}_{il}\}$$
$$ZTtW_{1il(s)} \geq tw_{1i} - tw_{1i}^U(1 - biT_{il(s)}) \quad \forall s \in \{1, 2, \dots, \bar{T}_{il}\}$$

$$0 \leq ZtT_{il(s)} \leq t_0^U biT_{il(s)} \quad \forall s \in \{1, 2, \dots, \bar{T}_{il}\} \tag{15}$$

$$ZtT_{il(s)} \leq t_0 \quad \forall s \in \{1, 2, \dots, \bar{T}_{il}\}$$
$$ZtT_{il(s)} \geq t_0 - t_0^U(1 - biT_{il(s)}) \quad \forall s \in \{1, 2, \dots, \bar{T}_{il}\}$$
$$TtW_{1il} = \sum_{s=1}^{\bar{T}_{il}} (T_{il(s)}ZTtW_{1il(s)}) \text{ and } Tt_{il} = \sum_{s=1}^{\bar{T}_{il}} (T_{il(s)}ZtT_{il(s)})$$

Parameter S_{il}: We first add a decision variable StX_{1il} and a set of constraints illustrated in (16) to replace the nonlinear term $S_{il}tx_{1l}$ in the model (12), then add a set of variables and constraints illustrated in (17) to replace the constraint $\xi \in \Xi$ for parameter S_{il} in the model (12) and finally add a variable StW_{2il} and St_{il} and a set of variables and constraints illustrated in (18) to replace the nonlinear terms $S_{il}tw_{2i}$ and $S_{il}t_0$ in the model (12), where tw_{2i}^U and tw_{2i}^L represent the upper bound and the lower bound of variable tw_{2i}, respectively.

$$StX_{1il} = \sum_{s=1}^{\bar{S}_{il}} S_{il(s)} ZStX_{1il(s)}$$
$$0 \le ZStX_{1il(s)} \le t_0 \quad \forall s \in \{1,2,\dots,\bar{S}_{il}\}$$

$$ZStX_{1il(s)} \le t_0^U biS_{il(s)} \quad \forall s \in \{1,2,\dots,\bar{S}_{il}\} \tag{16}$$

$$ZStX_{1il(s)} \le tx_{1i} \quad \forall s \in \{1,2,\dots,\bar{S}_{il}\}$$
$$ZStX_{1il(s)} \ge tx_{1i} - t_0^U(1 - biS_{il(s)}) \quad \forall s \in \{1,2,\dots,\bar{S}_{il}\}$$

$$S_{il} = \sum_{s=1}^{\bar{S}_{il}} S_{il(s)} biS_{il(s)}, \quad \sum_{s=1}^{\bar{S}_{il}} biS_{il(s)} = 1 \tag{17}$$

$$biS_{il(s)} \ge 0 \quad \forall s \in \{1,2,\dots,\bar{S}_{il}\} \text{ and } \cup_{\forall s}\{biS_{il(s)}\} \text{ is SOS1}$$

$$tw_{2i}^L biS_{il(s)} \le ZStW_{2il(s)} \le tw_{2i}^U biS_{il(s)} \quad \forall s \in \{1,2,\dots,\bar{S}_{il}\}$$
$$ZStW_{2il(s)} \le tw_{2i} - tw_{2i}^L(1 - biS_{il(s)}) \quad \forall s \in \{1,2,\dots,\bar{S}_{il}\}$$
$$ZStW_{2il(s)} \ge tw_{2i} - tw_{2i}^U(1 - biS_{il(s)}) \quad \forall s \in \{1,2,\dots,\bar{S}_{il}\}$$

$$0 \le ZSt_{il(s)} \le t_0^U biS_{il(s)} \quad \forall s \in \{1,2,\dots,\bar{S}_{il}\} \tag{18}$$

$$ZSt_{il(s)} \le t_0 \quad \forall s \in \{1,2,\dots,\bar{S}_{il}\}$$
$$ZSt_{il(s)} \ge t_0 - t_0^U(1 - biS_{il(s)}) \quad \forall s \in \{1,2,\dots,\bar{S}_{il}\}$$
$$StW_{2il} = \sum_{s=1}^{\bar{S}_{il}} (S_{il(s)} ZStW_{2il(s)}) \text{ and } St_{il} = \sum_{s=1}^{\bar{S}_{il}} (S_{il(s)} ZSt_{il(s)})$$

Parameter h_i and g_i: We first add a set of variables and constraints illustrated in (19) and (20) to replace constraints $\xi \in \Xi$ for parameter h_i and g_i, respectively, in the model (12). We then add variables HtW_{1i}, GtW_{2i} and a set of variables and constraints in (21) and (22) to replace the nonlinear terms $h_i tw_{1i}$ and $g_i tw_{2i}$, respectively, in the model (12). Finally, we add variables Ht_i, Gt_i and a set of variables and constraints in (23) and (24) to replace the nonlinear terms $h_i t_0$ and $g_i t_0$, respectively, in the model (12).

$$h_i = \sum_{s=1}^{\bar{h}_i} h_{i(s)} biH_{i(s)}, \quad \sum_{s=1}^{\bar{h}_i} biH_{i(s)} = 1 \tag{19}$$

$$biH_{i(s)} \ge 0 \quad \forall s \in \{1,2,\dots,\bar{h}_i\} \text{ and } \cup_{\forall s}\{biH_{i(s)}\} \text{ is SOS1}$$

$$g_i = \sum_{s=1}^{\bar{g}_i} g_{i(s)} biG_{i(s)}, \quad \sum_{s=1}^{\bar{g}_i} biG_{i(s)} = 1 \tag{20}$$

$$biG_{i(s)} \ge 0 \quad \forall s \in \{1,2,\dots,\bar{g}_i\} \text{ and } \cup_{\forall s}\{biG_{i(s)}\} \text{ is SOS1}$$

$$HtW_{1i} = \sum_{s=1}^{\bar{h}_i} h_{i(s)} ZHtW_{1i(s)}$$

$$tw_{1i}^L biH_{i(s)} \leq ZHtW_{1i(s)} \leq tw_{1i}^U biH_{i(s)} \quad \forall s \in \{1, 2, \ldots, \bar{h}_i\} \qquad (21)$$
$$ZHtW_{1i(s)} \leq tw_{1i} - tw_{1i}^L(1 - biH_{i(s)}) \quad \forall s \in \{1, 2, \ldots, \bar{h}_i\}$$
$$ZHtW_{1i(s)} \geq tw_{1i} - tw_{1i}^U(1 - biH_{i(s)}) \quad \forall s \in \{1, 2, \ldots, \bar{h}_i\}$$

$$GtW_{2i} = \sum_{s=1}^{\bar{g}_i} g_{i(s)} ZGtW_{2i(s)}$$

$$tw_{2i}^L biG_{i(s)} \leq ZGtW_{2i(s)} \leq tw_{2i}^U biG_{i(s)} \quad \forall s \in \{1, 2, \ldots, \bar{g}_i\} \qquad (22)$$
$$ZGtW_{2i(s)} \leq tw_{2i} - tw_{2i}^L(1 - biG_{i(s)}) \quad \forall s \in \{1, 2, \ldots, \bar{g}_i\}$$
$$ZGtW_{2i(s)} \geq tw_{2i} - tw_{2i}^U(1 - biG_{i(s)}) \quad \forall s \in \{1, 2, \ldots, \bar{g}_i\}$$

$$Ht_i = \sum_{s=1}^{\bar{h}_i} h_{i(s)} ZHt_{i(s)}$$

$$0 \leq ZHt_{i(s)} \leq t_0^U biH_{i(s)} \quad \forall s \in \{1, 2, \ldots, \bar{h}_i\} \qquad (23)$$
$$ZHt_{i(s)} \leq t_0 \quad \forall s \in \{1, 2, \ldots, \bar{h}_i\}$$
$$ZHt_{i(s)} \geq t_0 - t_0^U(1 - biH_{i(s)}) \quad \forall s \in \{1, 2, \ldots, \bar{h}_i\}$$
$$Gt_i = \sum_{s=1}^{\bar{g}_i} g_{i(s)} ZGt_{i(s)}$$

$$0 \leq ZGt_{i(s)} \leq t_0^U biG_{i(s)} \quad \forall s \in \{1, 2, \ldots, \bar{g}_i\} \qquad (24)$$
$$ZGt_{i(s)} \leq t_0 \quad \forall s \in \{1, 2, \ldots, \bar{g}_i\}$$
$$ZGt_{i(s)} \geq t_0 - t_0^U(1 - biG_{i(s)}) \quad \forall s \in \{1, 2, \ldots, \bar{g}_i\}$$

Parameter q_j: We first add a set of variables and constraints illustrated in (25) to replace the constraint $\xi \in \Xi$ for parameter q_j in the model (12). We then add decision variables QtY_{1j}, QtY_{2j}, Qt_j and a set of variables and constraints in (26) to replace the nonlinear terms $q_j ty_{1j}$, $q_j ty_{2j}$, and $q_j t_0$, respectively, in the model (12) where ty_{rj}^U and ty_{rj}^L represent the upper bound and the lower bound of variable ty_{rj}, respectively, for $r = 1$ and 2.

$$q_j = \sum_{s=1}^{\bar{q}_j} q_{j(s)} biQ_{j(s)}, \quad \sum_{s=1}^{\bar{q}_j} biQ_{j(s)} = 1 \qquad (25)$$

$$biQ_{j(s)} \geq 0 \quad \forall s \in \{1, 2, \ldots, \bar{q}_j\} \text{ and } \cup_{\forall s}\{biQ_{j(s)}\} \text{ is SOS1}$$

$$QtY_{1j} = \sum_{s=1}^{\bar{q}_j} q_{j(s)} ZQtY_{1j(s)}$$
$$ty_{1j}^L biQ_{j(s)} \leq ZQtY_{1j(s)} \leq ty_{1j}^U biQ_{j(s)} \quad \forall s \in \{1, 2, \ldots, \bar{q}_j\}$$
$$ZQtY_{1j(s)} \leq ty_{1j} - ty_{1j}^L(1 - biQ_{j(s)}) \quad \forall s \in \{1, 2, \ldots, \bar{q}_j\}$$
$$ZQtY_{1j(s)} \geq ty_{1j} - ty_{1j}^U(1 - biQ_{j(s)}) \quad \forall s \in \{1, 2, \ldots, \bar{q}_j\}$$
$$QtY_{2j} = \sum_{s=1}^{\bar{q}_j} q_{j(s)} ZQtY_{2j(s)}$$

$$ty_{2j}^{L}biQ_{j(s)} \leq ZQtY_{2j(s)} \leq ty_{2j}^{U}biQ_{j(s)} \quad \forall s \in \{1,2,\ldots,\bar{q}_j\} \qquad (26)$$

$$ZQtY_{2j(s)} \leq ty_{2j} - ty_{2j}^{L}(1 - biQ_{j(s)}) \quad \forall s \in \{1,2,\ldots,\bar{q}_j\}$$
$$ZQtY_{2j(s)} \geq ty_{2j} - ty_{2j}^{U}(1 - biQ_{j(s)}) \quad \forall s \in \{1,2,\ldots,\bar{q}_j\}$$
$$Qt_j = \sum_{s=1}^{\bar{q}_j} q_{j(s)} ZQt_{j(s)}$$
$$0 \leq ZQt_{j(s)} \leq t_0^{U}biQ_{j(s)}, \quad ZQt_{j(s)} \leq t_0, \quad \forall s \in \{1,2,\ldots,\bar{q}_j\}$$
$$ZQt_{j(s)} \geq t_0 - t_0^{U}(1 - biQ_{j(s)}) \quad \forall s \in \{1,2,\ldots,\bar{q}_j\}$$

Parameter W_{ij} and V_{ij}: We first add a set of variables and constraints illustrated in (27) and (28) to replace constraints $\xi \in \Xi$ for parameter W_{ij} and V_{ij}, respectively, in the model (12). We then add a set of variables and constraints illustrated in (29) and (30) together with variables WtY_{1ij}, VtY_{1ij}, WtY_{2ij}, VtY_{2ij}, WtW_{1ij}, and WtW_{2ij} to replace the nonlinear terms $W_{ij}ty_{1j}$, $V_{ij}ty_{1j}$, $W_{ij}ty_{2j}$, $V_{ij}ty_{2j}$, $W_{ij}tw_{1i}$, and $W_{ij}tw_{2i}$ in the model (12).

$$W_{ij} = \sum_{s=1}^{\bar{W}_{ij}} W_{ij(s)}biW_{ij(s)}, \quad \sum_{s=1}^{\bar{W}_{ij}} biW_{ij(s)} = 1 \qquad (27)$$

$$biW_{ij(s)} \geq 0 \quad \forall s \in \{1,2,\ldots,\bar{W}_{ij}\} \text{ and } \cup_{\forall s}\{biW_{ij(s)}\} \text{ is SOS1}$$

$$V_{ij} = \sum_{s=1}^{\bar{V}_{ij}} V_{ij(s)}biV_{ij(s)}, \quad \sum_{s=1}^{\bar{V}_{ij}} biV_{ij(s)} = 1 \qquad (28)$$

$$biV_{ij(s)} \geq 0 \quad \forall s \in \{1,2,\ldots,\bar{V}_{ij}\} \text{ and } \cup_{\forall s}\{biV_{ij(s)}\} \text{ is SOS1}$$

$$WtY_{1ij} = \sum_{s=1}^{\bar{W}_{ij}} W_{ij(s)}ZWtY_{1ij(s)}$$
$$ty_{1j}^{L}biW_{ij(s)} \leq ZWtY_{1ij(s)} \leq ty_{1j}^{U}biW_{ij(s)} \quad \forall s \in \{1,2,\ldots,\bar{W}_{ij}\}$$
$$ZWtY_{1ij(s)} \leq ty_{1j} - ty_{1j}^{L}(1 - biW_{ij(s)}) \quad \forall s \in \{1,2,\ldots,\bar{W}_{ij}\}$$
$$ZWtY_{1ij(s)} \geq ty_{1j} - ty_{1j}^{U}(1 - biW_{ij(s)}) \quad \forall s \in \{1,2,\ldots,\bar{W}_{ij}\}$$
$$WtY_{2ij} = \sum_{s=1}^{\bar{W}_{ij}} W_{ij(s)}ZWtY_{2ij(s)}$$

$$ty_{2j}^{L}biW_{ij(s)} \leq ZWtY_{2ij(s)} \leq ty_{2j}^{U}biW_{ij(s)} \quad \forall s \in \{1,2,\ldots,\bar{W}_{ij}\} \qquad (29)$$

$$ZWtY_{2ij(s)} \leq ty_{2j} - ty_{2j}^{L}(1 - biW_{ij(s)}) \quad \forall s \in \{1,2,\ldots,\bar{W}_{ij}\}$$
$$ZWtY_{2ij(s)} \geq ty_{2j} - ty_{2j}^{U}(1 - biW_{ij(s)}) \quad \forall s \in \{1,2,\ldots,\bar{W}_{ij}\}$$
$$WtW_{1ij} = \sum_{s=1}^{\bar{W}_{ij}} W_{ij(s)}ZWtW_{1ij(s)}$$
$$tw_{1i}^{L}biW_{ij(s)} \leq ZWtW_{1ij(s)} \leq tw_{1i}^{U}biW_{ij(s)} \quad \forall s \in \{1,2,\ldots,\bar{W}_{ij}\}$$
$$ZWtW_{1ij(s)} \leq tw_{1i} - tw_{1i}^{L}(1 - biW_{ij(s)}) \quad \forall s \in \{1,2,\ldots,\bar{W}_{ij}\}$$
$$ZWtW_{1ij(s)} \geq tw_{1i} - tw_{1i}^{U}(1 - biW_{ij(s)}) \quad \forall s \in \{1,2,\ldots,\bar{W}_{ij}\}$$

$$VtY_{1ij} = \sum_{s=1}^{\bar{V}_{ij}} V_{ij(s)} ZVtY_{1ij(s)}$$

$$ty_{1j}^L biV_{ij(s)} \le ZVtY_{1ij(s)} \le ty_{1j}^U biV_{ij(s)} \quad \forall s \in \{1, 2, \ldots, \bar{V}_{ij}\}$$

$$ZVtY_{1ij(s)} \le ty_{1j} - ty_{1j}^L(1 - biV_{ij(s)}) \quad \forall s \in \{1, 2, \ldots, \bar{V}_{ij}\}$$

$$ZVtY_{1ij(s)} \ge ty_{1j} - ty_{1j}^U(1 - biV_{ij(s)}) \quad \forall s \in \{1, 2, \ldots, \bar{V}_{ij}\}$$

$$VtY_{2ij} = \sum_{s=1}^{\bar{V}_{ij}} V_{ij(s)} ZVtY_{2ij(s)}$$

$$ty_{2j}^L biV_{ij(s)} \le ZVtY_{2ij(s)} \le ty_{2j}^U biV_{ij(s)} \quad \forall s \in \{1, 2, \ldots, \bar{V}_{ij}\} \qquad (30)$$

$$ZVtY_{2ij(s)} \le ty_{2j} - ty_{2j}^L(1 - biV_{ij(s)}) \quad \forall s \in \{1, 2, \ldots, \bar{V}_{ij}\}$$

$$ZVtY_{2ij(s)} \ge ty_{2j} - ty_{2j}^U(1 - biV_{ij(s)}) \quad \forall s \in \{1, 2, \ldots, \bar{V}_{ij}\}$$

$$VtW_{2ij} = \sum_{s=1}^{\bar{V}_{ij}} V_{ij(s)} ZVtW_{2ij(s)}$$

$$tw_{2i}^L biV_{ij(s)} \le ZVtW_{2ij(s)} \le tw_{2i}^U biV_{ij(s)} \quad \forall s \in \{1, 2, \ldots, \bar{V}_{ij}\}$$

$$ZVtW_{2ij(s)} \le tw_{2i} - tw_{2i}^L(1 - biV_{ij(s)}) \quad \forall s \in \{1, 2, \ldots, \bar{V}_{ij}\}$$

$$ZVtW_{2ij(s)} \ge tw_{2i} - tw_{2i}^U(1 - biV_{ij(s)}) \quad \forall s \in \{1, 2, \ldots, \bar{V}_{ij}\}$$

By applying these transformation steps, the model (12) can be transformed into its final formulation as a single level mixed integer linear programming problem as shown in the following model (31). In the model (31), T_{il}^* is the preprocessed value of T_{il} if T_{il} can be preprocessed and is equaled to zero otherwise. In addition, $IndT_{il}$ will take the value of one if T_{il} cannot be predetermined and will take the value of zero otherwise.

$$\max \{\Delta^{U^*} = \sum_j QtY_{2j} + \sum_l c_l x_{\Omega l} t_0 - \sum_j QtY_{1j} - \sum_l c_l tx_{1l}\}$$

$$\text{s.t.} \quad \sum_j WtY_{1ij} \ge Ht_i + \sum_{l|IndT_{il}=1} TtX_{1il} + \sum_{l|IndT_{il}=0} T_{il}^* tx_{1l} \quad \forall i \in L$$

$$\sum_j VtY_{1ij} = Gt_i + \sum_l StX_{1il} \quad \forall i \in E$$

$$\sum_j QtY_{1j} + \sum_l c_l tx_{1l} + \theta t_0 = 1$$

$$\sum_j WtY_{2ij} \ge Ht_i + \sum_{l|IndT_{il}=1} Tt_{il} x_{\Omega l} + \sum_{l|IndT_{il}=0} T_{il}^* x_{\Omega l} t_0 \quad \forall i \in L$$

$$\sum_j VtY_{2ij} = Gt_i + \sum_l St_{il} x_{\Omega l} \quad \forall i \in E$$

$$\sum_{i \in L} WtW_{1ij} + \sum_{i \in E} VtW_{2ij} \le Qt_j \quad \forall j \qquad (31)$$

$$\sum_j QtY_{2j} = \sum_{i \in L} HtW_{1i} + \sum_{i \in E} GtW_{2i} + \sum_{i \in E} \sum_l StW_{2il} x_{\Omega l}$$

$$+ \sum_{i \in L} \left(\sum_{l|IndT_{il}=1} TtW_{1il} x_{\Omega l} + \sum_{l|IndT_{il}=0} T_{il}^* tw_{1i} x_{\Omega l} \right)$$

$$0 \le tx_{1l} \le t_0 \quad \forall l, \quad tx_{1l} \le t_0^U x_{1l} \quad \forall l, \quad tx_{1l} \ge t_0 - t_0^U(1 - x_{1l}) \quad \forall l$$

$$ty_{1j} \ge 0 \quad \forall j, \quad ty_{2j} \ge 0 \quad \forall j, \quad tw_{1j} \ge 0 \quad \forall i \in L$$

$$t_0 \ge 0, \quad x_{1l} \in \{0, 1\} \quad \forall l$$

Condition for Constraints	Constraint Reference	Constraint Index Set
If $IndT_{il} = 1$	(13), (14), (15)	$\forall i \in L, \forall l$
Always	(16), (17), (18)	$\forall i \in E, \forall l$
Always	(19), (21), (23)	$\forall i \in L$
Always	(20), (22), (24)	$\forall i \in E$
Always	(25), (26)	$\forall j$
Always	(27), (29)	$\forall i \in L, \forall j$
Always	(28), (30)	$\forall i \in E, \forall j$

The optimal objective function value of the model (31), $\triangle^{U^*}$, is used to update the value of $\triangle^U$ by setting $\triangle^U$ to min $\{\triangle^{U^*}, \triangle^U\}$. The optimality condition is then checked. If the optimality condition is not satisfied, add scenario ω_2^* to the scenario set Ω and return to the first-stage algorithm where ω_2^* is the scenario generated by the optimal settings of $\xi = [\vec{\mathbf{c}}, \vec{\mathbf{q}}, \vec{\mathbf{h}}, \vec{\mathbf{g}}, \mathbf{T}, \mathbf{S}, \mathbf{W}, \mathbf{V}]$ from the model (31). Otherwise, the algorithm is terminated with an ϵ-optimal relative robust solution to the problem which is the discrete solution with the maximum relative regret value of $\triangle^U$, $\vec{\mathbf{x}}_{opt}$. The final formulation of the bi-level-2 model (model (31)) can be efficiently solved by using optimization techniques for the mixed integer linear program. The following Lemma 2 provides the important result that the algorithm always terminates at a globally ϵ-optimal relative robust solution in a finite number of algorithm steps.

Lemma 2. *The proposed three-stage algorithm terminates in a finite number of steps. After the algorithm terminates, it has either detected infeasibility or has found an ϵ-optimal relative robust solution to the problem.*

Proof. Notice that the relaxed model (2) is a relaxation of the original relative robust optimization problem and the feasible region of the model (1) contains the feasible region of the original problem. This has four important implications: (a) If the model (1) is infeasible, then the original problem is also infeasible, (b) if the relaxed model (2) is infeasible, then the original problem is also infeasible, (c) $\triangle^L \leq \min_{\vec{\mathbf{x}}} \max_{\omega \in \bar{\Omega}} \{(Z_\omega^*(\vec{\mathbf{x}}) - O_\omega^*)/O_\omega^*\}$ for all iterations, and (d) if $\vec{\mathbf{x}}^*$ is an optimal solution to the original problem, $\vec{\mathbf{x}}^*$ is a feasible solution to the relaxed model (2). From the first and second implications, it is clear that if the algorithm terminates because either the model (1) or the relaxed model (2) is infeasible then the original problem is infeasible. Suppose the algorithm terminates in Step 2 or Step 4 with $\triangle^L = \triangle^U$ and the first stage solution $\vec{\mathbf{x}}_{opt}$. Notice that we can go to Step 4 only if $\vec{\mathbf{x}}_{opt}$ is a feasible solution to the overall problem; then $\triangle^U = \max_{\omega \in \bar{\Omega}} \{(Z_\omega^*(\vec{\mathbf{x}}_{opt}) - O_\omega^*)/O_\omega^*\} \geq \min_{\vec{\mathbf{x}}} \max_{\omega \in \bar{\Omega}} \{(Z_\omega^*(\vec{\mathbf{x}}) - O_\omega^*)/O_\omega^*\}$. Therefore if $\triangle^L = \triangle^U$, then $\max_{\omega \in \bar{\Omega}} \{(Z_\omega^*(\vec{\mathbf{x}}_{opt}) - O_\omega^*)/O_\omega^*\} = \min_{\vec{\mathbf{x}}} \max_{\omega \in \bar{\Omega}} \{(Z_\omega^*(\vec{\mathbf{x}}) - O_\omega^*)/O_\omega^*\}$ or $\vec{\mathbf{x}}_{opt}$ is an optimal solution to the original problem. Because there is a finite number of possible combinations of and parameter settings, the proposed three-stage algorithm always terminates in a finite number of steps.

The following section demonstrates applications of the three-stage algorithm for scenario-based relative robust optimization problems with a mixed integer linear program (MILP) based model under full-factorial scenario design of data uncertainty. All results illustrate good performance of the algorithm.

4 Applications of the Three-Stage Algorithm

In this section, we apply the three-stage algorithm to three example problems. The first example problem illustrates the use of the algorithm on a simple MILP problem with ambiguity in model parameters' values. The second example problem demonstrates a counterexample that illustrates the inefficiency of the relative robust solution obtained by only considering a finite number of scenarios generated only by upper and lower bounds of each uncertain parameter. The third example problem demonstrates the application of the algorithm to make relative robust decisions for a facility location problem with an extremely large number of possible scenarios (1.215×10^{19}).

4.1 Applications of the Three-Stage Algorithm

This example problem is described by the following stochastic mixed integer linear programming problem. In this problem, our task is to find the robust settings of x_1 and x_2 that minimize that maximum relative regret value over all possible scenarios. In this small example problem, there are 243 possible scenarios (3^5).

$$
\begin{aligned}
\min \quad & 2y_1 + q\, y_2 + 8y_3 + 50x_1 + 350x_2 \\
\text{s.t.} \quad & -y_1 \geq -T_1\, x_1 \\
& -y_2 \geq -T_2\, x_2 \\
& 2y_1 + W\, y_2 + y_3 \geq h \\
& y_1, y_2, y_3 \geq 0 \quad x_1, x_2 \in \{0,1\} \\
& T_1 \in \{35, 40, 45\} \quad T_2 \in \{60, 70, 80\} \\
& h \in \{90, 120, 150\} \quad q \in \{1, 3, 4\} \quad W \in \{2, 3, 4\}
\end{aligned}
$$

In order to apply our algorithm to this problem, we start by considering four initial scenarios in the Ω set. Table 1 contains all parameters' settings,

Table 1. All parameter values and O^*_ω for four initial scenarios in Ω

Scenario	T_1	T_2	h	q	W	y_1^*	y_2^*	y_3^*	x_1^*	x_2^*	O^*_ω
1	35	60	150	1	2	15	60	0	1	1	490
2	45	80	90	1	2	45	0	0	1	0	140
3	35	60	90	4	4	35	0	20	1	0	280
4	45	80	150	4	4	0	37.5	0	0	1	500

Table 2. Solutions of the relaxed model (2) for four initial scenarios in Ω

Scenario	$y_{1\omega}$	$y_{2\omega}$	$y_{3\omega}$	$x_{1\omega}$	$x_{2\omega}$	O_ω^*	$Z_\omega^*(\vec{x}_\Omega)$	$(Z_\omega^*(\vec{x}_\Omega) - O_\omega^*)/O_\omega^*$
1	35	0	80	1	0	490	760	55.10%
2	45	0	0	1	0	140	140	0.00%
3	35	0	20	1	0	280	280	0.00%
4	45	0	60	1	0	500	620	24.00%

associated optimal variables settings, and the O_ω^* value (from the model (1)) for each scenario in Ω.

By using the information in Table 1, the relaxed model (2) considering only these four scenarios is solved and the solution is given in Table 2.

The candidate robust solution from the first stage is now $x_{1\Omega} = 1$ and $x_{2\Omega} = 0$ with the lower bound on the min-max relative regret value of 55.10%. At the second stage, after performing the preprocessing step, all model parameters can be fixed as follows: $T_1 = 35$, $T_2 = 60$, $h = 150$ and $W = 2$. Because this parameter setting is already considered in the scenario 1, the current candidate robust solution is already feasible for all possible scenarios. At the third stage, we are required to solve the model (31). By applying the parameter preprocessing step, T_2 can be fixed to the value of 80. Table 3 contains the new scenario generated by the model (31) with the upper bound on min-max relative regret value of 96.10%. This new scenario is added to the scenario set Ω and the model (1) and the relaxed model (2) are solved. The candidate robust solution from the first stage is now $x_{1\Omega} = 1$ and $x_{2\Omega} = 0$ with the lower bound on the min-max relative regret value of 96.10%. Because the difference between upper and lower bounds on min-max relative regret value is now zero, the algorithm is then terminated with the robust optimal solution of $x_1 = 1$ and $x_2 = 0$ with the maximum relative regret value of 96.10%. Note that by using the three-stage algorithm, we can identify the optimal relative robust solution by considering only five scenarios out of all 243 possible scenarios.

4.2 Comparison between Presented Algorithm and End Point Robust Solutions

In this subsection, we demonstrate a counterexample that illustrates the insufficiency of a relative robust solution obtained by considering only those scenarios generated by the upper and lower bound values for all uncertain

Table 3. The optimal solution for the model (31)

T_1	T_2	h	q	W	$\max_{\omega \in \bar{\Omega}} (Z_\omega^*(\vec{x}_\Omega) - O_\omega^*)/O_\omega^*$
35	80	150	1	4	96.10%

parameters. Let us consider the following MILP problem with uncertain parameter h with three possible values: 0, 5, and 10.

$$\begin{aligned}
\min \quad & -2y_1 - y_2 - y_3 + 20y_4 + x_2 \\
\text{s.t.} \quad & y_1 + y_2 \leq 10x_1 & y_1 \leq 5x_2 \\
& y_2 \leq h & y_1 \leq y_2 \\
& y_1 + y_3 \leq 5x_3 & y_4 \geq 1 \\
& y_1, y_2, y_3, y_4 \geq 0 & x_1, x_2, x_3 \in \{0,1\} \quad h \in \{0,5,10\}
\end{aligned}$$

The optimal robust solution to the problem where h can only take its values from its upper or lower bounds can be solved by using the relaxed model (2) with two scenarios ($h = 0$ and $h = 10$). The resulting optimal robust solution by considering only these two scenarios is $x_1 = 1$, $x_2 = 0$, and $x_3 = 1$ with the minimum-maximum relative regret value of 0%. In contrast, when applying the solution resulting from this approach ($x_1 = 1$, $x_2 = 0$, $x_3 = 1$) to the problem when h can take any values from the set $\{0, 5, 10\}$, the maximum relative regret of this solution is actually equal to 66.67% which happens when $h = 5$.

The proposed algorithm is now applied to solve this example problem by using two initial scenarios ($h = 0$ and $h = 10$). The first stage of the algorithm gives the candidate robust solution by setting $x_{1\Omega} = 1$, $x_{2\Omega} = 0$, and $x_{3\Omega} = 1$ with the lower bound of 0%. This candidate robust solution is then forwarded to the second stage of the algorithm for the feasibility check. After performing the preprocessing step, the h parameter can be fixed at 0, which has already been considered in the initial scenario. At the third stage, we are required to solve the model (31). The model (31) generates a scenario ($h = 5$) with the upper bound on min-max relative regret of 66.67%.

Because the current upper bound on the min-max relative regret value resulting from this model (31) is greater than the current lower bound, the algorithm forwards the setting of h at 5 (scenario 3) and the upper bound of 66.67% to the first stage of the algorithm. After one more iteration, the algorithm is then terminated when the upper bound and the lower bound are equal with the optimal robust solution of $x_1 = 1$, $x_2 = 1$ and $x_3 = 1$ with the maximum relative regret value of 20%. These results illustrate the superiority of the proposed algorithm over the use of a robust optimization algorithm that only considers each parameter at its boundaries, which can generate a less than optimal solution to the original problem.

4.3 Facility Location Example Problem with Extremely Large Number of Scenarios

In this subsection, we apply the algorithm to a hypothetical robust supply chain facility location problem with a large number of possible scenarios that rules out an explicit scheme. We consider a supply chain in which suppliers (sup) send material to factories (fac) that supply warehouses (wh) that supply

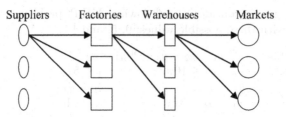

Fig. 2. Stages in the supply chain network ([18])

markets (mk) as shown in Figure 2 ([18]). Location and capacity allocation decisions have to be made for both factories and warehouses. Multiple warehouses may be used to satisfy demand at a market and multiple factories may be used to replenish warehouses. It is also assumed that units have been appropriately adjusted such that one unit of input from a supply source produces one unit of the finished product. In addition, each factory and each warehouse cannot operate at more than its capacity and a linear penalty cost is incurred for each unit of unsatisfied demand. The required parameters for the model are summarized in Table 4:

In the deterministic case, the goal is to identify factory and warehouse locations as well as quantities shipped between various points in the supply chain that minimize the total fixed and variable costs. In this case, the overall problem can be modeled as the mixed integer linear programming problem presented in the following model.

Table 4. Required parameters for the model

Parameter	Description
M	: Number of markets
N	: Number of potential factory locations
L	: Number of suppliers
T	: Number of potential warehouse locations
D_j	: Annual demand from customer j
K_i	: Potential capacity of factory site i
S_h	: Supply capacity at supplier h
W_e	: Potential warehouse capacity at site e
f_{1i}	: Fixed cost of locating a plant at site i
f_{2e}	: Fixed cost of locating a warehouse at site e
c_{1hi}	: Cost of shipping one unit from supplier h to factory i
c_{2ie}	: Cost of shipping one unit from factory i to warehouse e
c_{3ej}	: Cost of shipping one unit from warehouse e to market j
p_j	: Penalty cost per unit of unsatisfied demand at market j

$$\min \quad \sum_{i=1}^{N} f_{1i} y_i + \sum_{e=1}^{T} f_{2e} z_e + \sum_{h=1}^{L} \sum_{i=1}^{N} c_{1hi} x_{1hi}$$

$$+ \sum_{i=1}^{N} \sum_{e=1}^{T} c_{2ie} x_{2ie} + \sum_{e=1}^{T} \sum_{j=1}^{M} c_{3ej} x_{3ej} + \sum_{j=1}^{M} p_j s_j$$

$$\text{s.t.} \quad \sum_{i=1}^{N} x_{1hi} \le S_h \quad \forall h \in \{1, 2, \dots, L\}$$

$$\sum_{e=1}^{T} x_{2ie} \le K_i y_i \quad \forall i \in \{1, 2, \dots, N\}$$

$$\sum_{j=1}^{M} x_{3ej} \le W_e z_e \quad \forall e \in \{1, 2, \dots, T\}$$

$$\sum_{h=1}^{L} x_{1hi} - \sum_{e=1}^{T} x_{2ie} = 0 \quad \forall i \in \{1, 2, \dots, N\}$$

$$\sum_{i=1}^{N} x_{2ie} - \sum_{j=1}^{M} x_{3ej} = 0 \quad \forall e \in \{1, 2, \dots, T\}$$

$$\sum_{e=1}^{T} x_{3ej} + s_j = D_j \quad \forall j \in \{1, 2, \dots, M\}$$

$$x_{1hi} \ge 0 \ \ \forall h, i, \quad x_{2ie} \ge 0 \ \ \forall i, e \quad x_{3ej} \ge 0 \ \ \forall e, j$$

$$s_j \ge 0 \ \ \forall j, \quad y_i \in \{0, 1\} \ \ \forall i \quad z_e \in \{0, 1\} \ \ \forall e$$

When some parameters in this model are uncertain (ambiguous), the goal becomes to identify robust factory and warehouse locations (long-term decisions) with an objective utilizing the relative robust definition. Transportation decisions (short-term decisions) are now recourse decisions which can be made after all model parameters' values are realized. Table 5 summarizes information on approximated parameters' values associated with suppliers, factories, warehouses, and markets. A variable transportation cost of $0.01 per unit per mile is assumed in this example. The distance between each pair of locations is calculated based on the latitude and the longitude of each location.

The key uncertain parameters that we consider in this example are the supply quantity at each supplier, the potential capacity at each factory, the potential capacity at each warehouse, and finally the penalty cost per unit of unsatisfied demand at each market. We assume that each uncertain parameter can take its values from 80%, 100% and 120% of its approximated value

Table 5. Approximated parameters' information of the problem

sup(h)	S_h	fac(i)	f_{1i}	K_i	wh(e)	f_{2e}	W_e	mk(j)	D_i	p_j
San Diego	2500	Seattle	75000	2800	Sacramen.	37500	2500	Portland	800	125
Denver	3000	San Franc.	75000	2400	Oklahoma	37500	2600	LA	1050	140
Kansas C.	2000	Salt Lake	75000	2500	Lincoln	37500	2500	Phoenix	600	120
El Paso	1000	Wilmingt.	60000	2150	Nashville	40500	3000	Houston	1800	150
Cincinnati	800	Dallas	75000	2700	Cleveland	37500	2600	Miami	1500	125
Boise	500	Minneapo.	60000	2100	Fort W.	36000	2100	NY	1250	130
Austin	1100	Detroit	81000	3000	Eugene	37500	2300	St. Louis	1050	120
Charlotte	2000	Tampa	39000	1400	Santa Fe	34500	1700	Chicago	1750	140
Huntsville	700	Pittsburgh	75000	2500	Jacksonv.	37500	2400	Philadel.	750	125
Bismarck	600	Des Moines	45000	1600	Boston	37500	2600	Atlanta	1850	155

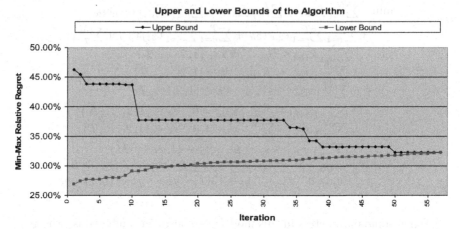

Fig. 3. Upper and lower bounds on min-max relative regret from the algorithm

reported in Table 5. In summary, this example contains 40 uncertain parameters with three possible values for each uncertain parameter. The total number of possible scenarios is 3^{40} or 1.2158×10^{19}. The case study is solved by our three-stage algorithm with $\epsilon = 0$ and the initial scenario set Ω containing 16 scenarios on a Windows XP-based Pentium(R) 4 CPU 3.60GHz personal computer with 2.00 GB RAM using a C++ program and CPLEX 10 for the optimization process. MS-Access is used for the case study input and output database. The algorithm terminates at an optimal robust solution within 57 iterations. The algorithm recommends opening production facilities at San Francisco, Dallas, Minneapolis, Detroit, Pittsburgh, and Des Moines and recommends opening warehouse facilities at Sacramento, Lincoln, Nashville, Cleveland, Fort Worth, and Jacksonville with the maximum relative regret value of 32.23%. Figure 3 illustrates the convergence of upper and lower bounds on the min-max relative regret value produced by the algorithm.

Finally, we compare the performance of the robust solution with the optimal solution from the deterministic model when each parameter is fixed at the value reported in Table 5. This deterministic solution recommends opening production facilities at San Francisco, Wilmington, Dallas, Minneapolis, and Detroit and recommends opening warehouse facilities at Sacramento, Nashville, Cleveland, Fort Worth, and Boston with the maximum relative regret value of 56.83%. These results illustrate the impact of ambiguity on the performance of the long-term decisions. They also illustrate the applicability and effectiveness of our three-stage algorithm for solving relative robust optimization problems that have a mixed integer (binary) linear programming base model with a full-factorial structure of data uncertainty.

5 Summary

This chapter develops a three-stage relative robust optimization algorithm for dealing with uncertainty in model parameter values of mixed integer linear programming problems under a full-factorial scenario structure of parametric uncertainty. The algorithm is an efficient approach to generate relative robust solutions when each uncertain parameter in a mixed integer (binary) linear programming problem independently takes its value from a finite set of real numbers with unknown probability distribution. The algorithm utilizes preprocessing steps, modeling techniques and problem transformation procedures to improve its computational tractability. The algorithm is proven to either terminate at an optimal relative robust solution or identify the nonexistence of the robust solution in a finite number of iterations. The proposed algorithm has been applied to solve the relative robust mixed integer linear programming problem under this structure of parametric uncertainty with the incredibly large number of possible scenarios (1.2158×10^{19}). All results illustrate good performance of the proposed algorithm under the problems considered.

References

1. Assavapokee, T., Realff, M., Ammons, J., and Hong, I. (2008), Scenario Relaxation Algorithm for Finite Scenario Based Min-Max Regret and Min-Max Relative Regret Robust Optimization, Computers and Operations Research, Vol. 35, No.6.
2. Assavapokee, T., Realff, M., and Ammons, J. (2008), A Min-Max Regret Robust Optimization Approach for Interval Data Uncertainty, Journal of Optimization Theory and Applications, Vol. 137, No. 2.
3. Averbakh, I. (2001), On the Complexity of a Class of Combinatorial Optimization Problems with Uncertainty, Mathematical Programming, 90, 263-272.
4. Averbakh, I. (2000), Minmax Regret Solutions for Minimax Optimization Problems with Uncertainty, Operations Research Letters, 27/2, 57-65.
5. Bai, D., Carpenter, T., Mulvey, J. (1997), Making a Case for Robust Optimization Models, Management Science, 43/7, 895-907.
6. Bajalinov, E.B. (2003), Linear Fractional Programming: Theory, Methods, Applications and Software, Applied Optimization, Kluwer Academic Publishers.
7. Bard, J.F. and Falk, J.E. (1982), An Explicit Solution to the Multi-level Programming Problem, Computers and Operations Research, Vol. 9/1, 77-100.
8. Bard, J.F. and Moore, J.T. (1990), A Branch and Bound Algorithm for the Bilevel Programming Problem, SIAM Journal of Scientific and Statistical Computing, Vol. 11/2, 281-292.
9. Bard, J.F. (1991), Some Properties of Bilevel Programming Problem, Journal of Optimization Theory and Application, Vol. 68/2, 371-378.
10. Bard, J.F. (1998), Practical Bilevel Optimization: Algorithms and Applications, Nonconvex Optimization and Its Applications, 30, Kluwer Academic Publishers.

11. Benders, J.F. (1962), Partitioning Procedures for Solving Mixed Variables Programming Problems, Numerische Mathematik, 4, 238-252.
12. Ben-Tal, A. and Nemirovski, A. (1998), Robust Convex Optimization, Mathematical Methods of Operations Research, 23, 769-805.
13. Ben-Tal, A. and Nemirovski, A. (1999), Robust Solutions to Uncertain Programs, Operations Research Letters, 25, 1-13.
14. Ben-Tal, A. and Nemirovski, A. (2000), Robust Solutions of Linear Programming Problems Contaminated with Uncertain Data, Mathematical Programming, 88, 411-424.
15. Ben-Tal, A., El-Ghaoui, L. and Nemirovski, A. (2000), Robust Semidefinite Programming, In: Saigal, R., Vandenberghe, L., Wolkowicz, H., (eds.), Semidefinite Programming and Applications, Kluwer Academic Publishers.
16. Bertsimas, D. and Sim, M. (2003), Robust Discrete Optimization and Network Flows, Mathematical Programming Series B, 98, 49-71.
17. Bertsimas, D. and Sim, M. (2004), The Price of Robustness, Operations Research, 52/1, 35-53.
18. Chopra, S. and Meindl, P. (2003), Supply Chain Management: Strategy, Planning, and Operations (2nd edition), Prentice Hall Publishers.
19. Hansen, P., Jaumard, B., and Savard, G. (1992), New Branch-and-Bound Rules for Linear Bilevel Programming, SIAM Journal of Scientific and Statistical Computing, Vol. 13/5, 1194-1217.
20. Huang, H.-X. and Pardalos, P.M. (2002), A Multivariate Partition Approach to Optimization Problems, Cybernetics and Systems Analysis, 38, 2, 265-275.
21. Kouvelis, P. and Yu, G. (1997), Robust Discrete Optimization and Its Applications, Kluwer Academic Publishers, Dordecht, The Netherlands.
22. Mausser, H.E. and Laguna, M. (1998), A New Mixed Integer Formulation for the Maximum Regret Problem, International Transactions in Operational Research, 5/5, 389-403.
23. Mausser, H.E. and Laguna, M. (1999), Minimizing the Maximum Relative Regret for Linear Programmes with Interval Objective Function Coefficients, Journal of the Operational Research Society, 50/10, 1063-1070.
24. Mausser, H.E. and Laguna, M. (1999), A Heuristic to Minimax Absolute Regret for Linear Programs with Interval Objective Function Coefficients, Europian Journal of Operational Research, 117, 157-174.
25. Migdalas, A., Pardalos, P.M., and Varbrand, P. (Editors) (1997), Multilevel Optimization: Algorithms and Applications, Kluwer Academic Publishers.
26. Mulvey, J., R. Vanderbei, S. Zenios (1995), Robust Optimization of Large-Scale Systems, Operations Research, 43, 264-281.
27. Terlaky, T. (1996), Interior Point Methods in Mathematical Programming, Kluwer Academic Publisher.
28. Von Stackelberg, H. (1943), Grundzuge der Theoretischen Volkswirtschaftslehre Stuttgart, Berlin: W. Kohlhammer.

Supply Chain and Logistics Design

An Enterprise Risk Management Model for Supply Chains

John M. Mulvey[1] and Hafize G. Erkan[2]

[1] Department of Operations Research and Financial Engineering
Princeton University, Princeton, NJ 08544
mulvey@princeton.edu
[2] Department of Operations Research and Financial Engineering
Princeton University, Princeton, NJ 08544
herkan@alumni.princeton.edu

Summary The design of an optimal supply chain rarely considers uncertainty within the modeling framework. This omission is due to several factors, including tradition, model size, and the difficulty in measuring the stochastic parameters. We show that a stochastic program provides an ideal framework for optimizing a large supply chain in the face of an uncertain future. The goal is to reduce disruptions and to minimize expected costs under a set of plausible scenarios. We illustrate the methodology with a global production problem possessing currency movements.

1 Introduction

Solving large-scale problems has been an important research focus for many years in which extensive research has been done and published (see [3]). Decentralized decision making has also been long recognized as an important decision-making problem. Decentralization among the divisions of a global firm requires a decomposition pattern that would also reduce the number of the steps required for the convergence to the centralized one (see [8],[9] for decentralized risk management theory and application in financial services industry). The comprehensive reviews published so far suggest that the large-scale models have proven to be extremely difficult to solve to optimality without the application of Benders' decomposition or factorization methods. The decentralization idea is often utilized in logistics problem or multiperiod production-distribution systems (see [4] for a discussion of a mixed integer programming formulation and an integrated design methodology based on primal (Benders') decomposition applied to the integrated design of strategic supply chain networks and to the determination of tactical production-distribution allocations in the case of customer demands with seasonal variations). The optimization framework embeds the inherent structure in which each decision

W. Chaovalitwongse et al. (eds.), *Optimization and Logistics Challenges in the Enterprise*, Springer Optimization and Its Applications 30,
DOI 10.1007/978-0-387-88617-6_5, © Springer Science+Business Media, LLC 2009

maker has a mathematical model that captures only the local dynamics and the associated interconnecting global constraints (see [1] for a description of a parallel implementation of the nested Benders' algorithm, which employs a farming technique to parallelize nodal subproblem solutions; see also [2] for such subproblems where the subproblems have a network or transportation problem structure and for ways of exploiting the structure of the subproblems and the combination of branch and bound with Benders' decomposition). Utilizing the extensive literature on stochastic decomposition (see [5] and [6]); we have unified the supply chain management problem with the asset liability management (ALM) and applied the nested Benders' decomposition method to formulate the subproblems associated with the divisions of an international production firm (see Figure 1).

During the decentralization, the subproblems are solved at every division and the only information passed to the headquarters is the price that the divisions are willing to charge for a possible flow of goods as depicted in Figure 2. The same set of scenarios are used for both of the divisions, and in that way through the possible flow among divisions the headquarters are allowed to hedge also the potential currency risk, which makes the supply chain management more robust.

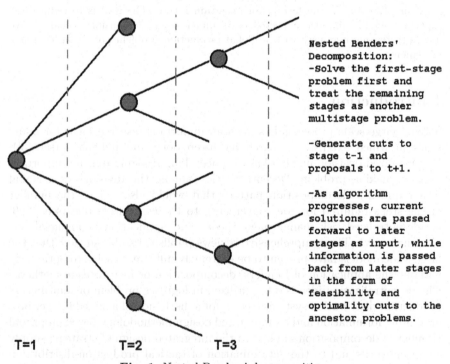

Nested Benders'
Decomposition:
-Solve the first-stage
problem first and
treat the remaining
stages as another
multistage problem.

-Generate cuts to
stage t-1 and
proposals to t+1.

-As algorithm
progresses, current
solutions are passed
forward to later
stages as input, while
information is passed
back from later stages
in the form of
feasibility and
optimality cuts to the
ancestor problems.

T=1   T=2   T=3

Fig. 1. Nested Benders' decomposition

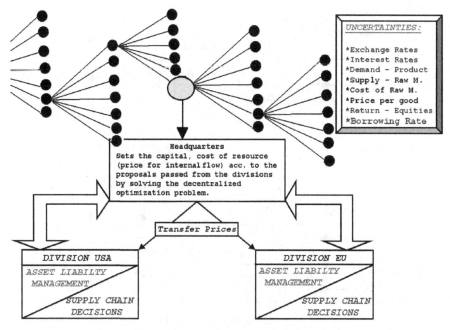

Fig. 2. Dantzig-Wolfe decomposition at a given stage

2 Asset Liability Management

The risk management of global companies is different from that of regional firms due to various reasons including exchange rate risk and different regulatory requirements in the international locations they operate. However, the targets for these multinational firms, which can span a wide range of spectrum from return benchmarks to capital allocation, are generally determined at the headquarter level. The decisions taken, on the other hand, are not the optimal choice for the regional subdivisions. Therefore, we define an optimization problem that emphasizes the exchange of goods among the subdivisions where timing is important and dependent on the market variables. We also consider dynamic borrowing rates as a function of leverage in our problem. Until now, the supply chain problem was analyzed by means of product flow and its efficiency. The problem of allocating capital to investing for production or to acquiring goods from other subdivisions is an important addendum we provide in our model.

The first step is to formulate the ALM model that incorporates the distinction between default and insolvency (see [7]). We will first define the variables and parameters and afterwards state the asset liability model which is presented in Figure 3.

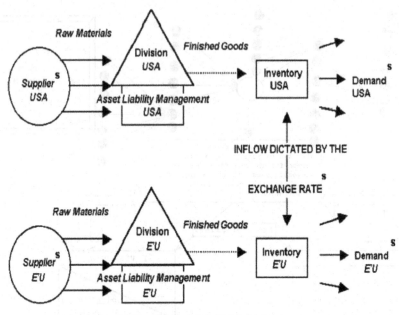

Fig. 3. Supply chain model

2.1 Indices

The decision variables are defined for each time period $t \in \mathcal{T}$, each scenario $s \in \mathcal{S}$, and each group $g \in \mathcal{G}$, where the groups represent the divisions of the firm; $i \in \mathcal{I}$ refers to different asset classes while $k \in \mathcal{K}$ represents the products; $m \in \mathcal{M}$ specifies the debt priority.

2.2 Parameters

$d_{g,k,t,s}$: demand (group, prod, time, scen)

$\rho^{\text{BOR}}_{g,t,t',s}$: % borrowing rate (group, time, time due, scen)

ρ^{TAX}_{g}: % tax rate (group)

$\rho^{\text{EXCH}}_{t,s}$: exchange rate (time, scen)

$r_{g,i,t,s}$: % return rate (group, asset, time, scen)

$\Psi_{t,s}$: capital (time, scen)

π_s: probabilities

$c^{\text{PROD}}_{g,k,t}$: product cost (group, prod, time)

$c^{\text{INV}}_{g,k,t}$: inventory cost (group, prod, time)

$c^{\text{SUP}}_{g,t,s}$: supplier cost (group, time, scen)

ξ: time periods (long-term debt)

θ_g^{DEBT}: debt ratio (borrowed/capital)

$\theta_g^{\mathrm{D/E}}$: debt/equity ratio

$\theta_g^{\mathrm{A/C}}$: asset/capital ratio

θ_g^{DIV}: dividend payout ratio

$p_{g,k,t,s}^{\mathrm{GOOD}}$: price/good (group, prod, time, scen)

h_k^{RAW}: usage (prod) - how much raw material used to produce one item

h_k^{FAC}: factory usage (prod)

$u_{g,t}^{\mathrm{FAC}}$: factory capacity (group, time)

$u_{g,t,s}^{\mathrm{SUP}}$: supplier capacity (group, time, scen)

$\overrightarrow{B}^{\mathrm{RATE}}$: vector of thresholds – e.g., as used in borrowing rate updates

$\overrightarrow{\gamma}_{g,t,s}^{\mathrm{RATE}}$: vector of bankruptcy-related parameters – e.g., as used in borrowing rate calculations (group, time, scen)

ρ^{RISK}: risk penalty/premium coefficient

λ: Lagrange multipliers

$f_t(\cdot)$: density function of a normal random variable at time t

2.3 Decision Variables

$x_{g,k,t,s}^{\mathrm{MAKE}}$: make (group, prod, time, scen)

$x_{g,k,t,s}^{\mathrm{INV}}$: inventory (group, prod, time, scen)

$x_{g,k,t,s}^{\mathrm{SELL}}$: sell (group, prod, time, scen)

$x_{g,k,t,s}^{\mathrm{IN}}$: inflow (group, prod, time, scen)

$x_{g,k,t,s}^{\mathrm{OUT}}$: outflow (group, prod, time, scen)

$x_{g,t,s}^{\mathrm{SUP}}$: supply (group, time, scen)

$\delta_{g,t,s}$: dividend – dollar amount (group, time, scen)

$y_{g,i,t,s}$: investment – dollar amount (group, asset, time, scen)

$w_{g,t,t',s,m}$: borrowed (group, time, time due, scen, priority)

$z_{g,t,s}^{\mathrm{NET}}$: net profit/loss (group, time, scen)

$z_{g,t,s}^{\mathrm{EARN}}$: earnings (group, time, scen)

$x_{g,t,s}^{\mathrm{TA}}$: total assets (group, time, scen)

$x_{g,t,s}^{\mathrm{TL}}$: total liabilities (group, time, scen)

$p_{g,k,t,s}^{\mathrm{TRAN}}$: transfer price (group, prod, time, scen)

$\nu_{g,t,s}$: valuation of cashflows (group, time, scen)

$\psi_{g,t,s}$: capital allocated to the groups (group, time, scen)

2.4 Model Formulation

The subsequent multistage stochastic programming model unifies the supply chain decisions with the ALM. Thus it enables a production firm operating globally to hedge against the exchange rate movements in the countries where the divisions are located. We formulate the problem for the case of two divisions which can be easily extended to cover more divisions.

The objective is to maximize the expected firm value which is a function of the earnings of the two divisions.

$$\max \quad \sum_s \pi_s f \left(z_{1,T,s}^{\text{EARN}}, z_{2,T,s}^{\text{EARN}} \right) \tag{1}$$

For each of the divisions, the amount sold plus the inventory to be held for the next period should be equal to the production amount plus the net product inflow and the previous inventory level in that stage.

$$x_{g,k,0,s}^{\text{INV}} = 0 \qquad \forall g,k,s \tag{2}$$

$$x_{g,k,t,s}^{\text{MAKE}} + x_{g,k,t,s}^{\text{IN}} - x_{g,k,t,s}^{\text{OUT}} + x_{g,k,t,s}^{\text{INV}}$$

$$= x_{g,k,t+1,s}^{\text{SELL}} + x_{g,k,t+1,s}^{\text{INV}} \qquad \forall g,k,s,t \in \{0,\ldots,T-1\} \tag{3}$$

$$x_{g,k,T,s}^{\text{INV}} = 0 \qquad \forall g,k,s \tag{4}$$

The amount sold in any division should be less than the corresponding market demand at that time.

$$x_{g,k,t,s}^{\text{SELL}} \leq d_{g,k,t,s} \qquad \forall g,k,s,t \in \{1,\ldots,T\} \tag{5}$$

The factory capacity is another constraint for both of the divisions.

$$\sum_k x_{g,k,t,s}^{\text{MAKE}} h_k^{\text{FAC}} \leq u_{g,t}^{\text{FAC}} \qquad \forall g,s,t \in \{0,\ldots,T-1\} \tag{6}$$

The amount of raw material purchased cannot exceed the supplier capacity at that time.

$$x_{g,t,s}^{\text{SUP}} \leq u_{g,t,s}^{\text{SUP}} \qquad \forall g,s,t \in \{0,\ldots,T-1\} \tag{7}$$

The raw material usage is restricted by the supplied quantity.

$$\sum_k x_{g,k,t,s}^{\text{MAKE}} h_k^{\text{RAW}} \leq x_{g,t,s}^{\text{SUP}} \qquad \forall g,s,t \in \{0,\ldots,T-1\} \tag{8}$$

The dividends are paid based on the net profit and the payout ratio.

$$\delta_{g,t,s} \leq \theta_g^{\text{DIV}} z_{g,t,s}^{\text{NET}} \qquad \forall g,s,t \in \{1,\ldots,T\} \tag{9}$$

The borrowing rate is adjusted based on the surplus the firm is holding at a given time point. For illustrative purpose, we assumed three different thresholds levels for setting the borrowing rate.

$$z_{g,t,s}^{\text{EARN}} - \sum_{t'=t+1}^{T} \sum_{m} w_{g,t,t',s,m} \leq B_1 + \gamma_{1,g,t,s}^{\text{RATE}} M \qquad \forall g, s, t \in \{1, \ldots, T-1\} \tag{10}$$

$$z_{g,t,s}^{\text{EARN}} - \sum_{t'=t+1}^{T} \sum_{m} w_{g,t,t',s,m} \leq B_2 + \gamma_{2,g,t,s}^{\text{RATE}} M \qquad \forall g, s, t \in \{1, \ldots, T-1\} \tag{11}$$

$$z_{g,t,s}^{\text{EARN}} - \sum_{t'=t+1}^{T} \sum_{m} w_{g,t,t',s,m} \leq B_3 + \gamma_{3,g,t,s}^{\text{RATE}} M \qquad \forall g, s, t \in \{1, \ldots, T-1\} \tag{12}$$

$$z_{g,t,s}^{\text{EARN}} - \sum_{t'=t+1}^{T} \sum_{m} w_{g,t,t',s,m} \geq B_1 - \left(1 - \gamma_{1,g,t,s}^{\text{RATE}}\right) M \qquad \forall g, s, t \in \{1, \ldots, T-1\} \tag{13}$$

$$z_{g,t,s}^{\text{EARN}} - \sum_{t'=t+1}^{T} \sum_{m} w_{g,t,t',s,m} \geq B_2 - \left(1 - \gamma_{2,g,t,s}^{\text{RATE}}\right) M \qquad \forall g, s, t \in \{1, \ldots, T-1\} \tag{14}$$

$$z_{g,t,s}^{\text{EARN}} - \sum_{t'=t+1}^{T} \sum_{m} w_{g,t,t',s,m} \geq B_3 - \left(1 - \gamma_{3,g,t,s}^{\text{RATE}}\right) M \qquad \forall g, s, t \in \{1, \ldots, T-1\} \tag{15}$$

The net profit is calculated by taking the return on equity, the interest paid on debt that was due at time t and the tax paid into account.

$$
\begin{aligned}
z_{1,t,s}^{\text{NET}} = \left(1 - \rho_1^{\text{TAX}}\right) \Bigg(& \sum_i y_{1,i+1,s} r_{1,i,t,s} \\
& - \sum_{t'=0}^{t-1} \sum_m \left(w_{1,t',t,s,m} \rho_{1,t',t,s}^{\text{BOR}} \left(3 - \gamma_{1,1,t,s}^{\text{RATE}} - \gamma_{2,1,t,s}^{\text{RATE}} - \gamma_{3,1,t,s}^{\text{RATE}}\right)\right) \\
& - \sum_k x_{1,k,t-1,s}^{\text{INV}} c_{1,k,t-1}^{\text{INV}} + \sum_k x_{1,k,t,s}^{\text{SELL}} p_{1,k,t,s}^{\text{GOOD}} - x_{1,t-1,s}^{\text{SUP}} c_{1,t-1,s}^{\text{SUP}} \\
& - \sum_k x_{1,k,t-1,s}^{\text{MAKE}} c_{1,k,t-1}^{\text{PROD}} - \sum_k x_{1,k,t-1,s}^{\text{IN}} p_{2,k,t,s}^{\text{TRAN}} \rho_{t,s}^{\text{EXCH}} \Bigg) \\
& \hspace{6cm} \forall s, t \in \{1, \ldots, T\} \tag{16}
\end{aligned}
$$

Due to the change in currencies, we have to consider the exchange rate when setting up the constraints for the other division.

$$
\begin{aligned}
z_{2,t,s}^{\text{NET}} = \left(1 - \rho_2^{\text{TAX}}\right) \Bigg(&\sum_i y_{2,i+1,s} r_{2,i,t,s} \\
&- \sum_{t'=0}^{t-1} \sum_m \left(w_{2,t',t,s,m} \rho_{2,t',t,s}^{\text{BOR}} \left(3 - \gamma_{1,2,t,s}^{\text{RATE}} - \gamma_{2,2,t,s}^{\text{RATE}} - \gamma_{3,2,t,s}^{\text{RATE}}\right)\right) \\
&- \sum_k x_{2,k,t-1,s}^{\text{INV}} c_{2,k,t-1,s}^{\text{INV}} + \sum_k x_{2,k,t,s}^{\text{SELL}} p_{2,k,t,s}^{\text{GOOD}} - x_{2,t-1,s}^{\text{SUP}} c_{2,t-1,s}^{\text{SUP}} \\
&- \sum_k x_{2,k,t-1,s}^{\text{MAKE}} c_{2,k,t-1,s}^{\text{PROD}} - \sum_k \left(x_{2,k,t-1,s}^{\text{IN}} p_{1,k,t,s}^{\text{TRAN}} \left(1/\rho_{t,s}^{\text{EXCH}}\right)\right) \Bigg)
\end{aligned}
$$
$$\forall s, t \in \{1,\ldots,T\} \quad (17)$$

The earnings of both divisions are calculated based on the previous earnings, the net profit/loss and the dividends paid.

$$
z_{g,t,s}^{\text{EARN}} = z_{g,t-1,s}^{\text{EARN}} + z_{g,t,s}^{\text{NET}} - \delta_{g,t,s} \qquad \forall g, s, t \in \{2,\ldots,T\} \qquad (18)
$$

$$
z_{g,1,s}^{\text{EARN}} = z_{g,1,s}^{\text{NET}} - \delta_{g,1,s} \qquad \forall g, s \qquad (19)
$$

In the following set of constraints, the risk and performance measures, constraint on the long-term debt (defined as debt due within τ time periods), and regulatory constraints are formulated for both of the divisions.

The total debt is bounded above by the division-specific debt ratio.

$$
\sum_{t'=t+1}^{T} \sum_m w_{g,t,t',s,m} \leq \theta_g^{\text{DEBT}} z_{g,t,s}^{\text{EARN}} \qquad \forall g, s, t \in \{1,\ldots,T-1\} \qquad (20)
$$

For the initial time period, the total debt is a function of the debt ratio as well as the starting capital allocated to the specific division.

$$
\sum_{t'=t+1}^{T} \sum_m w_{g,0,t',s,m} \leq \theta_g^{\text{DEBT}} \psi_{g,0,s} \qquad \forall g, s \qquad (21)
$$

The following is an additional constraint specifically restricting the long-term debt.

$$
\sum_{t'=t+\tau}^{T} \sum_m w_{g,t,t',s,m} \leq \theta_g^{\text{D/E}} z_{g,t,s}^{\text{EARN}} \qquad \forall g, s, t \in \{1,\ldots,T-1\} \qquad (22)
$$

The size of the investment portfolio is determined by the liquidity and asset/capital ratio of the division.

$$
\sum_i y_{g,i,t,s} \leq \theta_g^{\text{A/C}} y_{g,0,t,s} \qquad \forall g, s, t \in \{0,\ldots,T\} \qquad (23)
$$

Next, we consider the asset balance for each of the divisions at time zero where starting capital along with the debt issued account for the investments, production, supplies and inflow realized at the prevailing exchange rate.

$$\sum_i y_{1,i,0,s} + \sum_k x^{MAKE}_{1,k,0,s} c^{PROD}_{1,p,0} + x^{SUP}_{1,0,s} c^{SUP}_{1,0,s} + \sum_k x^{IN}_{1,k,0,s} p^{TRAN}_{2,k,1,s} \rho^{EXCH}_{1,s}$$

$$= \psi_{1,0,s} + \sum_{t'=1}^{T} \sum_m w_{1,0,t',s,m} \qquad \forall s \quad (24)$$

$$\sum_i y_{2,i,0,s} + \sum_k x^{MAKE}_{2,k,0,s} c^{PROD}_{2,p,0} + x^{SUP}_{2,0,s} c^{SUP}_{2,0,s} + \sum_k \left(x^{IN}_{2,k,0,s} p^{TRAN}_{1,k,1,s} (1/\rho^{EXCH}_{1,s}) \right)$$

$$= \psi_{2,0,s} + \sum_{t'=1}^{T} \sum_m w_{2,0,t',s,m} \qquad \forall s \quad (25)$$

The equations below define the asset balance for future stages. Different initial time period, we need to incorporate the sale revenues, inventory-related costs and interest expense paid at respective borrowing rates, which are a function of firm's surplus.

$$\sum_i y_{1,i,t+1,s} + \sum_k x^{MAKE}_{1,k,t+1,s} c^{PROD}_{1,k,t+1} + x^{SUP}_{1,t+1,s} c^{SUP}_{1,t+1,s}$$

$$+ \sum_k x^{IN}_{1,k,t+1,s} p^{TRAN}_{2,k,t+2,s} \rho^{EXCH}_{t+2,s} = \sum_{t'=t+2}^{T} \sum_m w_{1,t+1,t',s,m}$$

$$- \sum_{t''=0}^{t} \left(w_{1,t'',t+1,s} \left(1 + \left(3 - \gamma^{RATE}_{1,1,t,s} - \gamma^{RATE}_{2,1,t,s} - \gamma^{RATE}_{3,1,t,s} \right) \rho^{BOR}_{1,t'',t+1,s} \right) \right)$$

$$+ \sum_i y_{1,i,t,s} (1 + r_{1,i,t+1,s}) + \sum_k x^{SELL}_{1,k,t+1,s} p^{GOOD}_{1,k,t+1,s} - \sum_k x^{INV}_{1,k,t,s} c^{INV}_{1,k,t}$$

$$\forall s, t \in \{0, \dots, T-2\} \quad (26)$$

$$\sum_i y_{2,i,t+1,s} + \sum_k x^{MAKE}_{2,k,t+1,s} c^{PROD}_{2,k,t+1} + x^{SUP}_{2,t+1,s} c^{SUP}_{2,t+1,s}$$

$$+ \sum_k \left(x^{IN}_{2,k,t+1,s} p^{TRAN}_{1,k,t+2,s} (1/\rho^{EXCH}_{t+2,s}) \right) = \sum_{t'=t+2}^{T} \sum_m w_{2,t+1,t',s,m}$$

$$- \sum_{t''=0}^{t} \left(w_{2,t'',t+1,s} \left(1 + \left(3 - \gamma^{RATE}_{1,2,t,s} - \gamma^{RATE}_{2,2,t,s} - \gamma^{RATE}_{3,2,t,s} \right) \rho^{BOR}_{2,t'',t+1,s} \right) \right)$$

$$+ \sum_i y_{2,i,t,s} (1 + r_{2,i,t+1,s}) + \sum_k x^{SELL}_{2,k,t+1,s} p^{GOOD}_{2,k,t+1,s} - \sum_k x^{INV}_{2,k,t,s} c^{INV}_{2,k,t}$$

$$\forall s, t \in \{0, \dots, T-2\} \quad (27)$$

Lastly, the nonanticipativity constraints are defined for the stochastic programming problem.

3 Numerical Experimentation

Table 1 summarizes the comparison between the equivalent deterministic problem's and the decomposed problem's results via the nested Benders' decomposition algorithm. Via the decentralization, the only information passed to the headquarters consists of the prices that the divisions are willing to charge for a possible flow of goods. The rest of the optimization is handled at the division level where we pass the same set of scenarios to the divisions. We fix the following parameters: $\tau = 2$; debt ratio = USA 3 EU 1; debt/equity ratio = USA 2 EU 1; asset/capital ratio = USA 3 EU 2 ; dividend payout ratio = USA 0.05 EU 0.06; tax rate = USA 0.2 EU 0.1; risk penalty = USA 0.2 EU 0.4; starting capital = USA \$20,000 EU \$40,000.

In regard to the cross-divisional transfers/inflow, we observe that the inflow into the US division fluctuates with demand and the price per good in the USA as displayed in Figure 4. In this case study, we assumed a relatively flat exchange rate scenario over time in order to clearly reflect the interdivisional flows as a function of these two market variables.

In Figure 5, we see the effect of exchange rate and demand on the cross-divisional transfers. The fluctuations in demand over time dictate the amount of inflow into the US division, which in turn determines the production amount in that division.

In Figure 6, we change the borrowing rate thresholds, i.e., penalize the firm more as its surplus drops. The effect can be observed in the reduction of the expected terminal earnings when we incorporate more risk-averse borrowing behavior, which translates to a smaller probability of exceeding the target debt/equity ratio increase.

4 Conclusion

This chapter describes an integrated risk management approach for global supply chains. Enterprise risk management in this era requires managing all the risks embedded in production as well as global financial markets. Regulatory

Table 1. Comparison between Stochastic Benders' Decomposition (SBD) and the Equivalent Deterministic Problem (EDP) for Four Stages and Six Scenarios

Solution Method	Objective Function Value
SBD	\$139,014
EDP	\$138,785

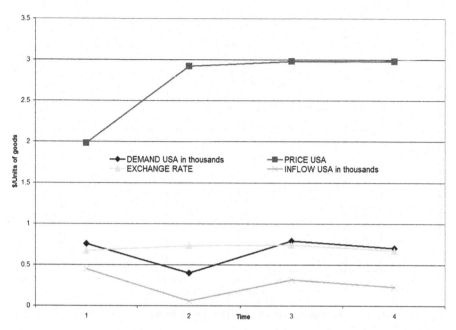

Fig. 4. Cross-divisional transfers/inflow

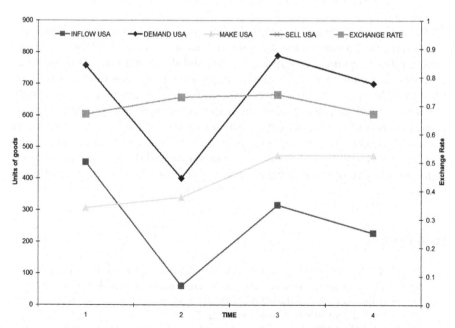

Fig. 5. The effect of demand and exchange rate on the inflow

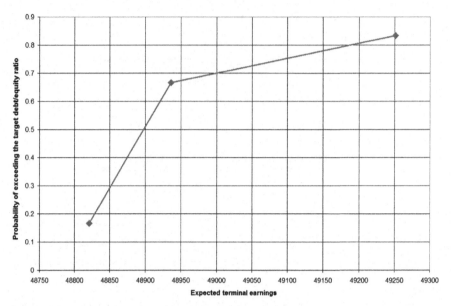

Fig. 6. Effect of borrowing rates on the probability of exceeding the target debt/equity ratio

constraints at the divisional as well as headquarters level introduce additional complexity.

In case of multinationals, proper valuation of expected long-term earnings and enterprise risk management requires a thorough modeling of dynamic borrowing rates as a function of firm's surplus and divisional tax and accounting practices. Exchange rates are also a major source of uncertainty, which we tried to address in our optimization model. Exchange rate movements as well as regional product pricing affect financing, investment and production decisions. In this context, use of foreign currency derivatives as hedges can enhance risk management. Any financial crisis exposes financial as well as production companies to large depreciations in exchange rate and decreased access of foreign capital. The optimal hedging strategy needs to be considered in future research.

References

1. Dempster MAH, Thompson RT (1998) Parallellization and aggregation of nested Benders decomposition. Annals of Operations Research 81:163–187
2. Dye S, Tomasgard A, Wallace SW (2000) Two-stage service provision by branch and bound. 35th ORSNZ Conference 221/7
3. Geoffrion AM (1969) An improved implicit enumeration approach for integer programming. Operations Research 17:437–454

4. Goetschalckx M, Vidal CJ, Dogan K (2002) Modeling and design of global logistics systems: a review of integrated strategic and tactical models and design algorithms. European Journal of Operational Research 143: 1–18

5. Higle JL, Rayco B, Sen S (2006) Stochastic scenario decomposition for multistage stochastic programs. To appear in Operations Research

6. Higle JL, Sen S (1996) Stochastic decomposition: a statistical method for large scale stochastic linear programming. Kluwer Academic Publishers, Dordrecht

7. Jarrow R, Purnanandam A (2004) The valuation of a firm's investment opportunities: a reduced form credit risk perspective. Working paper, Cornell University

8. Mulvey JM, Erkan HG (2005) Decentralized risk management for global property and casulty insurance companies. In: Wallace SW, Ziemba WT (eds) Applications of Stochastic Programming. SIAM Mathematical Series on Optimization 503–530

9. Mulvey JM, Erkan HG (2006) Applying CVaR for decentralized risk management of financial companies. Journal of Banking and Finance 30:627–644

Notes on using Optimization and DSS Techniques to Support Supply Chain and Logistics Operations

Tan Miller

Harper Professor of Global Supply Chain Management
College of Business Administration
Rider University, Lawrenceville, NJ 08648
tmiller@rider.edu, tanjean@verizon.net

Summary In this chapter, we offer recommendations on approaches that can further the rate of use of optimization and related methods in firms that may not have a rich history of utilizing these types of techniques. We begin by reviewing a framework for organizing operations from the strategic level to the daily operating level. Next, we consider traditional opportunities to employ optimization and related methods across this framework of activities. We then describe a general operating approach to facilitate the use of optimization and related decision support system (DSS) methods as a standard business practice. Finally, "lessons learned" from the author's implementation of DSS in several industries are discussed.

1 Introduction

There are certain industries that have historically used optimization and related operations research techniques as key components of their standard business practice. The airline and oil industries represent two examples that for decades have heavily utilized operations research (OR) techniques to support operations. There are also a wide range of firms that utilize optimization and OR techniques on a much more sporadic or one-off basis. In this article, we offer suggestions and recommendations on approaches that can further the rate of use of optimization and related methods in firms that may not have a rich history of utilizing these types of techniques.

The remainder of this chapter is organized as follows. We begin with a review of a framework for organizing operations from the strategic level to the daily operating level. This will provide a context for the balance of the chapter. We next consider traditional opportunities to employ optimization and related methods across this framework of activities. This section will also address "barriers and impediments" that exist in many organizations which lead to an underutilization of OR methods. Following this, we review an approach that

W. Chaovalitwongse et al. (eds.), *Optimization and Logistics Challenges in the Enterprise*, Springer Optimization and Its Applications 30, DOI 10.1007/978-0-387-88617-6_6, © Springer Science+Business Media, LLC 2009

the author has employed in industry practice to facilitate effective and regular use of optimization and related decision support system (DSS) methods. This discussion will include citations from several implementations. The chapter concludes with some "lessons learned" and final thoughts.

1.1 Hierarchical Framework for Organizing Supply Chain Operations

There are major organizational issues, systems and infrastructure considerations, methodology issues, and numerous other problem dimensions to evaluate in formulating a firm's distribution and supply chain network planning approach. From all perspectives, effective supply chain planning over multiple time horizons requires that a firm establish appropriate linkages across horizons and establish points of intersections among these horizons. To facilitate a planning system that possesses the appropriate linkages, a firm must have an overall framework that guides how different planning horizons and planning components fit together.

Figure 1 presents a general framework for hierarchical supply chain planning which defines three levels, namely the strategic, tactical, and operations level. As Figure 1 illustrates, strategic planning activities focus on a horizon of approximately two or more years into the future, while tactical and operational activities focus on plans and schedules for 12 to 24 months and one to 18 months in advance, respectively. At the strategic level, a firm must address such key issues as overall corporate objectives, market share and profitability goals, business and product mix targets, and so on. Planning decisions on overall corporate objectives drive strategic supply chain decisions. For example, market share and business or product mix objectives will strongly influence manufacturing capacity strategies.

At the strategic manufacturing planning level, the firm must address such issues as planned production capacity levels for the next three years and beyond, the number of facilities it plans to operate, their locations, the resources the firm will assign to its manufacturing operations, and numerous other important long-term decisions. Decisions made at the strategic level place constraints on the tactical planning level. At the tactical level, typical planning activities include the allocation of capacity and resources to product lines for the next 12 to 18 months, aggregate planning of workforce levels, the development or fine-tuning of distribution plans, and numerous other activities. Within the constraints of the firm's manufacturing and distribution infrastructure (an infrastructure determined by previous strategic decisions), managers make tactical (annual) planning decisions designed to optimize the use of the existing infrastructure. Planning decisions carried out at the tactical level impose constraints upon operational planning and scheduling decisions. At this level, activities such as distribution resource planning, rough cut capacity planning, master production scheduling, and shop floor control scheduling decisions occur.

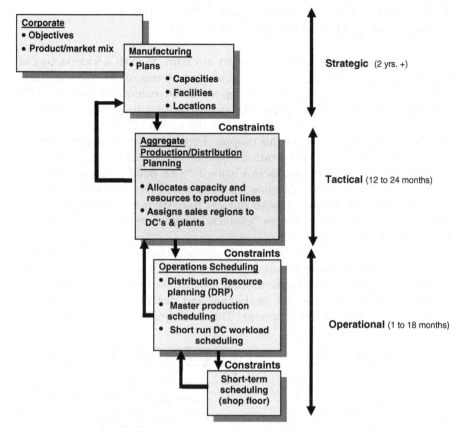

Fig. 1. A hierarchical supply chain planning framework

The feedback loops from the operational level to the tactical level and from the tactical level to the strategic level represent one of the most important characteristics of the supply chain planning system illustrated in Figure 1. To assure appropriate linkages and alignment between levels, a closed-loop system which employs a "top down" planning approach complemented by "bottom up" feedback loops is required. For example, production and distribution plans which appear feasible at an aggregate level can often contain hidden infeasibilities that only manifest themselves at lower, more disaggregated levels. Without proper feedback loops imbedded into its planning system, the danger that a firm will attempt to move forward with infeasible plans always exists. These infeasibilities often do not surface until a firm is in the midst of executing its operational plans and schedules. For additional detail on the importance of feedback loops, the reader is referred to hierarchical production planning literature (see, e.g., [3], [7]). Additionally, Appendix B presents an illustrative example of a feedback loop from the operational level to the tactical level.

2 The Role of Optimization and Simulation to Support a Hierarchical Framework

One can support the framework depicted in Figure 1[1] with a wide variety of optimization and simulation models at all three levels of the planning hierarchy. To illustrate these potential opportunities, consider Figure 2 which presents a generic tactical, operational (and strategic) supply chain planning system. We refer the reader to [7] for a lengthy review of the hierarchical system depicted. For purposes of this chapter, Figure 2 serves to illustrate where opportunities to employ optimization reside across the supply chain at a high level. Specifically, the rectangles in Figure 2 "with numbers in parentheses" represent activities where optimization models and related tools either play or should play a significant decision support role. We now briefly review these potential applications and discuss "enablers" and "impediments" that impact these opportunities.

Model (1) labeled "plant/DC/family optimization model" represents one of the most critical points in supply chain planning where firms should - but frequently do not - employ optimization models. As Figure 2 depicts, a firm should establish an integrated production and distribution plan as part of its tactical (annual) planning process. This need applies whether a particular

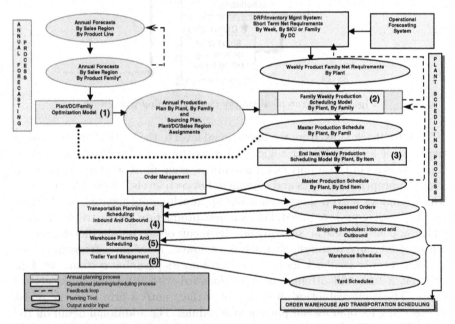

Fig. 2. Integrated production and distribution planning system (tactical/annual planning and operational scheduling)

---

[1] See Appendix A for a brief discussion of the components of this system.

firm has a large global manufacturing and distribution network or a more moderate-sized domestic network. As long as there are significant costs, service and/or capacity trade-offs that require evaluation; an optimization model has an important rule to play in the tactical planning process.

The integrated production and distribution tactical plan that model (1) creates should at the minimum provide the following:

- A production plan for each plant in the network. Typically this plan should be at a product family level or some other appropriate level of aggregation above the item level.

- A distribution plan for all distribution centers (DCs) that establishes network sourcing or flow relationships from plants to DCs to customers. Again this plan is generally developed at an appropriate level of product aggregation such as product family.

An optimization approach for this model assures that a firm can generate a cost minimizing or profit maximizing plan that accounts for all pertinent costs, capacities and service requirements across the network. Most critically, this approach facilitates an integrated plan which will align manufacturing and distribution activities.

Despite the seemingly self-evident reasons to employ optimization models at this point in the supply chain planning process, a number of barriers limit the implementation rate of truly cross-functional integrated optimization approaches. The following represent some of the key impediments to broader scale use of model (1) in practice.

- Many senior managers and decision makers responsible for manufacturing and distribution do not understand the potential benefits of optimization. It may seem hard to believe that after decades of use and availability, optimization methods remain a "black box" in many firms and industries. Managers caught up in the frenetic pace of the corporate world simply may not be willing or have the interest to invest their organization's time in understanding, developing and implementing large-scale optimization-based planning systems. This factor is an impediment to stand-alone manufacturing or distribution network implementations, as well as integrated manufacturing/distribution initiatives.

- There are often significant "organizational" barriers to the development of "integrated" optimization initiatives. Particularly in large firms, one frequently observes that the manufacturing and distribution organizations operate quite autonomously and each focus heavily on their own operating efficiencies. Even in firms with strong manufacturing/distribution linkages, one often still does not find the very high level of integration required to facilitate model (1). Specifically, for a firm to employ this model as a standard tactical planning process, its manufacturing and distribution functions must both adopt a holistic vision for their supply chain. This means each organization must be willing to sacrifice (or suboptimize) its own operating efficiency, if by doing so this will enhance the overall efficiency of the firm's supply chain. For example, manufacturing must be willing to accept higher operating costs, if by doing so it will facilitate distribution savings that will outweigh the incremental increased manufacturing costs. Similarly in the reverse situation distribution must also agree to accept higher incremental costs for the overall good of the firm. In many firms, senior functional leadership will not accept an approach which may lead to a denigration of its own operating efficiency.

- In large multinational or global firms, tax laws and benefits can heavily influence where a firm locates its plants and the firm's activity levels at different plants. In some firms this influence is so great that senior managers perceive that no need exists for an integrated planning methodology such as model (1). Instead managers believe that manufacturing location and production levels should be based primarily upon the analysis of the tax department. In such an environment it is extremely hard to implement model (1) - even using a version of this model that includes tax and transfer pricing (see [10]).

Model (2) labeled "family weekly production scheduling model" represents an opportunity to employ optimization to evaluate the trade-offs between production costs, changeover costs and inventory carrying costs while developing a product family production plan for a plant. This model can serve both a tactical and an operational role in a firm's hierarchical planning and scheduling process.

At the tactical (annual) level, model (2) can determine whether the production plans developed for a plant by model (1) remain feasible when evaluated in more detail. Examples of the greater level of detail typically found in model (2) compared to model (1) include evaluating production plans in weekly time buckets rather than in quarterly or annual buckets, and evaluating the impact of weekly changeovers on the feasibility of the annual production plan. Thus, at the tactical level, model (2) plays a critical role for each individual plant in validating whether a plant can operationally execute the schedule assigned to the plant by the centralized planning activity that model (1) supports. The feedback loop in Figure 2 from the output of model (2) back to model (1) represents the communication loop which must occur should model (2) identify that model (1) has assigned an infeasible annual plan to an individual plant. (See Appendix B for further discussion of this feedback loop.)

At the operational level, model (2) functions as a weekly product family production scheduling model for each individual plant in a network. This model generates a production plan by week, by product family, and by production line for a plant over a rolling planning horizon that typically spans 12 to 18 months. The reader is referred to [7] for a more detailed review of this model.

In contrast to model (1), optimization models are more commonly utilized by firms at this point in the supply chain planning process (i.e., at the model

(2) level). Nevertheless, when one considers the superior insights that optimization can provide for plant scheduling activities, the level of utilization across industries remains fairly low. Several reasons explain both the higher level of optimization utilization for model (2) vs. model (1) and the "relatively low" use of this methodology to support plant product family scheduling. These include the following:

- Model (2) has a relatively narrow functional role that does not span multiple functional areas within a firm. The manufacturing organization will utilize model (2) to schedule its plants. Thus, a manager who desires to employ optimization to schedule a plant does not typically have to obtain alignment across as broad a spectrum of organizational fiefdoms as would be necessary for model (1).
- In the most extreme isolated case, an individual manager or group of managers at one plant in a firm's network could choose to utilize model (2) just for their plant.[2] This could occur based upon just the knowledge or interest of a single influential manager.
- The increased availability of commercial software over the last 10 to 15 years which can perform plant scheduling problems at the line level with changeovers (see [7]) has further enhanced the probability that a firm will use optimization for model (2).
- A significant limiting factor, however, remains the hesitancy of managers and functional organizations to employ what are perceived as "black box" solutions and algorithms. Many managers and firms persist in their comfort level with simple spreadsheet scheduling tools that planners can easily understand and manipulate. One cannot underestimate the power that a firm's comfort level with alternative methodologies exerts on the selection of a decision support tool.

Model (3) labeled "end item weekly production scheduling model" represents a potentially similar opportunity to that of model (2) to employ an optimization-based methodology. This model establishes a production plan by week, by end item, and by production line for a plant over a rolling planning horizon that typically extends from several months to as many as 18 months into the future. The complexity of the cost and capacity trade-offs in this activity determines the potential benefit of using optimization at this point in the plant scheduling process. For example, in the case of a ceramic tile manufacturer, the changeover from one end item to another within a particular product family required only a 15-minute washup during a scheduled break for the workforce (see [5]). The significant capacity and changeover decisions occurred at the product family level for this manufacturer. Thus, while an optimization model could assist the firm at the model (3) level, similarly a well-designed spreadsheet model could also effectively translate the product family schedules generated by model (2) into an end item production schedule.

---

[2] The author has observed such cases in practice.

In contrast, for many firms and industries, changeover and capacity utilization alternatives at the end item level represent major decisions where the planner must evaluate critical choices. Production schedules for end items in the pharmaceutical industry represent such an example. The physical and/or regulatory requirements of a changeover often do not facilitate a rapid changeover over an employee work break. In such an environment, the potential contribution of an optimization methodology increases significantly.

The competing factors that facilitate and impede the use of optimization for model (3) are quite similar to those that influence model (2). Both models have applicability in a stand-alone plant environment with the primary difference between the two models being the level of product aggregation each respectively addresses.

Model (4) labeled "transportation planning and scheduling" represents an activity where private industry has employed optimization methods quite heavily. In fact, one would be hard-pressed to identify a general category of planning and scheduling activities where the use of optimization is more prevalent. A number of factors have contributed to this relatively high utilization rate include the following.

- Commercial user-friendly software for transportation scheduling has been widely available since the 1980s, and on a more limited basis even earlier.
- Further, commercial vendors effectively designed transportation scheduling software to integrate easily into a firm's existing supply chain IT systems (i.e., a transportation package typically requires more limited systems interfaces and fields of data than other supply chain systems).
- Transportation scheduling has a relatively narrow scope organizationally - it is usually the purview of a transportation group in a firm. This often allows transportation managers relatively more autonomy to employ niche products and methodologies for their function - without having to obtain "political" alignment across a broad spectrum of functions in an organization.
- The need for a method (i.e., optimization) that can mathematically determine how to schedule shipments is often more intuitively understood by non-technical managers than some other scheduling requirements for a firm. For example, this author's observation is that managers often have a difficult time comprehending how a mathematical tool "trades-off" changeover costs vs. variable production costs vs. inventory carrying costs in a production scheduling model - all while satisfying capacity constraints. In contrast, the concept that a mathematical tool can create a delivery schedule that consolidates orders/shipments with similar delivery dates and destinations is more easily understood. This "ease of understanding" facilitates a greater willingness and comfort level on the part of nontechnical managers to entrust the transportation scheduling activity to a mathematical optimization tool.

- In summary, the relatively narrow organizational domain of transportation scheduling, the relative simplicity of the activity, and the bountiful availability of good commercial user-friendly software have collectively contributed to a higher utilization of optimization in this functional area compared to other supply chain scheduling activities.

Model (5) displayed as "warehouse planning and scheduling" in Figure 2 represents another activity area that could benefit from increased utilization of optimization (and simulation). In fact, simulation perhaps even more than optimization has the potential to provide far greater contributions to warehouse operations in private industry. Although one can find occasional examples of a firm utilizing simulation or optimization to plan warehouse operations, the instances remain few and far between. Again, a number of factors have impeded the uptake of simulation or optimization for warehouse planning and scheduling including the following:

- Firms operating warehouses and even warehouse operations consulting firms in general have a strong confidence level in utilizing spreadsheet-based engineering models to design and operate warehouses. These spreadsheet models, fortified with good engineering rates on projected warehouse productivity measures, can provide good insights into warehouse size, labor and material handling equipment requirements (see e.g., [2]). The capability of spreadsheet models to evaluate effectively warehouse operations requirements and to predict the potential capacity limitations and bottlenecks of "sequential" warehouse operations has made these models very popular in practice. At the same time, this has reduced the need or drive of industry practitioners to invest more heavily in sophisticated techniques such as simulation.

- Further, the previously cited general comfort level that exists with easily understood spreadsheet models also dampens the use of simulation or optimization in warehouse operations, as does this factor in other functional areas.

The planning activity of warehouse design and layout has benefited the most to date in practice from simulation and optimization models. This typically represents more of a one-time long-run planning activity. Because of the perceived importance of optimizing warehouse operations flow, firms are in some cases more willing to invest the resources required to construct a sophisticated simulation model to assist in layout design. Nevertheless, based on this author's experience, the number of firms employing true simulation for strategic warehouse design remains a small minority. Further, when utilized, such models generally serve a one-time design planning purpose, but do not become incorporated into a standard business practice.

Model (6) noted as "trailer yard management" in Figure 2 is closely associated with model (5), warehouse planning and scheduling. In fact, commercial warehouse management systems (WMS) often include trailer yard management as a submodule within the overall WMS. We identify trailer yard management separately and explicitly because it represents a very important functional component of a firm's distribution network. However, because this scheduling area is often considered a subfunction of warehouse operations, it sometimes does not receive the attention it requires. An efficiently scheduled trailer yard can make the difference between a warehouse operation that successfully meets customer fulfillment requirements and an operation that experiences service failures.

Opportunities for the use of optimization in general are not as prevalent for warehouse trailer yard management as for the previous supply chain functions discussed (i.e., models 1 through 5). For extremely large warehouses trailer yards, sequencing issues can exist that optimization can better address than alternative methods. Certainly for large container rail yards optimization techniques can offer significant benefits. However, even for relatively large manufacturing and distribution warehouses (e.g., facilities of one million square feet or more), schedulers can usually generate effective unloading sequences using simple inventory prioritization logic in a database/spreadsheet application. Nevertheless, because selected opportunities do exist for optimization approaches to warehouse yard management and because of the often overlooked importance of this activity, we have included this functional area in this discussion.

3 An Approach to Promote the Use of Optimization and Related DSS Methods in Supply Chain Operations

Based upon implementations in a number of industries including pharmaceutical, consumer health care and ceramic tile among others, the author has employed a specific strategy to promote the use of optimization, simulation and related DSS techniques within a firm's operations. This section reviews this strategy. Additionally, the reader is referred to [2], [5], and [4] for detailed examples of DSS implementations based upon this approach. The following steps outline this strategy from the perspective of a manager seeking to develop and implement effective DSS tools such as optimization for his or her operation.

1. Evaluate your operation from the hierarchical, multiple planning period perspective previously described.
2. Select the planning activity or function within one's overall operation that can benefit most immediately from the introduction of an optimization (or simulation) based DSS. The factors that dictate this decision will vary by firm. Briefly, however, decision factors would naturally include: what facet of one's operation has the greatest need for enhanced DSS support, what project has the greatest likelihood of success, what project has the highest probability of stimulating further DSS enhancements and so on.
3. Staff the project with a colleague or colleagues who are employees of the firm, and ideally who are already established within the operation

where the implementation will occur (i.e., do not rely solely on third-party consultants to develop the DSS for your operation).

4. Utilize consultants to assist on the project only if necessary (e.g., if the skill sets and/or resources are not fully available internally). However, an internal employee must lead the project. Additionally, there must be one or more internal employees with the skill sets and knowledge to develop, implement and effectively utilize optimization and/or simulation. If there are internal resources who have the skill sets to utilize the applications, this will ensure continuity once the firm implements the planning methodology.

5. Establish the use of this DSS tool as a standard business practice once the initial DSS application is developed. The frequency with which the firm's operation will utilize the DSS tool will naturally vary depending upon the application. However, the critical point is that the role of this DSS tool in the operation's planning and scheduling activities be well defined.

6. Utilize the initial DSS applications to create a foundation from which to build and append additional related DSS planning tools and methodologies for the organization.

The strategy outlined in steps 1 through 6 sketches some basic high-level steps that have led to the successful implementations of DSS applications. Clearly there exist many more detailed steps necessary to facilitate a successful project and the reader is referred again to the citations at the beginning of this section for more detailed descriptions. These six steps, however, provide a general overview of the recommended approach.

In particular, once the DSS application has been developed and the initial implementation occurs, steps 5 and 6 assume critical importance. The manager(s) responsible for the DSS must ensure that the tool becomes an integral, standard component of the planning activity the DSS supports. Thus, the key question becomes how to assure this progression? Steps 3 and 4 provide guidance to this question.

Step 3 describes that one must select employees to manage the project who have existing roles and responsibilities within the firm. By having one or more key internal employees lead the development and implementation of the DSS project, this assures that there will be an in-house "advocate or champion" of the DSS on an ongoing basis. For this reason, if the sponsoring manager of the project can appoint two or more employees to major roles in the DSS project, this further enhances the ongoing support and advocacy for the system. Additionally, employees who hold key or central positions in the operation where the DSS will be implemented represent the best candidates to lead the DSS project—again because this will facilitate a very strong ongoing advocacy for the system. Similarly, should a sponsoring manager find it necessary to employ consultants to staff significant portions of the DSS project, it becomes important that the sponsor have a good exit strategy for the consultant(s). Specifically, an effective transition process must take place whereby internal employees must quickly assume ongoing operational responsibility for any portions of the system developed by the consultant(s). If this "hand-off" does not occur quickly, the probability significantly increases that the DSS will not become ingrained into the firm's operation. Instead the new system risks ending up unused "on the shelf" because it lacks ongoing supporters within the firm's operations.

With the appropriate advocacy established in steps 3 and 4, we return to steps 5 and 6. As noted, in step 5 the sponsoring manager must establish the use of the DSS tool as a standard business practice. Depending upon the level in the hierarchical planning framework (Figure 1) that the DSS supports, the operation may employ the tool as frequently as daily or as infrequently as quarterly to annually. Regardless of the appropriate frequency, the sponsoring manager and project advocates (i.e., the key employees who led the development project) must take the necessary steps to facilitate the DSS integration. This may represent a very simple or significant undertaking depending on the situation. If the project "advocates" plan to utilize the DSS in their own planning activities and area of responsibility, integrating the DSS as a standard business practice requires simply that these managers execute an implementation plan. Once implemented, the DSS becomes a core or standard business practice. However, if the sponsoring manager and DSS advocates must influence others to utilize the DSS to support planning activities that are either not their (the advocates') direct responsibility or are relatively remote (e.g., in distant plants), step 5 becomes far more difficult. For this reason, we highly recommend that in the initial DSS project selection process, the sponsoring manager choose an application that will support his or her area of responsibility directly. Further, we recommend that the manager select a DSS project that will support an operation or planning activity with which the manager has regular and direct contact. This will again facilitate a successful integration of the DSS as a standard business practice.

We consider now step 6 where the initial DSS implementation project becomes the foundation tool that stimulates the growth of an expanded system over time. Based on the implementations previously cited in this article, as well as other successful implementations, the author has observed a consistent pattern. Namely, a successful DSS implementation will "eventually" support several activities or operations in addition to the initial application. Several factors contribute to this.

An optimization project or similar large-scale DSS effort requires numerous data inputs. These data inputs can often support secondary DSS applications not originally envisioned when the initial system was planned. For example, the developers of the DSS reported in [2] originally constructed this system to provide general support for DC operations and customer logistics scorecards. However, when the operation that this DSS supported suddenly had significant storage capacity issues in the early to mid 2000s, the developers rapidly created another DSS to provide daily inventory deployment guidance for the overcapacitated network (see [8]. In developing this DSS, the in-house team that developed the second DSS relied heavily upon the data sets and existing data interfaces that supported the initial DSS. In some cases, the team simply augmented an existing interface from the firm's enterprise resource planning (ERP) to the existing DSS with additional data fields. These additional fields appended to the existing interface facilitated the rapid development of algorithms that formed the basis of a daily inventory deployment DSS. In this example, the fact that the firm already had one successful DSS in place - a "foundation" - paved the way for the rapid implementation of a second DSS.

A similar evolution occurred at a firm described in [4] and [9]. In this case, a firm developed and implemented a mathematical optimization model that generated integrated tactical manufacturing and distribution plans (i.e., model (1) in Figure 2). This DSS model quickly became a key component of the firm's standard tactical planning business process. This optimization model became the "foundation" tool that then spawned the development of several related DSS implementations at this firm over the next several years. In this case, the original optimization model generated very significant savings, as well as customer service level improvements for the firm. The success of this initial implementation created great interest at the firm in further improving the quality of key data inputs to this model. In particular, the firm next developed a new tactical forecasting system to provide product family demand projections to the integrated manufacturing and distribution model (see [9]). This integrated top-down/bottom-up forecasting system replaced a previous manual, anecdotal and judgment-based forecasting approach, and it dramatically improved the firm's forecast accuracy. Next, as the firm scrutinized its new optimization-based manufacturing and distribution planning system, it realized that it needed to improve the start of period inventory data inputs to its planning process. Specifically, the firm discerned that its aggregated inventory inputs to the production planning process were leading to production schedules that created poor customer service line item fill rates. Briefly, the aggregated inventory inputs did not properly recognize serious inventory imbalances that could exist at the end item and product family level. To correct this problem, the firm developed a DSS that evaluated its inventory simultaneously at multiple levels (i.e., at the end item and product family levels). This DSS application became a preprocessing step that the firm utilized to develop its inventory inputs to the tactical production planning process (see [6] for additional details).

The examples cited in this section illustrate how an initial DSS implementation stimulates over time the development of additional DSS applications. As noted a well-implemented DSS typically provides secondary and tertiary benefits and decision support for additional problems beyond those which the developers of such systems explicitly designed the original application. Additionally, once an initial optimization model or similar DSS provides significant contributions to a firm, this usually spawns further interest and inquiries from other managers in an organization. Specifically, managers who observe a DSS system improving the effectiveness of decision making in a periphery function to their own become interested in utilizing similar methodologies to support their own areas of functional responsibility. This can lead to either new DSS projects or the expansion of existing DSS implementations. In either case, a key point remains that the original DSS project creates a foundation from which to build other applications to benefit a firm. This further illustrates the importance for a manager of carefully identifying and implementing an initial DSS project.

3.1 Lessons Learned and Final Thoughts

The hierarchical planning framework reviewed at the outset of this chapter provides a unifying perspective with which to view existing and potential DSS applications in a firm. An organization with strong capabilities will have DSS systems to support planning activities at the strategic, tactical and operational levels (i.e., support for all planning horizons). For the manager or firm seeking to enhance internal DSS capabilities, a review of the firm's current decision support systems across all planning horizons represents a good starting point. This review will serve to help identify those planning activities that could benefit the most from additional support.

Once the sponsoring manager has selected a DSS application for development and implementation, it is critical to keep in mind that optimization and similar operations research techniques remain unknown, "black box" methods for many firms and managers. Thus, the success of a DSS application may hinge upon whether the sponsoring manager can install one or more internal project leads who have both strong operational knowledge and a good comfort level with technical operations research type methodologies. Finally, once the DSS application is developed and implemented, the sponsoring manager must quickly incorporate this new capability into the firm's standard business planning process.

In summary, there is no one "right way" to implement optimization and other DSS techniques into a firm's core planning processes. However, the approaches outlined in this chapter have facilitated a number of successful implementations in different firms and industries. It is hoped that the approaches described herein may offer some useful insights for other potential applications.

A Brief Review of Key Components of the Integrated Production and Distribution Planning System Depicted in Figure 2

A.1 Annual Forecasts by Sales Region by Product Line

- A high level forecast (e.g., at a division or major product line) typically initiates the tactical planning process. The "sales region" would be a very large region such as a country in a global planning model.

A.2 Annual Forecasts by Sales Region by Product Family

- A manufacturing and distribution optimization model generally requires forecasts at the product family level (i.e., at a more disaggregated level than the initial high level forecast). A product family is defined as a group of individual end items that have similar manufacturing and distribution characteristics (e.g., items that can be produced on the same production lines at the same plants and at similar costs and rates).

- Additionally, the sales or geographic regions in this product family forecast should be more disaggregated than in the initial high level forecast. For example in a U.S. domestic model, the sales regions defined in the optimization model may consist of 100 to several hundred metropolitan areas and surrounding areas. In contrast the initial high level forecast may be a total U.S. forecast by product line.

A.3 Plant/DC/Family Model

- This optimization model generally includes the following components:
  1. All plants and other major finished goods supply points (e.g., contract manufacturers).
  2. All major finished goods production lines at all plants.
  3. Definitions (and explicit modeling) of all major product families produced at plants/supply points. This includes variables and constraints in the model that define the rate and cost of production of each product family on each production line.
  4. The freight cost to transport each product family from its supply point origin(s) to the sales regions defined in the model where there exists demand for the product family. Note that this also includes the freight costs to transport product families across intermediate origin-destination pairs between the original supply point and final point of demand.
  5. The cost and capacity constraints (or throughput rates) associated with product families moving through intermediate facilities between the original product family supply point(s) and the final demand point(s). A distribution center (DC) is the most typical example of an intermediate facility.
  6. A demand forecast for each product family at each defined sales region (final demand point) in the optimization model.

- The plant/DC/family model often contains many more components such as inventory carrying costs, duty and customs costs, etc. The reader interested in a more comprehensive review is referred to [4] and [7]. Note also that this type of model frequently contains a mix of fixed and variable costs. Thus, model design decisions typically include an evaluation of whether to utilize a mixed integer or linear programming approach.

A.4 Annual Production Plan by Plant, by Family and Sourcing Plan

- An integrated networkwide production and distribution plan produced by the plant/DC/family optimization model.

A.5 Operational Forecasting System

- A system that generates demand forecasts for all stockkeeping units (skus). A sku is defined as an individual finished good end item that is stored at a defined geographic location (e.g., a DC or retail store) to satisfy customer orders (i.e., meet demand).

A.6 DRP/Inventory Management System

- A system that projects inventory requirements at finished goods inventory stocking locations (e.g., DCs). This system utilizes sku forecasts, sku service level targets, and current sku inventory levels at a stocking location to project net inventory requirements by sku.

A.7 Weekly Product Family Net Requirements by Plant

- A projection of the net inventory and production requirements for each product family that a plant produces. The product family net requirements are developed by aggregating the net requirements of each individual end item within a product family.

A.8 Family Weekly Production Scheduling Model by Plant, by Family

- See description of model (2) in text.

A.9 Master Production Schedule by Plant, by Family

- The planned production schedule and related information (e.g., total demand) generated by the weekly production-scheduling model for each product family that a plant produces. (See APICS dictionary for additional descriptions[1].)

A.10 End Item Weekly Production Scheduling Model

- See description of model (3) in text.

A.11 Master Production Schedule by Plant, by End Item

- The "planned" production schedule and related information generated by the weekly production-scheduling model for each end item that a plant produces.

A.12 Order Management, Transportation Planning and Scheduling, Warehouse Planning and Scheduling, and Trailer Yard Management

- For a comprehensive review of order, transportation, warehouse and trailer yard processes and management, the reader is referred to [11].

B Illustration of a Feedback Loop from the Operational Level to the Tactical Level

Feedback loops from the operational level to the tactical level and from the tactical level to the strategic level represent a key and "defining attribute" of any hierarchical supply chain planning system. To provide additional perspective of what a feedback loop is, in this appendix we review an illustrative feedback loop from model (2) to model (1) in Figure 2.

As previously discussed, in the tactical planning process, model (1) generates a 12- to 18-month production plan at the product family level for each plant in a network. The model also creates an integrated distribution plan that identifies which plants supply which DCs and which DCs serve which customers, again at the product family level. Figure 3 displays a networkwide

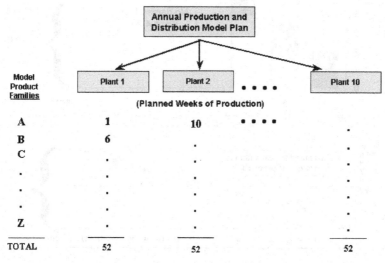

Fig. 3. Illustrative tactical production plan created by model (1)

annual production plan that for illustrative purposes we will assume model (1) has created. This plan displays the weeks of production of each product family that each plant will manufacture over a 12-month planning horizon. (For simplicity the figure primarily shows "dots" rather than all 26 product families and their assignments.)

For illustration we now focus on model (1)'s assignment that plant 1 should produce one week of product family A. We will also assume that product family A has the following attributes:

1. It contains 20 finished good end items, and
2. Each of the 20 end items has a minimum production run length of a 1/2 day (i.e., if the plant has to produce an item, it must produce the item for a minimum of 1/2 of a day).

Briefly end items are aggregated into product families for tactical planning based upon their respective similar characteristics. For example, assume that this is a ceramic tile manufacturing network and that the 20 end items in product family A are different color 2" × 2" wall tile end items (e.g., blue, green, yellow, etc.). Each end item can be produced on the same production lines at the same plants and at very similar costs per unit and at similar output rates. These similar end items would be planned as one product family in model (1) at the tactical planning level.

Now let's consider Figure 4 which depicts two very different scenarios (case 1 and case 2) under which model (1) could generate an initial assignment of one week of production for product family A at plant 1.

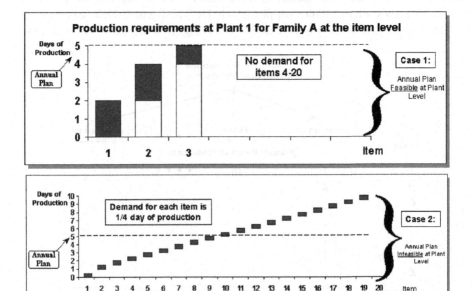

Fig. 4. Two scenarios for product family A at plant 1

The total demand for product family A consists of the sum of the demand for the 20 end items that comprise this product family. (For simplicity we will also define production requirements as equal to total demand in this example.) Now consider case 1 and case 2 in Figure 4.

Case 1 • The total demand (and production requirements) for product family A at plant 1 is in three end items (1, 2, and 3).
- There is no demand for end items 4 through 20 (i.e., demand = 0)
- Thus, as Figure 3 depicts, to satisfy the demand for product family A at plant 1 will require 2 days production of item 1, 2 days of item 2 and 1 day of item 3. No production of items 4 through 20 is required.
- Therefore, plant 1 can feasibly produce the production assignment from model (1) of 1 week of family A. (Note that we define 5 business days as 1 week in this example.)

Case 2 • The total demand for product family A at plant 1 consists of $1/4$ of a day's production for each of its 20 end items.
- $20 \times 1/4 = 5$ business days total demand, or 1 week of demand (and production) - the assignment of model (1) to plant 1 for family A.
- Recall, however, that plant 1 has a minimum production run length of $1/2$ day for any item.
- Therefore, for plant 1 to produce all 20 items in family A will require $20 \times 1/2 = 10$ business days of production.
- Thus, the production assignment from model (1) to plant 1 for product family A is not feasible.

B.1 How Can This Infeasible Production Assignment Occur

At the networkwide tactical planning level, models and planners generally do not evaluate very detailed issues such as the minimum run length of individual end items and individual plants. The purpose and objectives of 12 to 18 months planning exercises at the tactical level necessitate that planning/modeling be conducted at more aggregated levels (e.g., product families rather than end items). This allows the possibility that plans developed at the tactical level may in some cases be infeasible to implement at the operational level. Case 2 illustrates how these infeasibilities may arise.

In practice, "feedback loops" from lower planning and scheduling levels to higher levels take on great importance because of the type of situation illustrated in case 2. As plans cascade down from one level to the next lower level (e.g., networkwide to individual plant), managers at the lower level must evaluate these plans and communicate back any infeasibilities identified. This becomes an iterative process whereby tactical plans should be revised based on feedback loop communications, and then revised tactical plans are re-evaluated at the operational level. This process continues until a feasible plan, at all levels, has been developed. The reader is referred to [7] for additional discussion of feedback loops and hierarchical systems.

References

1. James F. Cox and John H. Blackstone. *American Production and Inventory Control Society (APICS) Dictionary.* APICS Educational Society for Resource Manage, eleventh edition, 2004.
2. Vijay Gupta, Emmanuel Peters, Tan Miller, and Kelvin Blyden. Implementing a distribution network decision support system at Pfizer/Warner-Lambert. *Interfaces*, 32(4), 2002.
3. Arnoldo Hax and Harlan Meal. Hierarchical integration of production planning and scheduling. In M.A. Geisler, editor, *TIMS Studies in Management Science, Vol. 1, Logistics.* Elsevier, 1975.
4. Matthew Liberatore and Tan Miller. A hierarchical production planning system. *Interfaces*, 15(4), 1985.
5. Renato De Matta and Tan Miller. A note on the growth of a production planning system: A case study in evolution. *Interfaces*, 23(4), 1993.
6. Tan Miller. A note on integrating current end item inventory conditions into optimization based long run aggregate production and distribution planning activities. *Production and Inventory Management Journal*, 32(4), 1991.
7. Tan Miller. *Hierarchical Operations and Supply Chain Management.* Springer-Verlag Press, second edition, 2002.
8. Tan Miller. An applied logistics decision support system. In *INFORMS Annual Meeting, Pittsburg, PA*, 2006.
9. Tan Miller and Matthew Liberatore. Seasonal exponential smoothing with damped trends: an application for production planning. *International Journal of Forecasting*, 9(4):509–515, 1993.
10. Tan Miller and Renato De Matta. A global supply chain profit maximization and transfer pricing model. *Journal of Business Logistics*, 29(1):175–200, 2008.
11. James R. Stock and Douglas Lambert. *Strategic Logistics Management.* Irwin/McGraw-Hill, fourth edition, 2001.

On the Quadratic Programming
Approach for Hub Location Problems

Xiaozheng He[1], Anthony Chen[2], Wanpracha Art Chaovalitwongse[3], and Henry Liu[4]

[1] Department of Civil Engineering
University of Minnesota, Minnesota, MN 55455
hexxx069@umn.edu
[2] Department of Civil Engineering
Utah State University, Logan, UT 84322
achen@engineering.usu.edu
[3] Department of Industrial and Systems Engineering
Rutgers University, Piscataway, NJ 08854
wchaoval@rci.rutgers.edu
[4] Department of Civil Engineering
University of Minnesota, Minnesota, MN 55455
henryliu@umn.edu

Summary Hub networks play an important role in many real-life network systems such as transportation and telecommunication networks. Hub location problem is concerned with identifying appropriate hub locations in a network and connecting an efficient hub-and-spoke network that minimizes the flow-weighted costs across the network. This chapter is focused on the uncapacitated single allocation p-hub median problem (USApHMP), which arises in many real-world hub networks of logistics operations. There have been many approaches used to solve this problem. We herein focus on a quadratic programming approach, which has been proven very effective and efficient. This approach incorporates the use of the linearization for 0-1 quadratic programs. In this chapter, we give a brief review of the linearization techniques for 0-1 quadratic programs and compare the performance of several existing linearization techniques for USApHMP. Toward the end, we discuss some properties, comments and possible developments of these linearization techniques in the real-life USApHMP.

1 Introduction

Hub networks arise in many real life transportation and telecommunication problems. Hub location problems, traditionally viewed as facility location or network design problems, are concerned with identifying appropriate hub locations and connecting an efficient hub-and-spoke network that minimizes the flow-weighted costs throughout the network. There are two types of

W. Chaovalitwongse et al. (eds.), *Optimization and Logistics Challenges in the Enterprise*, Springer Optimization and Its Applications 30, DOI 10.1007/978-0-387-88617-6_7, © Springer Science+Business Media, LLC 2009

arcs/connections in the network in a hub location problem: (1) hub arcs between two hubs and (2) access arcs between a hub and an origin/destination. The hub arcs are used to connect nodes selected to be the hub nodes in the network, and the hub nodes are fully connected. The access arcs are used to connect the nonhub nodes (origins/destinations) to hubs. The p-hub median location problem is a fundamental discrete hub facility location and hub network design problem analogous to the p-median problem [10]. The p-hub median location problem can be formally defined as follows. Given a number p of hubs to be selected, the objective is to select/locate p appropriate nodes in a network to be hubs in a hub-and-spoke network and to allocate all nonhub nodes to hubs such that the flow-weighted (e.g., transportation) costs across the network are minimized. The cost per unit flow between two hubs is usually smaller than the cost per unit flow between an origin/destination node and a hub to reflect the economies of scale, resulting from aggregation of flows on the interhub arcs. This hub-and-spoke network is practically efficient and commonly arises in telecommunication and transportation networks. Instead of setting up a full-connected network for every node, some nodes are selected to be hubs and the flows from nonhub nodes are assigned to be transferred to the hubs. Since all hubs are interconnected, only one transmission is needed between the collection of transportation flows from origin and the redistribution to their destination. Therefore, the flows in the hub-and-spoke network can be treated as a transshipment problem and the transportation procedure can be separated as three parts: collection, transmission and redistribution. The hub-and-spoke network may typically be illustrated as in Figure 1.

The location problems in hub networks arise in a wide variety of real-life applications including airline management, postage delivery network, and telecommunication networks. These networks provide connections between

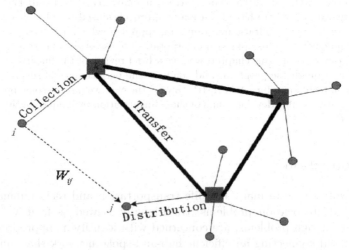

Fig. 1. Hub-and-spoke network with 14 nodes and 3 hubs

many origin nodes and destination nodes via hub nodes that may serve as switching, sorting, connecting, rerouting, and consolidation points for flows of freight, passengers, or information. The consolidation of flows on relatively few arcs in hub networks is very beneficial and gives the economies of scale as it reduces a great number of arcs required to connect all nodes. Generally, the design of a hub network involves locating hub locations/facilities and determining the arcs to connect origins, destinations, and hubs. In airline systems, the hub location problem was traditionally focused on the passenger's flight times. A passenger having a trip from one origin to one destination is firstly transferred to some hub, then to another hub if necessary, and finally to the final destination. The passengers traveling to different destinations are amalgamated when they are traveling between hubs. Due to a highly efficient subnetwork of the hubs and shorter distances between hubs and non-hub nodes, the whole network runs in an efficient condition and the total transportation costs decrease. Later, the transient times spent at hubs for unloading, loading, and sorting operations have been considered. The transient times may constitute a significant portion of the passenger's travel times or the total delivery time for cargo delivery systems. Several studies in the literature are focused on the minimization of the arrival time of the last arrived item in cargo delivery systems. Many nonlinear and linear integer programs to compute the arrival times by taking into account both the flight times and the transient times have been developed. Another hub location problem with aircraft application is to consider several different types of aircraft, which can be viewed as different types of arcs in the network, to serve specified origin-destination flows via hub locations. The objective of this problem is to select the number of aircrafts of each type on each link to minimize the total aircraft utilization cost of flying. This problem is also similar to the modeling of light rail connections in urban transport planning. In addition to transportation context, similar logistics issues also arise in parcel delivery operations and telecommunication networks. In telecommunication network design, the hub network consists of two different kinds of nodes: access nodes and transit nodes. Access nodes represent source and destination of traffic demands but cannot be directly connected while transit nodes, which are typically fully connected, do not generate or attract traffic demands but collect traffic from access nodes and route them through the network. Given a set of access nodes and a set of potential locations for the transit nodes, the hub-location problem is to decide the number and locations of the transit nodes to guarantee that all access nodes are allocated to a transit node such that traffic capacity constraints are satisfied. Specifically, the objective is to minimize the total cost of the network, which is the sum of connection costs and fixed hub costs.

Hub-location problem can be viewed as the $\mathcal{NP}$-hard location-allocation problem, which is extremely difficult to solve. Even in the case that the hub locations are determined *a priori*, this problem is still $\mathcal{NP}$-hard [27]. Because of its difficulties and practical benefits, there has been an increasing interest in investigating the formulations and the methodologies for hub-location problem since the mid 1980s [28]. This problem can be divided into two classes based on different allocation modes. The first one is single allocation hub median problems, in which each nonhub node is incident with exactly one access arc, i.e., all the flows to and from nonhub nodes are forced onto a single access arc. The p-hub median problem is a single allocation problem, whose objective is to connect the network with $p(p-1)/2$ undirected arcs among hubs for all the hub pairs and to connect each of the nonhub nodes with an access arc closest to a hub. The second one is multiple allocation hub median problems, in which nonhub nodes may be incident with more than one access arc, i.e., all the flows to and from each nonhub node may be divided among the incident access arcs to more than one hubs. Both single and multiple allocation hub median problems have been extended in a variety of directions including direct origin-destination arcs [5, 6], fixed costs for hubs [1, 5, 6, 26, 29], and capacities [14, 19]. For a more complete survey on hub location problems, see [11]. One of the most widely studied versions is the uncapacitated single allocation p-hub median problem (USApHMP), which is the main focus of this chapter.

The remainder of this chapter is organized as follows. Section 2 presents the definition and mathematical formulation of the USApHMP. Section 3 describes the linearization techniques, some of which can be used to solve the USApHMP. Section 4 presents some computational results which demonstrate the efficacy of linearization techniques. Section 5 presents the conclusion and a discussion of some future research directions.

2 Uncapacitated Single Allocation p-Hub Median Problem (USApHMP)

The mathematical formulation of USApHMP was first introduced in the late 1980s [28]. The formulation is in a form of quadratic integer program with nonconvex objective, which is straightforward from the description of the hub-location problem. The hub-location problem has a close connection with the quadratic assignment problem and the formulation proposed in [28] is a special case of the 0-1 quadratic problem. There are many well-developed solution approaches for 0-1 quadratic programs that can be directly applied to the hub-location problem. There have been many studies focused on the efficient methodologies to solve the quadratic integer programming model of USApHMP. Thorough reviews on the p-hub location problems and their solution approaches can be found in [8, 9]. Considering the p-hub location problems can be treated as discrete mathematical programming problems, they have the same fundamental properties as all other integer programming (IP) problems. Several methodologies developed for general IP could be altered or combined with additional constraints in order to be applied to the p-hub location problems. Those approaches include full enumeration [28], branch-and-bound [16], simulated annealing [24], Tabu search [33], shortest path search [18], Lagrangian decomposition [13], and linearization technique [7]. The linearization approach was originally developed for solving 0-1 quadratic programming. There have been many studies investigating the use of linearizations for solving USApHMP [10, 16, 34] and empirical studies of linearizations for UMApHMP can be found in [7].

The mathematical formulation for USApHMP was first introduced in [28], which can be described as follows. Suppose we have a USApHMP with a given network $G(N, A)$, where $N = 1, 2, ..., n$ is the node set and $A = N \times N$ is the arc set. We use notations $i, j, k, m \in N$ as the indices of nodes. Let W_{ij} represent the flow demands from node i to j, and C_{ij} represent the transportation costs on arc (i, j). In general, $C_{ij} = C_{ji}$. Some papers also assumed that $W_{ij} = W_{ji}$ for convenience, but it is not necessary. Let the 0-1 decision variables $Z_{ik} \in \{0, 1\}$ be 1 if node i is assigned to be connected to node k and 0 otherwise. And then, $Z_{kk} = 1$ implies that node k is selected as a hub. In a USApHMP, every node i is connected to only one hub. It means that we have the constraints $\sum_k Z_{ik} = 1, \forall i \in N$. Due to the budget constraint, only p hubs may be located in the network, i.e., $\sum_k Z_{kk} = p$. For all nodes, the demand flows can only travel to selected hub nodes, i.e., $Z_{ik} \leq Z_{kk}, \forall i, k \in N$. The objective of USApHMP is to minimize the total transportation costs under the above-mentioned constraints. The transportation costs are determined by the connection between nodes and hubs. Then the 0-1 quadratic programming formulation in [28] is given by:

$$\min \sum_{i \in N} \sum_{k \in N} \sum_{j \in N} \sum_{m \in N} W_{ij}(\chi C_{ik} Z_{ik} + \alpha C_{km} Z_{ik} Z_{jm} + \delta C_{jm} Z_{jm}) \quad (1)$$

$$\text{s.t.} \sum_{k \in N} Z_{kk} = p \quad (2)$$

$$\sum_{k \in N} Z_{ik} = 1, \qquad \forall i \in N \quad (3)$$

$$Z_{ik} \leq Z_{kk}, \qquad \forall i, k \in N \quad (4)$$

$$Z_{ik} \in \{0, 1\}, \qquad \forall i, k \in N. \quad (5)$$

The transportation costs of each flow traveling from node i to node j consist of three parts: "collection cost," "transfer cost" and "distribution cost" as shown in the objective function in Eq. (1). The coefficients χ, α and δ represent the weights on different transportation parts. In general, $\chi = \delta = 1$ which means that we have "collection cost" equal to "distribution cost," and $0 < \alpha \leq 1$ that is because of the interhub transportation discount. In practice, α is much smaller than χ and δ. However, in other applications, χ and δ may be different. For example, the Australia post (AP) data which were given in some papers as a test problem have $\chi = 3$, $\delta = 2$, and $\alpha = 0.75$. We shall call the above formulation USApHMP-Q, which is a 0-1 quadratic program. This formulation has been shown to be very hard to solve. Many approaches have been presented for solving this type 0-1 quadratic program, in which the linearizaion technique is one of the most efficient strategies.

3 Linearization Techniques

Linearization technique was firstly developed for solving 0-1 polynomial programs. Early linearizations were presented in Zangwill [36] and Watters [35]. Based on their pioneering works, other researchers have developed some alternative linearizations. Among those are Glover and Woolsey [22, 23], Adams and Sherali [4], and most recently Chaovalitwongse et al. [12]. The performance comparison of these linearizations with other techniques for solving 0-1 quadratic program was presented in [15]. The application of these linearizations to real-world problems was demonstrated in [31]. In this chapter, we mainly apply one of the above-mentioned linearizations to the USApHMP. We give a brief review of existing linearizations for 0-1 quadratic program in this section and for USApHMP in next section.

Consider a particular 0-1 quadratic programming (QP) problem given by:

$$\min \sum_{i \in N} q_i x_i + \sum_{i \in N} \sum_{j \in N} q_{ij} x_i x_j \tag{6}$$

$$\text{s.t.} \sum_{j \in N} a_{ij} x_j \leq b_i \qquad \forall i \in N \tag{7}$$

$$x_i \in \{0, 1\}, \qquad \forall i \in N, \tag{8}$$

where x_i are binary decision variables and a_{ij}, b_i, q_{ij} are coefficients.

Zangwill [36] and Watters [35] were the first to propose a linearization approach by introducing additional binary variables w_{ij} to replace the quadratic terms $x_i x_j$ in the objective function Eq. (6). Additional constraints, $x_i + x_j - w_{ij} \leq 1$ and $x_i + x_j \geq 2w_{ij}, \forall i, j$, are necessary to be introduced into this reformulation. A large number of additional binary decision variables and additional constraints is the main disadvantage of this linearization technique. Subsequently, Glover and Woolsey [22] gave an improved linearization that avoided the necessity to introducing new binary variables. Their additional cut constraints, $x_i \geq w_{ij}$ and $x_j \geq w_{ij}, \forall i, j$ enforce the additional variables w_{ij} to be binary. However, this linearization doubles the number of additional constraints added. This drawback could be problematic in the case when the size of original QP becomes very large. Their following improvements of linearization that used alternative constraints $(n - i)x_j \geq \sum_{j \in N} w_{ij}, \forall i$ to enforce the additional variables to be binary were presented in [23]. Using similar technique, Glover [21] provided a concise representation which is further extended in recent papers [2, 3]. This technique yields a much smaller number of additional constraints. However, since the first set of constraints $x_i + x_j - w_{ij} \leq 1, \forall i, j$ must be reserved in this formulation, the size of the reformulation increases quadratically. Although increasing the size of the reformulation may handicap the application of the linearization, Adams and

Sherali [4] proposed their reformulation technique that is tighter, in LP relaxation sense, than other linearizations although the size of their reformulation is larger than other linearizations. The most recent and interesting linearization technique developed for 0-1 quadratic program was given by Chaovalitwongse et al. [12]. They have shown that when the coefficient matrix $Q = \{q_{ij}\}$ is nonnegative (or $q_{ij} \geq 0, \forall i, j$) the 0-1 QP in Eqs. (6-8) is equivalent to the following formulation:

$$\min \sum_{i \in N} s_i \tag{9}$$

$$\text{s.t.} \sum_{j \in N} a_{ij} x_j \leq b_i \qquad \forall i \in N \tag{10}$$

$$\sum_{j \in N} q_{ij} x_j - y_i - s_i = 0 \qquad \forall i \in N \tag{11}$$

$$y_i \leq \mu(1 - x_i) \qquad \forall i \in N \tag{12}$$

$$x_i \in \{0, 1\}, y_i \geq 0, s_i \geq 0 \quad \forall i \in N, \tag{13}$$

where $\mu = \max_i \sum_j |q_{ij}| = \|Q\|_\infty$ and y_i, s_i are additional real variables. Based on the structure of (9-13), Sherali and Smith [32] improved the tightness by strengthening the bounds on y and s.

With only $2n$ additional variables and constraints, this linearization technique has a much smaller size compared with other linearizations. The above reformulation for QP needs the assumption of a nonnegative objective coefficient matrix Q. However, for a general case of matrix Q, similar proof was also proposed in [12]. It is important to note that, in all distance matrices, the coefficients of the Q matrix satisfy the nonnegativity requirement. Therefore, this linearization can be applied to the USApHMP.

4 Linearization Techniques for USApHMP

After the QP for USApHMP in Eqs. (1-5) was proposed in [28], O'Kelly et al. [30] reformulated the constraints in Eq. (4) using the summation of i as

$$\sum_{i \in N} Z_{ik} \leq (n - p + 1) Z_{kk} \qquad \forall k \in N \tag{14}$$

in order to reduce the number of additional constraints. This strategy is similar to the one given by Glover and Woolsey [23] of using fewer constraints $\sum_{j \in N} w_{ij} \leq (n - i) x_j, \forall i$ to replace $w_{ij} \leq x_i$ and $w_{ij} \leq x_j, \forall i, j$ for 0-1 QP problems. This technique could be applied to the following linearizations to reduce the number of additional constraints as well.

Campbell [9] elaborated on the formulation of hub location problems and was the first to propose a mixed integer linear programming (MILP) formulation for general p-hub median problem, which is given by

$$\min \sum_{i\in N}\sum_{k\in N}\sum_{j\in N}\sum_{m\in N} W_{ij}C_{ikjm}X_{ikjm} \tag{15}$$

$$\text{s.t.} \sum_{k\in N} Y_k = p \tag{16}$$

$$\sum_{k\in N}\sum_{m\in N} X_{ikjm} = 1, \quad \forall i,j \in N \tag{17}$$

$$X_{ikjm} \leq Y_k, \qquad \forall i,k,j,m \in N \tag{18}$$

$$X_{ikjm} \leq Y_m, \qquad \forall i,k,j,m \in N \tag{19}$$

$$0 \leq X_{ikjm} \leq 1, \qquad \forall i,k,j,m \in N \tag{20}$$

$$Y_k \in \{0,1\}, \qquad \forall k \in N, \tag{21}$$

where Y_k are decision variables indicating whether node k is selected to be a hub and X_{ikjm} are decision variables indicating whether the demand flow W_{ij} travels from the origin node i, via hubs k and m, to the destination node j. The cost coefficients C_{ikjm} is equivalent to $(\chi C_{ik} + \alpha C_{km} + \delta C_{jm})$ in Eq. (1) representing the path costs from node i to node j via hubs k and m.

4.1 Tight Linearization Technique (T-LT)

Based on the above MILP formulation, Skorin-Kapov et al. [34] developed a tight linearization technique for USApHMP. Their linear relaxation, called T-LT in this chapter, is given by

$$\min \sum_{i\in N}\sum_{k\in N}\sum_{j\in N}\sum_{m\in N} W_{ij}(\chi C_{ik} + \alpha C_{km} + \delta C_{jm})X_{ikjm} \tag{22}$$

$$\text{s.t.} \sum_{k\in N} Z_{kk} = p \tag{23}$$

$$\sum_{k\in N} Z_{ik} = 1, \qquad \forall i \in N \tag{24}$$

$$Z_{ik} \leq Z_{kk}, \qquad \forall i,k \in N \tag{25}$$

$$\sum_{m\in N} X_{ikjm} = Z_{ik}, \qquad \forall i,k,j \in N \tag{26}$$

$$\sum_{k\in N} X_{ikjm} = Z_{jm}, \qquad \forall i,j,m \in N \tag{27}$$

$$X_{ikjm} \geq 0, \qquad \forall i,k,j,m \in N \tag{28}$$

$$Z_{ik} \in \{0,1\}, \qquad \forall i,k \in N, \tag{29}$$

where X_{ikjm} are additional variables with the same meaning as in Campbell's formulation. This formulation introduces X_{ikjm} to replace the quadratic terms in Eq. (1). This idea could be traced to the linearization for quadratic assignment problem (QAP) published in [20]. Their formulations involve a huge number of variables and constraints. Considering a small network with only 10 nodes, the reformulation induces 10,000 additional variables and 2,000 additional constraints compared with O'Kelly's original QP formulation. Although this formulation has been shown to be very tight, it becomes much harder to solve due to the fast-growing number of variables and constraints as the problem size increases.

4.2 Multicommodity Flow Linearization Technique (MC-LT)

Ernst and Krishnamoorthy [16] proposed a formulation with fewer variables and constraints by introducing additional variables Y_{km}^i to represent the flow transported between hubs k and m generated from node i. They modeled the USApHMP as a multicommodity flow problem, which is given by (with $\chi = \delta = 1$)

$$\min \sum_{i \in N} \sum_{k \in N} C_{ik}(O_i + D_i)Z_{ik} + \sum_{i \in N} \sum_{k \in N} \sum_{m \in N} \alpha C_{km} Y_{km}^i \tag{30}$$

$$\text{s.t.} \sum_k Z_{kk} = p \tag{31}$$

$$\sum_{k \in N} Z_{ik} = 1, \qquad\qquad \forall i \in N \tag{32}$$

$$Z_{ik} \leq Z_{kk}, \qquad\qquad \forall i, k \in N \tag{33}$$

$$\sum_{m \in N} Y_{km}^i - \sum_{m \in N} Y_{mk}^i = O_i Z_{ik} - \sum_{j \in N} W_{ij} Z_{jk}, \quad \forall i, k \in N \tag{34}$$

$$Z_{ik} \in \{0, 1\}, \qquad\qquad \forall i, k \in N \tag{35}$$

$$Y_{km}^i \geq 0, \qquad\qquad \forall i, k, m \in N, \tag{36}$$

where $O_i = \sum_j W_{ij}$ denotes the amount of flow emanating from node i, and $D_j = \sum_i W_{ij}$ denotes the amount of flow going into node j. The additional constraints in Eq. (34) represent the flow divergence at node k. This formulation, called MC-LT here, involves n^3 additional variables and n^2 additional constraints, which are much less than the T-LT formulation for USApHMP.

4.3 Quadratic Assignment Linearization Technique (QA-LT)

Referring to the study by Kaufman and Broeckx [25], we observe that the linearization formulation generated for QAP, which may involve the least number of additional variables and constraints, could be applied to the USApHMP. We call this linearization formulation QA-LT, which is given by

$$\min \sum_{i \in N} \sum_{k \in N} \{C_{ik}(O_i + D_i)Z_{ik} + \alpha Y_{km}\} \tag{37}$$

$$\text{s.t.} \sum_k Z_{kk} = p \tag{38}$$

$$\sum_{k \in N} Z_{ik} = 1, \qquad\qquad \forall i \in N \tag{39}$$

$$Z_{ik} \leq Z_{kk}, \qquad\qquad \forall i, k \in N \tag{40}$$

$$\sum_{j \in N} \sum_{m \in N} W_{ij}C_{km}(Z_{ik} - 1)$$
$$+ \sum_{j \in N} \sum_{m \in N} W_{ij}C_{km}Z_{jm} \leq Y_{ik}, \quad \forall i, k \in N \tag{41}$$

$$Z_{ik} \in \{0, 1\}, \qquad\qquad \forall i, k \in N \tag{42}$$

$$Y_{ik} \geq 0, \qquad\qquad \forall i, k \in N, \tag{43}$$

where the coefficient O_i and D_i have the same meaning as the MC-LT formulation. Note that only n^2 additional real variables for Y_{ik} and only n^2 additional constraints are involved into this formulation. This linearization is perhaps the smallest in terms of problem size.

4.4 Quadratic Programming Linearization Technique (QP-LT)

Considering the linearization for 0-1 QP in Eqs. (9-13) given by Chaovalit-wongse et al. [12], we apply this linearization to the USApHMP in Eqs. (1-5) by replacing the i and k indices by a single index (say j) for the n^2 node-hub pairs (i, k). Consequently, we obtain exactly the same objective function as in Eq. (6). Then we can directly apply the linearization in Eqs. (9-13) to the USApHMP. This linearization technique, called QP-LT, is presented by

$$\min \sum_{i \in N} \sum_{k \in N} S_{ik} \tag{44}$$

$$\text{s.t.} \sum_k Z_{kk} = p \tag{45}$$

$$\sum_{k \in N} Z_{ik} = 1, \qquad\qquad \forall i \in N \tag{46}$$

$$Z_{ik} \leq Z_{kk}, \qquad\qquad \forall i, k \in N \tag{47}$$

$$\sum_{j \in N} \sum_{m \in N} q_{ikjm}Z_{jm} = Y_{ik} + S_{ik}, \qquad \forall i, k \in N \tag{48}$$

$$Y_{ik} \leq \mu(1 - Z_{ik}), \qquad\qquad \forall i, k \in N \tag{49}$$

$$Z_{ik} \in \{0, 1\}, \qquad\qquad \forall i, k \in N \tag{50}$$

$$Y_{ik}, S_{ik} \geq 0, \qquad\qquad \forall i, k \in N, \tag{51}$$

where Y_{ik} and S_{ik} are new variables, and the coefficients q_{ikjm} in Eq. (48) are the demand-weighted transportation costs $q_{ikjm} = W_{ij}(\chi C_{ik} + \alpha C_{km} + \delta C_{jm})$, and $\mu = \max_{i,k} \sum_j \sum_m |q_{ikjm}|$ in constraints (49). Due to a nice problem structure of USApIIMP, we could modify the above formulation as follows.

Theorem 1. *The QP formulation in Eqs. (1-5) is equivalent to following minimization problem:*

$$\min \sum_{i \in N} \sum_{k \in N} (S_{ik} - \sigma Z_{ik}) \tag{52}$$

$$s.t. \sum_k Z_{kk} = p \tag{53}$$

$$\sum_{k \in N} Z_{ik} = 1, \qquad \forall i \in N \tag{54}$$

$$\sum_{i \in N} Z_{ik} \leq (n - p + 1) Z_{kk}, \qquad \forall k \in N \tag{55}$$

$$\sigma + \sum_{j \in N} \sum_{m \in N} q_{ikjm} Z_{jm} = Y_{ik} + S_{ik}, \quad \forall i, k \in N \tag{56}$$

$$Y_{ik} \leq (\mu + \sigma)(1 - Z_{ik}), \qquad \forall i, k \in N \tag{57}$$

$$Z_{ik} \in \{0, 1\}, \qquad \forall i, k \in N \tag{58}$$

$$Y_{ik}, S_{ik} \geq 0, \qquad \forall i, k \in N, \tag{59}$$

where $q_{ikjm} = W_{ij}(\chi C_{ik} + \alpha C_{km} + \delta C_{jm})$, $\mu = \max_{i,k} \sum_j \sum_m |q_{ikjm}|$ *and* $\sigma \geq 0$.

Proof. Necessity: From [30], $\sum_{i \in N} Z_{ik} \leq (n - p + 1) Z_{kk}, \forall k$ has been shown to be equivalent to $Z_{ik} \leq Z_{kk}, \forall i, k$. The main difference between the revised QP-LT formulation in Eqs. (52-59) and the QP-LT in Eqs. (44-51) is at the constraints in Eqs. (56-57). Similar to the proof of Theorem 2 in [12], if we multiply Z_{ik} to both sides of the constraints in Eq. (56) and sum up Eqs. (56) by all $i \in N$ and $k \in N$, we obtain $\sum_{i \in N} \sum_{k \in N} \sum_{j \in N} \sum_{m \in N} q_{ikjm} Z_{ik} Z_{jm} = \sum_{i \in N} \sum_{k \in N} Y_{ik} Z_{ik} + \sum_{i \in N} \sum_{k \in N} S_{ik} Z_{ik} - \sum_{i \in N} \sum_{k \in N} \sigma Z_{ik}$. From Eq. (57) and the binary property of Z_{ik}, we can derive $Y_{ik} Z_{ik} = 0, \forall i, k \in N$. We can show that $\sum_{i \in N} \sum_{k \in N} S_{ik} Z_{ik} = \sum_{i \in N} \sum_{k \in N} S_{ik}$. Assuming (Z^*, Y^*, S^*) is an optimal solution to Eqs. (52-59), the above equation is equivalent to showing that, for any i, k, if $Z_{ik}^* = 0$, then $S_{ik}^* = 0$. By exactly the same contradiction method as a result of Theorem 1 in [12], assuming for certain i, k, we have $Z_{ik}^* = 0$ and $S_{ik}^* > 0$. Then we could let $\hat{Y}_{ik} = Y_{ik}^* + S_{ik}^*$, and $\hat{S}_{ik} = 0$. For all other $j \neq i$, or $m \neq k$, let $\hat{Y}_{jm} = Y_{jm}^*$ and $\hat{S}_{jm} = S_{jm}^*$. Then $(Z^*, \hat{Y}, \hat{S})$ is a feasible solution to Eqs. (52-59) as well. However, $\sum_{i \in N} \sum_{k \in N} \hat{S}_{ik} < \sum_{i \in N} \sum_{k \in N} S_{ik}$, which contradicts with the assumption that (Z^*, Y^*, S^*) minimizes the objective function. Therefore, we conclude that

$$\sum_{i \in N} \sum_{k \in N} \sum_{j \in N} \sum_{m \in N} q_{ikjm} Z_{ik} Z_{jm} = \sum_{i \in N} \sum_{k \in N} (S_{ik} - \sigma Z_{ik}).$$

From Eq. (56) and the necessity of the complementarity property of $Y_{ik}Z_{ik} = 0$, Eq. (57) is now proven.

Sufficiency: The proof is similar.

From the above theorem, we notice the following remarks.

- Since USApHMP offers the linkage constraints $\sum_{k \in N} Z_{ik} = 1, \forall i \in N$, such that $\sum_{i \in N} \sum_{k \in N} \sigma Z_{ik} = n\sigma$, the objective function of the above formulation can always be simplified as $\min \sum_{i \in N} \sum_{k \in N} S_{ik}$, which is *not* equivalent to the objective of the original QP formulation for USApHMP, when $\sigma > 0$.
- The QP-LT formulation for USApHMP in Eqs. (44-51) is a specific case of the revised QP-LT formulation for USApHMP in Eqs. (52-59) when $\sigma = 0$.
- The linearization formulation for general cases in Theorem 4 in [12] is another specific of the revised QP-LT formulation for USApHMP in Eqs. (52-59) when $\sigma = \mu$.

We note that, in practice, the constraints in Eq. (57) are very loose for USApHMP and might not give good bounds. Therefore, we tighten Eq. (57) using the following corollary.

Corollary 1. *Assume for any fixed j where $Z_{jm} = 1, \forall m \in N$. We obtain*
$$\sum_{j \in N} \sum_{m \in N} q_{ikjm} Z_{ik} \leq \sum_{j \in N} \max_{m \in N} q_{ikjm} \leq \mu \forall i, k \in N.\ \textit{From this inequality, we}$$
let $\bar{\mu}_{ik} = \sum_{j \in N} \max_{m \in N} q_{ikjm}, \forall i, k \in N$. Then the constraints in Eq. (57) can be formulated as $Y_{ik} \leq (\bar{\mu}_{ik} + \sigma)(1 - Z_{ik}) \forall i, k \in N$.

Using the above corollary, we can further tighten these constraints by eliminating the relaxation parameter σ and reformulate Eq. (57) as

$$Y_{ik} \leq \bar{\mu}_{ik}(1 - Z_{ik}), \qquad \forall i, k \in N. \tag{60}$$

Although these alternative constraints give better computational performances, we cannot guarantee the optimality with this strategy.

4.5 Space Complexity Comparison of Linearization Techniques for USApHMP

In order to give a theoretical comparison between all the above-mentioned linearization techniques for USApHMP, we investigate the numbers of additional constraints and variables needed to be introduced by each technique given in Table 1.

From Table 1, the QA-LT yields the least numbers of variables and constraints while QP-LT also has the same polynomial degree. It is worth noting that the quadratic programming based linearizations are more compact, in terms of space complexity, than other linearization techniques.

Table 1. Comparison between different formulations for USApHMP

Formulations	QP	T-LT	MC-LT	QP-LT	QA-LT
# of new constraints	0	$2n^3$	n^3	$2n^2$	n^2
# of new real variables	0	n^4	n^3	$2n^2$	n^2
Total # of constraints	$n^2 + 1$	$2n^3 + n^2 + 1$	$n^3 + n^2 + 1$	$3n^2 + 1$	$2n^2 + 1$
Total # of variables	n^2	$n^4 + n^2$	$n^3 + n^2$	$3n^2$	$2n^2$

5 Empirical Study

In order to evaluate the performance of the two quadratic programming based linearizations discussed in the previous section, empirical studies on real USApHMPs are investigated and discussed in this section. Generally, additional variables and constraints from linearizations increase the size of the linear integer programming problems. In many cases where the original QP problem is large, the number of additional variables and constraints could be very large with the polynomial growth in dimension. Due to the $\mathcal{NP}$-hard nature of integer program, this makes the problem much harder to solve. In this section, we investigate the computational performances of QP-LT and QA-LT. After USApHMPs were reformulated as mixed integer linear programming problems, CPLEX solvers were used to solve these MIP problems. All the numerical tests were carried out on a desktop computer with 3.2 GHz processor and 2 GB of RAM. The USApHMPs were modeled in MATLAB and solved by calling the CPLEX solver in the TOMLAB software. In order to get a reasonable comparison, we used several parameter settings in CPLEX. The stopping criterion used here is the relative MIP gap (cpxControl.EPGAP in CPLEX), which is set to be smaller than 0.01%, and the maximum CPU running time, which is set to 1,000 seconds. In this study, we used the Civil Aeronautics Board (CAB) data as our test case. This data set was given in [28] and has been widely studied as a baseline test problem. The CAB data contains 25 nodes and we assumed that all of them were potential hub nodes and used $p = 2, 3, 4$. The sketch of this data set is presented in Figure 2.

In the first trial when we directly used the QP-LT formulation in Eqs. (44-51) to solve the USApHMP with CAB data set, the performance of this QP-LT is very poor. At the stopping criteria after 1,000 seconds, the gap was in the range of 65%. The reason could be that the value of coefficient μ in Eq. (49) is too large and the lower-bound heuristic of CPLEX failed to give a good bound. Consequently, we applied the QP-LT formulation in Eqs. (52-59) and the revised constraints in Eq. (60). We set $\sigma = \mu$ and eliminated σ in Eq. (57). In other words, we used the revised constraints, $\mu + \sum_{j \in N} \sum_{m \in N} q_{ikjm} Z_{jm} = Y_{ik} + S_{ik} \forall i, k \in N$. After μ was added to Eq. (48), the constraints in Eq. (49) were tightened by the range of Y_{ik}. In this study, we show only one test scenario with the relaxation parameter $\sigma = \mu$. Note that since parameter μ is a relatively large number, the probability of

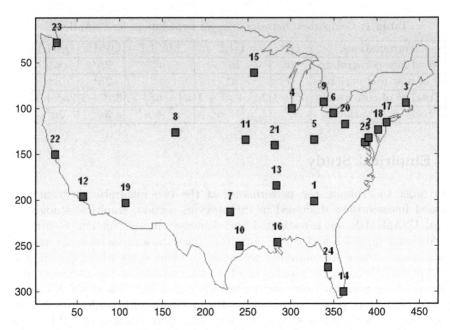

Fig. 2. Civil Aeronautics Board data with 25 nodes

cutting out optimal solutions increases while the CPLEX performs faster. It is possible to investigate the range of parameter μ to get good bounds. In this test, this setting of σ made the convergence much faster while it increased a probability of not getting the real optimal solution. We tested and compared the performances of QP-LT and QA-LT under different settings of α, $\alpha = 0.2, 0.4, 0.6, 0.8$. The computational results are shown in Table 2.

Table 2. Computational results for USApHMP

Problem		QA-LT				QP-LT			
p	α	Nodes	Gap(%)	CPU(s)	Costs	Nodes	Gap(%)	CPU(s)	Costs
2	0.2	916	0.01	49.1	8.5478e5	535	0.01	0.85	8.5478e5
2	0.4	824	0.01	53.4	9.4079e5	444	0.01	0.75	9.4079e5
2	0.6	2450	0.01	97.0	1.0258e6	438	0.01	0.80	1.0410e6
2	0.8	5833	0.01	128.4	1.1051e6	638	0.01	0.9	1.1566e6
3	0.2	4330	0.01	164.2	6.5532e5	1978	0.01	2.0	6.6126e5
3	0.4	14424	0.01	349.0	7.7005e5	1596	0.01	1.7	7.8900e5
3	0.6	54755	0.01	790.8	8.8266e5	1252	0.01	2.1	9.1157e5
3	0.8	35356	17.9	> 1000	9.8964e5	1503	0.01	2.1	1.0456e6
4	0.2	9150	0.01	334.0	5.3771e5	2391	0.01	1.7	5.4211e5
4	0.4	25468	15.3	> 1000	6.7884e5	1669	0.01	1.5	6.8903e5
4	0.6	30255	34.1	> 1000	8.0208e5	2983	0.01	2.3	8.0371e5
4	0.8	40035	41.5	> 1000	9.2886e5	3714	0.01	3.0	9.4007e5

From our computational results, we observe that the QP-LT by far out-performed the QA-LT in all test cases. The computational time of QP-LT is drastically smaller than that of QA-LT in every case. Also, the QP-LT is converged by the solution gap-stopping criteria. Note that the final objective function cost in QP-LT is higher than that in QA-LT. As mentioned earlier, the addition of coefficient σ to Eq. (48) could overconstrain the problem and cut out the true optimal solution off the feasible region. Although the objective function cost of QP-LT is higher than that of QA-LT, the difference in value is very minimal ($< 6\%$). One can also consider different values of parameter σ, which can be fine-tuned to get the best trade-off between the solution quality and computational time.

6 Conclusion

In this chapter, we discussed various mathematical programming formulations of USApHMP. We theoretically and empirically compared current linearization techniques for USApHMP. Based on our preliminary study, we have found that the QP-LT in Eqs. (44-51) shows a great potential for solving large-scale problems, especially problems with quadratic terms. Although we may get good bounds by fine-tuning some coefficients in QP-LT, we still consider the bounding strategy as a heuristic and we cannot theoretically guarantee the optimality. Meanwhile, the choice of relaxation parameter σ needs to be further investigated, e.g., what is a good rule of thumb for choosing the relaxation parameter σ? In addition, we only tested the QP-LT on a relative small problem, only 25 nodes. In the future, it will be very interesting to apply the QP-LT to large-scale USApHMPs. One potential USApHMP is the Australian post data, which contain 200 nodes shown in Figure 3. We expect that the QP-LT will yield a better performance than other linearizations on larger problems. However, there are also some disadvantages of this QP-LT for USApHMP. First, the coefficients appearing in the constraints in Eqs. (48,49) need extra computational time. When the problem size becomes larger, it may take a long time to solve the problem. The linearizations are just a technique to reformulate nonlinear optimization problems. Since they are not a solution approach, the performance of linearizations highly depends on the choice of solution methods for MIP used by CPLEX. Therefore, some specific approach may be developed for linearizations, e.g., specific branch-and-bound approaches, special cuts, and other bounding heuristics. Also, we note that the lower bounding procedure is very crucial to the performance of linearizations.

The coefficient matrix of constraints in Eq. (48) is very vulnerable to input data. In some practical problems, the coefficient matrix could be very dense; therefore, the memory requirement of this matrix could be up to n^4. This high memory requirement heavily handicaps the efficiency of linearizations in practice. It may be possible to relax this set of constraints in order to get a new set of constraints with a sparse coefficient matrix. But the size of the problem

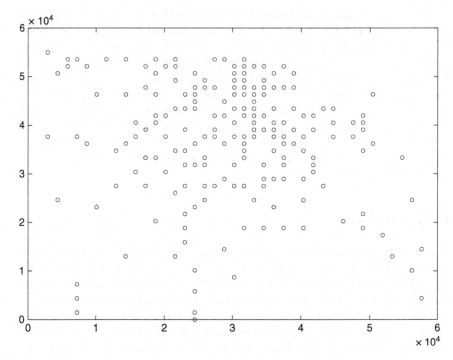

Fig. 3. Australia post data with 200 nodes

should be kept at the same level, which is $O(n^2)$ additional variables and constraints. However, we always have the linkage constraints in USApHMPs, which makes the relaxation very difficult.

We notice that the uncapacitated multiple allocation p-hub median problem (UMApHMP) [17] has a similar problem structure and problem fomulation as the USApHMP. Because the number of binary decision variables in UMApHMP is only n and other decision variables are relaxed to be real values of the interval $[0, 1]$, the UMApHMP is much easier than USApHMP. Therefore, all the considerations concerning the linearization techniques for USApHMP can be easily applied to UMApHMP.

References

1. S. Abdinnour-Helm and M. A. Venkataramanan. Solution approaches to hub location problems. *Annals of Operations Research*, 78:31-50, 1998.
2. W.P. Adams and R.J. Forrester. A simple recipe for concise mixed 0-1 linearizations. *Operations Research Letters*, 33:55-61, 2005.
3. W.P. Adams, R.J. Forrester, and F.W. Glover. Comparison and enhancement strategies for linearizing mixed 0-1 quadratic programs. *Discrete Optimization*, 1:99-120, 2004.

4. W.P. Adams and H.D. Sherali. A tight linearization and an algorithm for zero-one quadratic programming problems. *Management Science*, 32:1274-1290, 1986.
5. T. Aykin. Lagrangian relaxation based approaches to capacitated hub-and-spoke network design problem. *European Journal of Operational Research*, 79:501-523, 1994.
6. T. Aykin. Networking policies for hub-and-spoke systems with applications to the air transportation system. *Transportation Science*, 29(3):201-221, 1995.
7. N. Boland, A. Ernst, M. Krishnamoorthy, and J. Ebery. Preprocessing and cutting methods for multiple allocation hub location problems. *European Journal of Operational Research*, 155:638-653, 2004.
8. D.L. Bryan and M.E. O'Kelly. Hub-and-spoke network in air transportation: an analytical revew. *Journal of Regional Science*, 39:275-295, 1999.
9. J.F. Campbell. Integer programming formulations of discrete hub location problems. *European Journal of Operational Research*, 72:387-405, 1994.
10. J.F. Campbell. Hub location and the p-hub median problem. *Operations Research*, 44:923-935, 1996.
11. J.F. Campbell, A. Ernst, and M. Krishnamoorthy. Hub location problems. In H. Hamacher and Z. Drezner, editors. *Location Theory: Applications and Theory*, pp. 373-406, Springer-Verlag, New York, 2001.
12. W. Chaovalitwongse, P.M. Pardalos, and O.A. Prokoyev. A new linearization technique for multi-quadratic 0-1 programming problems. *Operations Research Letter*, 32:517-522, 2004.
13. P. Chardaire and A. Sutter. A decomposition method for quadratic zero-one programming. *Management Science*, 41:704-712, 1995.
14. J. Ebery, M. Krishnamoorthy, A. Ernst, and N. Boland. The capacitated multiple allocation hub location problem: Formulations and algorithms. *European Journal of Operational Research*, 120:614-631, 2000.
15. S. Elloumi, A. Faye, and E. Soutif. Decomposition and linearization for 0-1 quadratic programming. *Annals of Operations Research*, 99:79-93, 2000.
16. A.T. Ernst and M. Krishnamoorthy. Efficient algorithms for the uncapacitated single allocation p-hub median problem. *Location Science*, 4:139-154, 1996.
17. A.T. Ernst and M. Krishnamoorthy. Exact and heuristic algorithm for the uncapacitated multiple allocation p-hub median problem. *European Journal of Operational Research*, 104:100-112, 1998.
18. A.T. Ernst and M. Krishnamoorthy. An exact solution approach based on shortest-paths for p-hub median problem. *INFORMS Journal on Computing*, 10:149-162, 1998.
19. A.T. Ernst and M. Krishnamoorthy. Solution algorithms for the capacitated single allocation hub location problem. *Annals of Operations Research*, 86:141-159, 1999.
20. A. Frieze and J. Yadeger. On the quadratic assignment problem. *Discrete Applied Mathematic*, 5:89-98, 1983.
21. F. Glover. Improved linear integer programming formulations of nonlinear integer programs. *Management Science*, 22:455-460, 1975.
22. F. Glover and E. Woolsey. Further reduction of zero-one polynomial programming problems to zero-one linear programming problems. *Operations Research*, 21:156-161, 1973.

23. F. Glover and E. Woolsey. Converting the 0-1 polynomial programming problem to a 0-1 linear program. IRMIS working paper, 9304, School of Business, Indiana University, Bollomington, Indiana, 1974.

24. S.A. Helm and M.A. Venkataramanan. Using simulated annealing to solve the p-hub location problem. *Operations Research*, 22:180-182, 1993.

25. L. Kaufman and F. Broeckx. An algorithm for the quadratic assignment problem using Benders' decomposition. *European Journal of Operational Research*, 2:204-211, 1978.

26. J.G. Klincewicz. Dual algorithms for the uncapacitated hub location problem. *Location Science*, 4:173-184, 1996.

27. R.F. Love, J.G. Morris, and G.O. Wesolowsky. *Facilities Location*, North Holland, Amsterdam, 1988.

28. M. O'Kelly. A quadratic integer program for the location of interacting hub facilities. *European Journal of Operational Research*, 32:393-404, 1987.

29. M. O'Kelly. Hub facility location with fixed costs. *Papers in Regional Science: The Journal of the RSAI*, 71:293-306, 1992.

30. M. O'Kelly, D. Skorin-Kapov, and J. Skorin-Kapov. Lower bounds for the hub location problem. *Management Science*, 41:713-721, 1995.

31. H.D. Sherali, J. Desai, and H. Rakha. A discrete optimization approach for locating automatic vehicle identification readers for the provision of roadway travel times. *Transportation Research Part B: Methodological*, 40:857-871, 2006.

32. H.D. Sherali and J.C. Smith. An improved linearization strategy for zero-one quadratic programming problems. *Optimization Letters*, 1:33-47, 2007.

33. D. Skorin-Kapov and J. Skorin-Kapov. On tabu search for the location of interacting hub facilities. *European Journal of Operational Research*, 73:502-509, 1994.

34. D. Skorin-Kapov, J. Skorin-Kapov, and M. O'Kelly. Tight linear programming relaxations of uncapacitated p-hub median problems. *European Journal of Operational Research*, 94:582-593, 1996.

35. L. Watters. Reduction of integer polynomial programming problems to zero-one linear programming problems. *Operations Research*, 15:1171-1174, 1967.

36. W.I. Zangwill. Media selection by decision programming. *Journal of Advertising Research*, 5:30-36, 1965.

Nested Partitions and Its Applications to the Intermodal Hub Location Problem

Weiwei Chen[1], Liang Pi[2], and Leyuan Shi[2]

[1] Department of Industrial and Systems Engineering
University of Wisconsin-Madison, Madison, WI 53706
wchen26@wisc.edu
[2] Department of Industrial and Systems Engineering
University of Wisconsin-Madison, Madison, WI 53706
lpi@wisc.edu
[3] Department of Industrial and Systems Engineering
University of Wisconsin-Madison, Madison, WI 53706
leyuan@engr.wisc.edu

Summary The nested partitions (NP) method has been proven to be a useful framework for effectively solving large-scale discrete optimization problems. In this chapter, we provide a brief review of the NP method and its applications. We then present a hybrid algorithm that integrates mathematical programming with the NP framework. The efficiency of the hybrid algorithm is demonstrated by the intermodal hub location problem (IHLP), a class of discrete facility location problems. Computational results show that the hybrid approach is superior to the integer programming approach and the Lagrangian relaxation method.

1 Discrete Optimization

Many key business investment decisions are large-scale discrete optimization problems casting as designing underlying dynamic systems, at the lowest possible cost, to achieve desired service levels. These problems are challenging and are notoriously difficult to solve. There are two principal technologies for large-scale discrete optimization problems: (1) exact algorithms that are guaranteed to find optimal solutions; and (2) heuristic algorithms that quickly find acceptable solutions.

Exact solution methods are grounded in mathematical programming theories. Such methods have been studied for decades. Significant breakthroughs in the ability to solve large-scale discrete problems using mathematical programming have been achieved. Generally, branching methods and decomposition methods are two primary classes of mathematical programming methods used to solve discrete optimization problems [32, 33]. Lagrangian relaxation can be thought of as a decomposition method with respect to the constraints

W. Chaovalitwongse et al. (eds.), *Optimization and Logistics Challenges in the Enterprise*, Springer Optimization and Its Applications 30,
DOI 10.1007/978-0-387-88617-6_8, © Springer Science+Business Media, LLC 2009

since it moves one or more constraints into the objective function. Relaxation methods play a key role in the use of mathematical programming for solving discrete optimization problems [10, 18]. The Lagrangian problem is easy to solve since the complicated constraints are no longer present. Furthermore, it often produces a fairly tight and hence useful bound.

Nonexact solution methods include two classes: approximation algorithms and heuristic algorithms. Approximation algorithms are often based on similar mathematical programming theories as exact solution methods and are usually applied when an efficient exact algorithm can hardly be found. Approximation algorithms can guarantee that the solution lies within a certain range of the optimal solution. They can also usually provide provable runtime bounds. A typical example for an approximation algorithm is that for the covering problems [4, 15].

Unlike approximation algorithms, heuristic algorithms aim to find reasonably good solutions within an acceptable time frame, without making a performance guarantee. Due to its efficiency and effectiveness, heuristic algorithms have recently drawn much attention. Many efficient heuristic methods exist, ranging from the simplest heuristic such as the greedy algorithm to more sophisticated or randomized heuristics such as simulated annealing (SA) [16, 31], genetic algorithm (GA) [13, 19], tabu search [7, 12], and ant colony optimization [8, 9].

Introduced by Shi and Ólafsson [26], the nested partitions (NP) method is a metaheuristic framework, which is best suited for solving large-scale discrete optimization problems, though it is also applicable to solving continuous optimization problems. The NP method is a partitioning and sampling based strategy that focuses computational effort on the most promising region of the solution space while maintaining a global perspective on the problem. Therefore, it is particularly efficient for problems where the feasible region can be partitioned such that good solutions tend to be clustered together and the corresponding regions are hence natural candidates for concentrating the computation effort. The NP method has been successfully applied in many areas, such as planning and scheduling [24, 28, 29], logistics and transportation [22], supply chain design [25], data mining [21], and health care [28].

In this chapter, we first provide a brief review of the NP method. We then present a detailed implementation of the NP method for the intermodal hub location problem (IHLP), an emerging research area that has drawn great attention from both academics and industrial sources. The IHLP can be viewed as a class of discrete facility location problems (DFLPs) that has certain unique and difficult constraints such as the concave transportation cost function. To effectively solve the problem, we develop hybrid algorithms that utilize the linear programming (LP) lower bound for guiding the sampling procedure of the NP method. Numerical results show that the hybrid algorithms can outperform the mathematical programming and Lagrangian relaxation approaches.

The rest of the chapter is organized as follows. Section 2 provides the review of the NP framework and some successful applications. In Section 3, IHLP is formulated and hybrid algorithms are developed. Numerical results are presented in Section 4. Section 5 concludes the chapter with a discussion about future research.

2 Nested Partitions and Applications

In this section, we first introduce the basic methodology and procedure of the NP framework, then some successful applications of the NP and hybrid NP algorithms are reviewed.

2.1 The NP Framework

Consider the following optimization problem:

$$\min_{x \in X} f\{x\} \tag{1}$$

Problem (1) can be a combinatorial optimization problem or a mixed integer program where multiple local optima exist. The feasible region is denoted as X, and an objective function $f : X \to \mathbb{R}$ which can be linear or nonlinear is defined on this set.

For these types of problems, the only known method to guarantee global optimum is to enumerate all possible solutions throughout the whole region X and compare their performance values to find the best one. However, in most practical applications, the feasible region is too large to be enumerated. This type of problem is what the NP method targets.

In each iteration of the NP algorithm, we assume that there is a region (subset) of X that is considered the *most promising*. We partition this most promising region into a fixed number of M *subregions* and aggregate the entire *complementary region* (also called surrounding region) into one region, that is, all the feasible solutions that are not in the most promising region. Therefore we consider $M+1$ subsets that are a partition of the feasible region X, namely they are disjointed and their union is equal to X. Each of these $M+1$ regions is sampled using some random sampling scheme to generate feasible solutions that belong to that region. The performance values (objective values) of the randomly generated samples are used to calculate the *promising index* for each region. This index determines which region is the most promising region in the next iteration. If one of the subregions is found to be the best, this region becomes the most promising region. The next most promising region is thus nested within the last. If the complementary region is found to be the best, then the algorithm *backtracks* to a larger region that contains the previous most promising region. This larger region becomes the new most promising region, and is then partitioned and sampled in the same fashion.

If region η is a subregion of region σ, we call σ the *superregion* of η. Let $\sigma(k)$ denote the most promising region in the kth iteration. We further denote the *depth* of $\sigma(k)$ as $d(k)$. The feasible region X has depth 0, the subregions of X have depth 1, and so forth. When X is finite, eventually there will be regions that contain only a single solution. We call such singleton regions regions of *maximum depth*. If the problem is infinite, we define the maximum depth to correspond to the smallest desired sets. The maximum depth is denoted as d^*. With this notation, we describe the *generic nested partitions algorithm*, distinguished from hybrid NP algorithms, in the following paragraph [28]. Notice that the special cases of being at minimum or maximum depth are considered separately.

Generic Nested Partitions Algorithm ($0 < d(k) < d^*$)

1. **Partitioning.** Partition the most promising region $\sigma(k)$ into M subregions $\sigma_1(k), \ldots, \sigma_M(k)$, and aggregate the complementary region $X \backslash \sigma(k)$ into one region $\sigma_{M+1}(k)$.
2. **Random Sampling.** Randomly generate N_j sample solutions from each of the regions $\sigma_j(k)$, $j = 1, 2, \ldots, M + 1$:

$$x_1^j, x_2^j, \ldots, x_{N_j}^j, \ j = 1, 2, \ldots, M + 1.$$

Calculate the corresponding performance values:

$$f(x_1^j), f(x_2^j), \ldots, f(x_{N_j}^j), \ j = 1, 2, \ldots, M + 1.$$

3. **Calculate Promising Index.** For each region σ_j, $j = 1, 2, \ldots, M + 1$, calculate the promising index as the best performance value within the region:

$$I(\sigma_j) = \min_{i \in \{1, 2, \ldots, N_j\}} f(x_i^j), \ j = 1, 2, \ldots, M + 1.$$

4. **Move.** Calculate the index of the region with the best performance value.

$$\hat{j}_k = \arg \min_{j \in \{1, \ldots, M+1\}} I(\sigma_j).$$

If more than one region is equally promising, the tie can be broken arbitrarily. If this index corresponds to a region that is a subregion of $\sigma(k)$, that is $\hat{j}_k \le M$, then let this be the most promising region in the next iteration:

$$\sigma(k + 1) = \sigma_{\hat{j}_k}(k).$$

Otherwise, if the index corresponds to the complementary region, that is $\hat{j}_k = M + 1$, backtrack to the superregion of the current most promising region:

$$\sigma(k + 1) = \sigma(k - 1).$$

For the special case of $d(k) = 0$, the steps are identical except there is no complementary region. The algorithm hence generates feasible sample solutions from the subregions and in the next iteration moves to the subregion with the best promising index. For the special case of $d(k) = d^*$, there are no subregions. The algorithm therefore generates feasible sample solutions from the complementary region and either backtracks or stays in the current most promising region.

The above procedure of the NP method gives us a framework that guides the search and enables convergence analysis [28]. The NP method is also applicable to problems where the objective function is noisy, for example, when it can only be evaluated as a realization of some random variables. The NP method for stochastic optimization can still be proved to converge to the global optimum [27]. The NP framework provides the flexibility to incorporate domain knowledge or local search into the search procedure. According to our computational experience, hybrid NP algorithms with local heuristics almost always outperform local heuristics alone. We will provide a detailed implementation example in Section 3.

2.2 Applications

The NP method is a global metaheuristic found to be effective for difficult large-scale combinatorial optimization problems. The NP framework allows us to incorporate many efficient heuristics such as the tabu search and genetic algorithm into its procedure. The resulting hybrid algorithms are more efficient than either generic NP or the heuristics alone. Domain knowledge and the special structure of the problem can be utilized in the NP procedures, such as biased sampling. The NP method can also be combined with mathematical programming to produce more efficient search algorithms when the mathematical programming approach alone fails to solve the problem. A few successful NP applications are summarized here.

Product Design

Some heuristic algorithms are successful at solving certain types of discrete optimization problems. Product design problems occur when designing new products to satisfy the preferences of expected customers. An important problem is discerning how to use the preferences of potential customers to design a new product such that the market share of the new product is maximized. This problem is very difficult to solve, especially as the product complexity increases and more attributes are introduced. In fact, it belongs to the class of NP-hard problems. In literature, the greedy search (GS) heuristic and dynamic programming (DP) heuristic have been applied [17], and a genetic algorithm (GA) approach has also been introduced [2, 3].

By incorporating these heuristics into the NP framework, new NP/GS, NP/DP, NP/GA, and NP/GA/GS algorithms are designed in [29]. The greedy search and dynamic programming heuristic are used to bias the sampling distribution in the NP method, and the GA is utilized to improve the initial population and the promising index for each region. Numerical examples are used to compare the new optimization framework with existing heuristics, and the results indicate that the new method is able to produce higher-quality product profiles. Furthermore, these performance improvements were found to increase with increased problem size. This finding indicates that the NP optimization framework is an important addition to the product design and development process, and will be particularly useful for designing complex products that have a large number of important attributes. Detail numerical comparisons are presented in [29].

Buffer Allocation

Buffer allocation problems exist in various kinds of manufacturing systems. In the design of production lines, people are interested in the issue of how to optimally allocate a given buffer capacity between a number of stations such that a desired system performance is achieved. There are two factors that make this problem difficult. First, due to the inherent complexity of the buffer allocation problems, we expect that most objective functions cannot be defined by an analytical expression. When this is the case, analytic approximation or simulation can be used for performance evaluation. Secondly, the total number of feasible solutions grows exponentially when the total number of machines and the total buffer capacity increase. It is usually impossible to search through the whole solution space by enumeration. Heuristics or local search methods are usually used to obtain the near-optimal solutions. By combining the NP method with tabu search (TS), a hybrid NP/TS algorithm is proposed in [24]. Tabu search attempts to avoid some solutions from being visited repeatedly. The NP method guarantees that more effort is expended in good regions that are most likely to contain the best solution. The hybrid NP/TS algorithm exploits both the advantage of NP and tabu search, and better results have been reported than those using tabu search without NP framework, as shown in Figure 1.

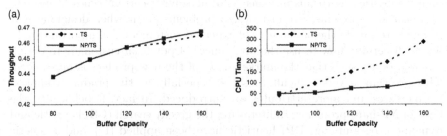

Fig. 1. (a) The throughput as a function of buffer capacity for the TS and NP/TS algorithms; and (b) the CPU time as a function of buffer capacity for the TS and NP/TS algorithms

Supply Chain Optimization

In recent years, supply chain optimization has become very popular among the world's leading manufacturers, distributors and retailers. Solving the strategic-level warehouse location problem can lead to significant savings for the company by determining the lowest cost or most efficient strategy. With advances in technology and information systems, these problems become more and more complicated in terms of problem size and uncertainty. From a computational point of view, multicommodity capacitated facility location problems are quite challenging because they are combinatorially explosive – some of these optimization models have millions of variables and millions of constraints. They are NP-hard and generally intractable with respect to standard mixed integer programming (MIP) tools, such as the direct application of general purpose branch-and-cut commercial solvers (e.g., CPLEX).

In [25], a hybrid NP/CPLEX approach is introduced for solving large-scale multicommodity facility location problems. The algorithm is applied to 17 hard problem instances. These problems have various values for five design parameters: numbers of plants, warehouses, open warehouses, customers, and products resulting in problems of varying size and difficulty. The new NP/CPLEX algorithm is significantly faster, and better feasible solutions are achieved in 14 cases compared to CPLEX and specialized approaches based on Lagrangian relaxation [25]. This example illustrates that the NP framework can effectively combine problem-specific heuristics with MIP tools (such as AMPL/CPLEX) in implementations.

Feature Selection

The subject of data mining has recently enjoyed enormous interests from both academia and industry. One of the problems that must usually be solved as a part of practical data mining projects is the feature selection problem, which involves selecting a good subset of variables to be used by subsequent inductive data mining algorithms. From an optimization point of view, feature selection can clearly be formulated as a binary combinatorial optimization problem where the decision variables determine if a feature (variable) is included or excluded. The problem is generally difficult to solve. The number of possible feature subsets is 2^n, where n is the number of features, and evaluating every possible subset is therefore prohibitively expensive unless n is very small. There is generally no structure present that allows for an efficient search through this large space, and a heuristic approach that sacrifices optimality for efficiency is typically applied in practice. Another difficulty in solving this problem is that there is no simple method for dealing an objective function, and the quality of the solution (feature subset) is often measured by its eventual performance when used with a learning algorithm. Thus, the objective function becomes the estimated performance of a complex learning algorithm. Since the NP method can deal effectively with such complex objective functions, it becomes an attractive method for solving this problem.

Feature selection methods are generally classified as either filtering methods, which produce a ranking of all features before the learning algorithm is applied, or wrapper methods, which use the learning algorithm to evaluate subsets of features. Depending on the method that is used to evaluate sample subsets, the NP method can be implemented as either a filter or a wrapper algorithm, but it always searches through the space of feature subsets by evaluating all subsets. On the other hand, it can also incorporate methods that evaluate individual features into intelligent partitioning to impose a structure that speeds the search. Based on the above idea, the NP-wrapper and NP-filter algorithms are developed and numerically proven to have good performance [21].

Resource Constrained Project Scheduling

In the above applications, general purpose metaheuristics and problem-specific local search methods are incorporated into the NP framework to improve both the partitioning and the generation of feasible solutions. The efficiency and effectiveness of the NP method can be further improved by incorporating expert domain knowledge in a similar manner, resulting in a knowledge-based NP algorithm. This provides us with a possibility to exploit domain knowledge, while exact methods cannot take advantage of this since this would make the problem more complicated and therefore intractable.

A resource-constrained project scheduling problem in many manufacturing and service applications can be described as follows. A project consists of a set of tasks to be performed and given precedence requirements between some of the tasks. The project scheduling problem involves finding the starting time of each task so that the overall completion time of the project is minimized. One or more resources is required to complete each task. The resources are limited so if a set of tasks requires more than the available resources, they cannot be performed concurrently. The problem now becomes NP-hard and cannot be solved efficiently to optimality using any traditional methods. In practice, constraints that are difficult for optimization methods such as mathematical programming are sometimes easily addressed by incorporating domain knowledge. For example, a domain expert may easily be able to specify priorities among tasks requiring the same resource(s) in the resource-constrained project scheduling problems. The domain expert therefore, with some assistance perhaps from an interactive decision support system, can specify some priority rules to convert a very complex problem into an easy-to-solve problem. The NP method can effectively incorporate such domain knowledge into the optimization framework by using the priority rules when generating feasible solutions, as well as when partitioning intelligently. An example on the optimal process planning is presented in [28], and the knowledge-based NP algorithm for process planning is effective and capable of producing high-quality solutions rapidly.

Radiation Treatment Planning

Health care delivery is an area of immense importance where optimization techniques have increasingly been used in recent years. Radiation treatment planning is an important example of this. The intensity modulated radiation therapy (IMRT) is a recently developed complex technology for such treatment. Because of its complexity, the treatment planning problem is generally divided into several subproblems. The first is termed the beam angle selection (BAS) problem. In essence, beam angle selection requires the determination of roughly 4-9 angles from 360 possible angles subject to various spacing and opposition constraints. It is computationally intense to solve this selection problem. In modern clinics, the rotational angles of a treatment couch are also considered as another set of decision variables. This adds even more complexity to the problem. Because of these reasons, currently the angles are selected manually by clinicians based on their experiences.

By applying the NP framework for automating beam angle selection, high-quality solutions have been reported [28]. Relative to good quality beam angle sets constructed via expert clinical judgment and other approaches, the beam sets generated via NP showed significant reduction (up to 32%) in radiation delivered to noncancerous organs-at-risk near the tumors. Thus, in addition to providing a method for automating beam angle selection, the NP framework yields higher quality beam sets that significantly reduce radiation damage to critical organs.

3 Intermodal Hub Location Problems

In this section, we provide a detailed implementation of the NP method to the intermodal hub location problem (IHLP). In particular, we will focus on three important steps of the NP method: partitioning, sampling and backtracking.

3.1 Problem Formulation

Due to its economic impact, the IHLP has drawn great attention by researchers [5, 20, 23]. The intermodal transportation via truck and rail is an alternative to single mode, truckload carriage. Intermodal operations utilize a sealed container or truck trailer that is mechanically moved between modes (truck, rail) in a seamless fashion. An intermodal terminal has equipments suitable for transferring the containers and trailers between modes. For distances over a certain threshold, rail transportation becomes more efficient than truck transportation and results in savings in time, operating costs and labor. Typically an IHLP consists of three connected movements: two local ones that involve truck movements and an intermediate one that involves rail movement. The IHLP aims to minimize the costs of the total intermodal transportation system. These costs consist of operation costs of the opened hubs and the routing costs of the intermodal movements.

The IHLP can be viewed as a type of discrete location facility problems. Before we formulate the problem, we introduce the following notations.

Sets:

- $I = \{I_1, \ldots, I_{|I|}\}$: set of origin/destination terminal locations.
- $R = \{R_1, \ldots, R_{|R|}\}$: set of intermodal hub locations.
- $F = \{F_1, \ldots, F_{|F|}\}$: set of demanding flows, i.e., movement demand from certain origin to certain destination.

Parameters:

- $O_f, f \in F$: origin terminal of flow f.
- $D_f, f \in F$: destination terminal of flow f.
- $W_f, f \in F$: amount of flow f.
- $C1_r, r \in R$: operating cost of hub r, if r is opened.
- $C2_{ab}, (a, b) \in I \times R \cup R \times R \cup R \times I$: transportation cost function of the flow between location a to location b. Due to the scale economy which is a crucial consideration in transportation industry, $\forall (a, b) \in I \times R \cup R \times R \cup R \times I$, we can assume $C2_{ab}$ to be a nondecreasing concave function of the amount of the flow from location a to location b. Furthermore, in the formulation, we assume that these functions are piecewise linear, and the values of these functions are given.
- $C3_f, f \in F$: cost rate per unit amount if flow f is not moved or moved by other more expensive method such as pure truck movement.
- $M_{ab}, (a, b) \in I \times R \cup R \times I$: $= 1$, if movement from a to b are allowed; $= 0$, otherwise.

Then, we define the decision variables of this problem as follows:

- $x_{fkm}, f \in F, k \in R, m \in R, \geq 0$: the amount of flow f moved through intermodal rail line (k, m).
- $y_{ab}, (a, b) \in I \times R \cup R \times R \cup R \times I, \geq 0$: the amount of flow from location a to location b.
- $z_r, r \in R$: 0-1 facility location variables. $\forall r \in R, z_r = 1$ if hub r is opened; $= 0$, otherwise.
- $u_f, f \in F, \geq 0$: the amount of flow f that is not moved through the intermodal operations.

Then, the formulation of the problem is described as follows:
Objective: Minimize $Q =$

$$\sum_{r \in R} C1_r \cdot z_r \tag{2}$$

$$+ \sum_{(a,b) \in I \times R \cup R \times R \cup R \times I} C2_{ab} \cdot y_{ab} \tag{3}$$

$$+ \sum_{f \in F} C3_f \cdot u_f \tag{4}$$

Subject to:

$$u_f + \sum_{k \in R, m \in R} x_{fkm} = W_f \quad \forall f \in F, \tag{5}$$

$$\sum_{m \in R} x_{fkm} \leq W_f \cdot z_k \quad \forall f \in F, k \in R, \tag{6}$$

$$\sum_{k \in R} x_{fkm} \leq W_f \cdot z_m \quad \forall f \in F, m \in R, \tag{7}$$

$$\sum_{f \in F, m \in R: O_f = i} x_{fkm} \leq y_{ik} \quad \forall i \in I, k \in R, \tag{8}$$

$$\sum_{f \in F} x_{fkm} \leq y_{km} \quad \forall k \in R, m \in R, \tag{9}$$

$$\sum_{f \in F, k \in I: D_f = j} x_{fkm} \leq y_{mj} \quad \forall m \in R, j \in I, \tag{10}$$

$$x_{fkm} = 0 \quad \forall f \in F, k \in I, m \in I : M_{(O_f)k} \cdot M_{m(D_f)} = 0 \tag{11}$$

Here, in the objective function, (2) are the costs of hub operations, (3) are the costs of flows moved by the intermodal operations, and (4) are the costs of flows not moved by the intermodal operations. Constraints (5) are the requirements that all the flows should be covered. Constraints (6) and (7) require that each flow can only be routed via opened hub in a intermodal movement. Constraints (8), (9) and (10) are the relationships between variables x's and variables y's. Constraints (11) are the restrictions of the movements between terminals and hubs. In the formulation above, term (3) includes concave piecewise linear functions, leading to one major difficulty of solving the problem. There are standard procedures to linearized term (3) [6]; however, the problem size will be increased due to the new variables and constraints introduced.

3.2 Hybrid NP and Mathematical Programming Algorithm

To effectively solve the IHLP, in this subsection, we developed a hybrid NP and mathematical programming (HNP-MP) approach. The effectiveness of the hybrid algorithm is demonstrated through a set of test problems that have been widely used in the area of logistics.

In most of NP implementations, complete samples (or solutions) are generated in each sampling step. To be able to integrate the LP/MIP techniques into the NP framework, we introduce a concept of *a partial sample*. A partial sample is a set of solutions that is generated by sampling only a part of the variables in a given region. For example, assume that the solution space can be denoted as: $(X_1, \ldots, X_n)$ and the sampling region is $(X_1^*, \ldots, X_k^*, X_{k+1}, \ldots, X_n)$ where X_1^*, ..., X_k^* are fixed. By sampling X_{k+1}

to X_{k+j}, $1 \leq j < n - k$, we have a partial sample of the following form: $(X_1^*, \ldots, X_k^*, \bar{X}_{k+1}, \ldots, \bar{X}_{k+j}, X_{k+j+1}, \ldots, X_n)$. The purpose of introducing partial samples is that by reducing the number of variables of the original optimization problem, the subproblem with only variables of $(X_{k+j+1}, \ldots, X_n)$ could be solved effectively and quickly using mathematical programming techniques.

HNP-MP Algorithm

S0: Set the initially most promising region as the overall solution space. Set the initial surrounding region as ϕ. Go to S1.

S1: If stopping conditions hold, restart (go to S0) or stop (the best sample obtained so far is returned); otherwise, go to S2;

S2: Obtain LP solution for current most promising region. Do LP biased sampling (see below) over the most promising region and surrounding region to generate partial samples $(X_1^*, \ldots, X_k^*, \bar{X}_{k+1}, \ldots, \bar{X}_{k+j}, X_{k+j+1}, \ldots, X_n)$. Go to S3.

S3: Evaluate partial samples $(X_1^*, \ldots, X_k^*, \bar{X}_{k+1}, \ldots, \bar{X}_{k+j}, X_{k+j+1}, \ldots, X_n)$ by solving the embedded problems. Calculate the promise indices for both the most promising region and the surrounding region. If the most promising region is more promising, go to S4; otherwise, go to S5.

S4: Perform the partitioning and get a new most promising region. (LP solution-based partitioning can be used.) Go to S1.

S5: Carry out backtracking. The resulting region is set as the next most promising region. Go to S1.

There are several important elements in the HNP-MP algorithm that are described in the following.

Sampling

The first step in applying the HNP-MP approach to IHLP problems is to determine a proper form of partial samples such that we can fully leverage the capability of IP/MIP solvers such as CPLEX or specialized algorithms to efficiently solve the small-scale subproblems associated with the partial samples. For the IHLP, we define partial samples as feasible solutions to the problem in the form of letting a set of hubs be closed (some z variables are fixed to 0), and no flow can move through these closed hubs. Figure 2 shows an example with seven hubs $r_1, \ldots, r_7$, and four flows $f_1, \ldots, f_4$; and in a partial solution, hub r_1, r_4 and r_6 are closed.

Biased sampling techniques can be used to obtain partial samples that contain high-quality samples. We develop a sampling procedure, called the linear programming (LP) solution-based sampling, which will be applied to our problem. The detailed procedure of *LP biased sampling* is given below:

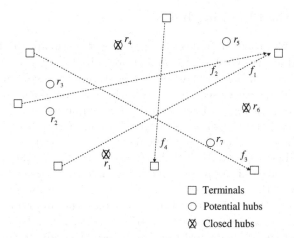

Fig. 2. IHLP: A partial solution

1. Obtain the LP solution. Denote the z part by $\mathbf{z}^*$.

2. Calculate the sampling weights of variable $\mathbf{z}$, based on the value of $\mathbf{z}^*$. For $\forall h \in H$, the sampling weight is positively correlated to the value of z_h^*.

3. Based on the sampling weights, a partial sample can be generated: randomly select M $(< |R|)$ variables from all $z's$ [34] and fix the remaining $(|R| - M)$ z variables to zero. M is controlled to make the subproblems of partial samples easy to solve.

In this chapter, a linear function of z^* is used to calculate the sampling weights. For $\forall k \in R$, we define $W_k' = z_k^* + \epsilon$ (ϵ is a very small nonnegative number, $\epsilon = 0.01$ is used in our tests). Then after the normalization step, we can obtain the sampling weights as: $\forall k \in R, W$ (hub k is open) $= W_k = W_k' / \sum_{k' \in R} W_{k'}'$.

An example with seven hubs $(r_1, \ldots, r_7)$ is shown in Table 1. Based on these weights, we can sample a set of hubs to be potentially opened, and let all other hubs to be closed. One of the possible sampling results using the weights in Table 1 can be hubs r_1, r_4 and r_6 are closed and others are potentially opened (as shown in Fig. 2).

Table 1. An example of the sampling weights ($\epsilon = 0.01$)

Hub (k)	LP solution (z_k^*)	Weight (W_k')	Normalized weight (W_k)
r_1	0	0.01	1/227
r_2	0.3	0.31	31/227
r_3	0.7	0.71	71/227
r_4	0	0.01	1/227
r_5	0.5	0.51	51/227
r_6	0.1	0.11	11/227
r_7	0.6	0.61	61/227

Calculating the Promising Index

To calculate the promising index, we first need to evaluate the partial samples generated in the sampling step and obtain a best (or good) sample within each partial sample using standard integer programming algorithms. In Fig. 2, hub r_1, r_4 and r_6 are closed. A sample which is contained by the partial solution is shown in Fig. 3. Each sample is a complete solution of the original problem, and the top samples obtained in the partial sample evaluation step will be used to calculate the promising index and guide the partitioning/backtracking step. For the IHLP, if only a fraction of the z variables is fixed (to 0) in a partial sample, the partial sample corresponds to a relatively small problem with the same structure as the original problem; if the partial samples are in the form of fixing all the z variables, the subproblem associated with each partial sample is a pure routing problem. For both cases, standard integer programming algorithms can be used to evaluate partial samples efficiently.

Partitioning, Backtracking and Stopping

If the most promising region needs to be further partitioned, we keep the current best sample in the next most promising region, which provides a set of partitioning variables. For the IHLP, each available partitioning attribute opens a certain hub, and can be potentially used to partition a current most promising region into two subregions (one partition with this hub open which is selected as the next most promising region, one with this hub closed which is aggregated into the surrounding region). Also, we can use several attributes for partitioning. For example, when an attribute with hub k open

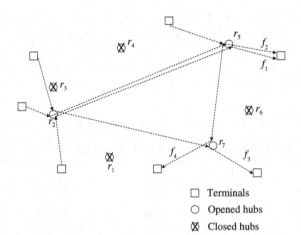

Fig. 3. IHLP: A sample

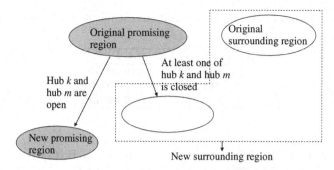

Fig. 4. An example of effective partitioning

and an attribute with hub m open are used for partitioning, then one partition can be both hub k and hub m are open (selected as the next most promising region) and the other partition is where at least one of the two hubs is closed (aggregated into the surrounding region). This example can be described in Fig. 4. Then, we can use the LP solution-based partitioning to select the partitioning attribute(s) from all available ones. Basically, we use the LP solution on the current most promising region as the partitioning index, called the LP solution-based partitioning index. For the IHLP, the value of z_k^* ($k \in R$) is used as the LP solution-based partitioning index for the attribute that hub k is open. We can select one attribute (or several attributes) with the best (or top) LP solution-based partitioning index value for partitioning.

If backtracking is performed, then some constraints on the promising region are relaxed. For the IHLP, by dropping some cuts which let certain hub(s) open on the current most promising region, the backtracking area (the most promising region in the next iteration) will include the current most promising region and the best sample obtained so far. The algorithm usually stops when the computational resource (e.g., time) reaches a predefined value. Other problem-depend stopping criteria can also be designed.

Table 2. DFLP: Scale settings

Index	$|I|$	$|R|$	$|F|$
a	60	30	300
b	30	30	300
c	60	40	200
d	80	40	200
e	50	50	150
f	40	50	150
g	40	40	200

Table 3. DFLP: Transportation cost function settings

Index	S1	S2	S3	S4	P1	P2
1	0.07	0.09	0.14	0.16	30	40
2	0.1	0.11	0.15	0.16	20	20
3	0.07	0.11	0.12	0.16	10	10

Table 4. DFLP: CPLEX results and LR results vs. HNP-MP results

Ins	CPub	CPlb	CPgap(%)	LRup	LRgap(%)	NP-CPub	NP-CPgap(%)
1a	604171	487561	23.9	518017	6.2	494705	1.5
1b	461617	448997	2.8	505079	12.5	454880	1.3
1c	403153	320345	25.8	337503	5.4	324313	1.2
1d	385607	313337	23.1	329818	5.3	317509	1.3
1e	302019	238229	26.8	256319	7.6	241663	1.4
1f	253196	232600	8.9	234998	1.0	234898	1.0
1g	350581	312283	12.3	329982	5.7	315783	1.1
2a	604171	553196	9.2	567793	2.6	565163	2.2
2b	570222	523013	9.0	555238	6.2	528691	1.1
2c	401353	365622	9.8	371864	1.7	372036	1.8
2d	385607	354859	8.7	361218	1.8	359298	1.3
2e	302020	275713	9.5	281922	2.3	275793	0.0
2f	300074	267871	12.0	274028	2.3	269229	0.5
2g	395128	353558	11.8	367467	3.9	358710	1.5
3a	604171	477633	26.5	485076	1.6	481618	0.8
3b	455698	447052	1.9	484173	8.3	454129	1.6
3c	401353	305522	31.4	314423	2.9	312196	2.2
3d	385607	295429	30.5	312687	5.8	301513	2.1
3e	302020	229240	31.7	237010	3.4	233042	1.7
3f	300074	222205	35.0	248134	11.7	226639	2.0
3g	395128	299684	31.8	312503	4.3	305789	2.0
Ave.			18.2		4.8		1.4

4 Computational Results

In this subsection, we report our computational experience on applying HNP-MP to the IHLP.

Testing Instances

We randomly generated a set of instances with typical settings to test our solution approach. The experiment settings are described as follows:

Map and Locations: We generate a rectangle map of 500×500 square miles. $|I|$ terminal locations and $|R|$ ramp locations are randomly generated

Table 5. Scale setting

Index	$|I|$	$|R|$	$|F|$
a	60	30	300
b	30	30	300
c	60	40	200
d	80	40	200
e	50	50	150
f	40	50	150
g	40	40	200
h	100	30	400
i	80	30	500
j	60	40	300

Table 6. Transportation cost function setting

Index	S1	S2	S3	S4	P1	P2
1	0.07	0.09	0.14	0.16	40	30
2	0.1	0.11	0.15	0.16	20	20
3	0.07	0.11	0.12	0.16	10	10

over the map. For each location pair, the distance between the two locations is the Euclidean distance on the map.

Flows: $|F|$ flows are randomly generated on the origin-destination location pairs with distance larger than 300 miles. $\forall f \in F, W_f$ is randomly generated by Uniform(10,50).

Cost Functions: The cost of open a certain hub r $(r \in R)$ is set to $|L| \cdot Uniform(30, 40)$. the cost rate per unit amount of unmoved flows is $0.18 \cdot (direct \ distance \ of \ the \ flow)$. The transportation cost function (cost rate per mile) between a location pair is set to a four piece linear concave function: for each location pair with truck movement, the cost rate (the slop for each piece of the cost function) is randomly generated over the range (S1, S2), and the three nondifferentiable points are set to $P1 \cdot |L|/|I|$, $2 \cdot P1 \cdot |L|/|I|$ and $3 \cdot P1 \cdot |L|/|I|$; for each location pair with truck movement, the cost rate is randomly generated over the range (S3, S4), and the three nondifferentiable points are set to $0.01 \cdot P2 \cdot |L|$, $0.02 \cdot P2 \cdot |L|$ and $0.03 \cdot P2 \cdot |L|$.

Routing: When the distance between location a and location b is less than a predetermined parameter N, $M_{ab} = 1$; otherwise $M_{ab} = 0$. This constraint on the possible movement, adopted by many big transportation companies nowadays, actually reduces the problem size and thus the solution time of the problem. Here, N is set to 200 miles.

Overall, we generated 30 testing instances of common properties and scales in real applications, with ten different groups of scale setting (as shown in Table 5) and three different groups of parameter setting (as shown in Table 6).

Algorithm Settings

We first test all instances through CPLEX 9.1 with default CPLEX parameter settings in our computer with P4 2.8G CPU and 1G memory. The reason we used CPLEX is that integer programming is one widely used approach to solve DFLPs in the literature. For all instances, we set a time limit of 2 hours. Then, we test the HNP-MP approach on these instances. Some algorithm setups are described as follows:

- Dual simplex method is used to solve the LP problem on the promising region in each iteration.
- CPLEX is used to evaluate each partial solution: we apply the value of the current best solution as the feasible bound, set the MIP tolerance gap to 0.005, and set the computation time limit to 2 minutes.
- Depending on the scale of the problem, in each iteration, take 8-20 partial solutions and fix two-three hubs open if backtracking is not needed.
- Stop the algorithm when the stopping conditions are met and no restart is used.
- To show the generality of our approach, no domain knowledge is used in the tests. (In applications, we may combine special knowledge to further improve the performance of our approach.)

For comparison purpose, we also test the Lagrangian relaxation (LR) approach [11] which is widely used to solve DFLPs in the literature. The settings of the LR approach are briefly described as follows:

- Constraints (8) (9) and (10) are relaxed in the LR subproblem, and multiplers corresponding to these constraints are added to the objective function.
- Subgradient algorithm is used to update the multiplers in each iteration of the LR procedure.
- In each iteration, after obtaining the integer solution of the LR subproblem, fix the y's variables and solve the original problem to get the feasible integer solution. The best feasible solution is reported when the algorithm ends.

Simplified LP Solution

If solving the LP problem becomes the bottleneck for the hybrid HNP-MP algorithm, we will use simplified LP problems to replace the original LP problems. Our numerical results show that solving the simplified LP problem requires much less computation time, and the LP solution of the simplified LP problem, which we call the simplified LP solution, can still be used to guide the sampling and partitioning steps. In the partial sample evaluation step, we refer to the original problem to get complete samples.

Four kinds of LP simplification techniques which are easily applicable to IHLPs are provided here. Each of these techniques can be applied alone and their combination might result in better performance in some cases.

Flow Reduction (FR): Flow reduction uses a part of, instead of all of, the flows in the formulation. Denote the set of selected flows to be $L' \subset L$, then the basic problem here is how to select the set L'. After selecting the set L', we also need to change the transportation cost function $C2$ accordingly. Particularly, we tend to change the scale of the cost function. The following procedure is used to calculate the adjusted cost function $C2'$: denote $S = \sum_{l \in L'} M_l / \sum_{l \in L} M_l$, then, $\forall (a,b) \in \{I,R\} \cup \{R,R\} \cup \{R,I\}, g \in \Re^+$, define $C2'_{ab}(g) = C2_{ab}(g/S)$. In the test examples, we uniformly select set L', and 60-70% of all the flows are selected in the simplified LP problem.

Route Restriction (RR): We try to set some restriction on the possible routes each flow can take to reduce the LP problem size, i.e., to fix some x's variables to be 0. In the original problem, each flow f ($\in F$) can be moved through a set of possible routes when all hubs are open, denoted as U_f. For each flow $f \in F$, we select a set of unpromising routes $U'_f \subset U_f$, and forbid all routes in U'_f for flow f. In the test examples, we use a simple heuristic to select forbidden routes: for each flow f and a route $u \in U_f$ with rail movement from k to m, if *distance between O_f and k > G*, or *distance between m and D_f > G* ($G < N$ is a predetermined parameter), the route u is forbidden for flow f. G is set to 130-150 miles.

Hub Division (HD): We divide the original hub set R into some subsets (two equal size subsets are used in the test examples), denoted as R_e, $e \in E$. $\prod_{e \in E} R_e = R$, and in a certain iteration, if a hub k is open for the promising region, we have $\forall e \in E, k \in R_e$. We can get $|E|$ subproblems by replacing the set R with set R_e in the original problem for each $e \in E$. Then the biased sampling weight and the LP solution-based partitioning index of a particular hub h ($\in R$) can be calculated according to the LP solutions of the subproblems. Denote LP solution of the z variable(s) in subproblem R_e ($e \in E$) to be z_e, then the LP solution z^* used in the HNP-MP approach can be calculated as follows: $\forall k \in R$. $z^*_k = (\sum_{e \in E: k \in R_e} z_e)/(\sum_{e \in E: k \in R_e} 1)$.

Cost Simplification (CS): In many real situations, the complicated transportation cost functions associated with each location pair increase the computational requirement of the LP problem dramatically. The CS technique used in this chapter is called individual cost function simplification. By reducing the number of pieces in the functions $C2$'s, we can largely reduce the difficulty in solving the LP problem. In the test examples, a two-piece linear function is used to approximate the original four-piece linear function.

Testing Results

For 21 out of 30 test instances, the LP relaxed problem can be solved relatively easily, and the computational results are shown in Table 7. (Here, CPub is the

Table 7. Computational results on instances with small LP time

Ins	CPub	CPlb	CPgap(%)	LRup	LRgap(%)	NP-MPub	NP-MPgap(%)
1a	604171	487561	23.9	518017	6.2	494705	1.5
1b	461617	448997	2.8	505079	12.5	454880	1.3
1c	403153	320345	25.8	337503	5.4	324313	1.2
1d	385607	313337	23.1	329818	5.3	317509	1.3
1e	302019	238229	26.8	256319	7.6	241663	1.4
1f	253196	232600	8.9	234998	1.0	234898	1.0
1g	350581	312283	12.3	329982	5.7	315783	1.1
2a	604171	553196	9.2	567793	2.6	565163	2.2
2b	570222	523013	9.0	555238	6.2	528691	1.1
2c	401353	365622	9.8	371864	1.7	372036	1.8
2d	385607	354859	8.7	361218	1.8	359298	1.3
2e	302020	275713	9.5	281922	2.3	275793	0.0
2f	300074	267871	12.0	274028	2.3	269229	0.5
2g	395128	353558	11.8	367467	3.9	358710	1.5
3a	604171	477633	26.5	485076	1.6	481618	0.8
3b	455698	447052	1.9	484173	8.3	454129	1.6
3c	401353	305522	31.4	314423	2.9	312196	2.2
3d	385607	295429	30.5	312687	5.8	301513	2.1
3e	302020	229240	31.7	237010	3.4	233042	1.7
3f	300074	222205	35.0	248134	11.7	226639	2.0
3g	395128	299684	31.8	312503	4.3	305789	2.0
Ave.			18.2		4.8		1.4

upper bound from CPLEX; CPlb is the lower bound from CPLEX MIP solver; CPgap is the gap between CPub and CPlb; LRub is the upper bound of the solution reported by the LR approach; LRgap is the gap between LRub and CPlb; NP-MPub is the upper bound from HNP-MP algorithm; NP-MPgap is the gap between NP-MPub and CPlb.)

For the rest of nine instances where the LP solving time of the original problem is large, the LP simplification techniques described in the previous section are tested. The computational results are summarized in Table 8. (FRub/RRub/HDub/CSub is the bound of the solution obtained by the HNP-MP approach with each kind of LP simplification technique, respectively, FRgap/RRgap/HDgap/CSgap is the gap between FRub/RRub/HDub/CSub and CPlb, respectively.)

Overall, the HNP-MP approach outperforms CPLEX and LR significantly in terms of the solution quality within the time limit.

5 Conclusion

In this chapter, we present a review of the nested partitions method and its successful applications. These applications show that the NP method provides

Table 8. Computational results on instances with large LP time

Ins	CPub	CPlb	CPgap(%)	LRub	LRgap(%)
1h	802648	665327	20.6	717057	7.8
1i	1025920	834316	23.0	910059	9.1
1j	601230	480885	25.0	522441	8.6
2h	802648	747940	7.3	777700	4.0
2i	1025920	943455	8.7	986426	4.6
2j	601230	547922	9.7	566821	3.4
3h	802648	655665	22.4	687713	4.9
3i	1025920	842773	21.7	854783	1.4
3j	601230	472398	27.3	488013	3.3
Ave.			18.43		5.23

Ins	FRub	FRgap(%)	RRub	RRgap(%)	HDub	HDgap(%)	CSub	CSgap(%)
1a	673163	1.2	673142	1.2	673904	1.3	675075	1.5
1b	842734	1.0	845456	1.3	847589	1.6	842691	1.0
1c	488195	1.5	488373	1.6	494604	2.9	488243	1.5
2a	763505	2.1	760147	1.6	760808	1.7	764629	2.3
2b	963960	2.2	966587	2.5	970019	2.8	965599	2.3
2c	562097	2.6	560644	2.3	560378	2.3	560510	2.3
3a	666734	1.7	668592	2.0	670286	2.2	666942	1.7
3b	850293	0.9	850228	0.9	857518	1.7	848557	0.7
3c	481328	1.9	481328	1.9	482605	2.2	480691	1.8
Ave.		1.67		1.69		2.08		1.67

us a flexible framework for solving discrete optimization problems effectively. We then develop a hybrid HNP-MP algorithm for the intermodal hub location problem (IHLP). IHLP is an important class of discrete facility location problems that has drawn great attention recently by researchers from both academia and industry. The computational results show that the HNP-MP approach is superior to the integer programming approach and the Lagrangian relaxation approach which are two conventional approaches for many facility location problems. The HNP-MP approach is also applicable for other discrete facility location problems.

Further research on improving the efficiency of HNP-MP approach to solve large-scale IHLPs and other discrete facility location problems may include the following aspects:

• In the standard HNP-MP approach, for evaluating each partial sample, we use the MIP solvers. For a specific problem, it is possible to apply or develop some efficient heuristics [1, 14] to replace the MIP solvers and improve the efficiency of the HNP-MP approach.

- For some problems, some local search algorithms (based on the top samples) can also be combined into the HNP-MP approach to achieve even better performance, i.e., to use the results of the local search procedure to guide the partitioning/backtracking step in each iteration.
- For multilevel discrete facility location problems [35], each step of the HNP-MP approach may need a more complicated design and deserves dedicated research work.
- As we considered a deterministic version of IHLPs in this chapter, we are also interested in the stochastic version of IHLPs [30]. Potentially, the NP method can be even more beneficial for stochastic problems.

References

1. Sue Abdinnour-Helm and M. A. Venkataramanan. Solution approaches to hub location problems. *Annals of Operations Research*, 78(0):31–50, 1998.
2. P. V. Balakrishnan and V. S. Jacob. Triangulation in decision support systems: Algorithms for product design. *Decision Support Systems*, 14:313–327, 1995.
3. P. V. Balakrishnan and V. S. Jacob. Genetic algorithms for product design. *Management Science*, 42:1105–1117, 1996.
4. Dimitris Bertsimas and Chung-Piaw Teo. From valid inequalities to heuristics: a unified view of primal-dual approximation algorithms in covering problems. *Operations Research*, 46(4):503–514, 1998.
5. J. F. Campbell. Integer programming formulations of discrete hub location problems. *European Journal of Operation Research*, 72:387–405, 1994.
6. K. L. Croxton, B. Gendron, and T. L. Magnanti. A comparison of mixed-integer programming models for nonconvex piecewise linear cost minimization problems. *Management Science*, 49(9):1268–1279, 2003.
7. Djurdje Cvijović and Jacek Klinowski. Taboo search: An approach to the multiple minima problem. *Science*, 267(5198):664–666, 1995.
8. Marco Dorigo. *Optimization, Learning and Natural Algorithms*. PhD thesis, Politecnico di Milano, Italy, 1992.
9. Marco Dorigo, Gianni Di Caro, and Luca M. Gambardella. Ant algorithms for discrete optimization. *Artificial Life*, 5(2):137–172, 1999.
10. Marshall L. Fisher. The lagrangian relaxation method for solving integer programming problems. *Management Science*, 50(12):1861–1871, 2004.
11. A. Geoffrion. Lagrangian relaxation for integer programming. *Mathematic Programming Study 2*, 2:82–114, 1974.
12. Fred W. Glover and Manuel Laguna. *Tabu Search*. Kluwer Academic Publishers, Boston, MA, 1997.
13. D. E. Goldberg. *Genetic Algorithm in Search, Optimization and Machine Learning*. Addison-Wesley, Reading, MA, 1989.
14. Trevor S. Hale and Christopher R. Moberg. Location science research: A review. *Annals of Operations Research*, 123:21–35, 2003.
15. N. G. Hall and D. S. Hochbaum. A fast approximation algorithm for the multicovering problem. *Discrete Applied Mathematics*, 15(1):35–40, 1986.
16. S. Kirkpatrick, C. D. Gelatt, and M. P. Vecchi. Optimization by simulated annealing. *Science*, 220(4598):671–680, 1983.

17. R. Kohli and R. Krishnamurti. Optimal product design using conjoint analysis: Computational complexity and algorithms. *European Journal of Operations Research*, 40:186–195, 1989.
18. Claude Lemaréchal. Lagrangian relaxation In Michael Jünger and Denis Nad def, editors, *Computation Combanatorial Optimization*, volume 2241 of *Lecture Notes in Computer Science*, pages 112–156. Springer-Verlag, Berlin Heidelberg, 2001.
19. Melanie Mitchell. *An Introduction to Genetic Algorithms*. MIT Press, Cambridge, MA, 1998.
20. M. E. O'kelly and D. L. Bryan. Hub location with flow economies of scale. *Transportation Research Part B*, 32(8):605–616, 1998.
21. Sigurdur Ólafsson and Jaekyung Yang. Intelligent partitioning for feature selection. *INFORMS Journal on Computing*, 17(3):339–355, 2005.
22. Liang Pi, Yunpeng Pan, and Leyuan Shi. Hybrid nested partitions and mathematical programming approach and its applications. *IEEE Transactions on Automation Science & Engineering*, 5(4): 573–586, 2008.
23. Illia Racunica and Laura Wynter. Optimal location of intermodal freight hubs. *Transportation Research Part B*, 39:453–477, 2005.
24. Leyuan Shi and Shuli Men. Optimal buffer allocation in production lines. *IIE Transactions*, 35:1–10, 2003.
25. Leyuan Shi, Robert R. Meyer, Mehmet Bozbay, and Andrew J. Miller. A nested partitions framework for solving large-scale multicommodity facility location problems. *Journal of Systems Science and Systems Engineering*, 13(2):158–179, 2004.
26. Leyuan Shi and Sigurdur Ólafsson. Nested partitions method for global optimization. *Operations Research*, 48(3):390–407, 2000.
27. Leyuan Shi and Sigurdur Ólafsson. Nested partitions method for stochastic optimization. *Methodology and Computing in Applied Probability*, 2(3):271–291, 2000.
28. Leyuan Shi and Sigurdur Ólafsson. *Nested Partitions Optimization: Methodology and Applications*, volume 109 of *International Series in Operations Research & Management Science*. Springer, 2007.
29. Leyuan Shi, Sigurdur Ólafsson, and Qun Chen. An optimization framework for product design. *Management Science*, 47(12):1681–1692, 2001.
30. Lawrence V. Snyder. Facility location under uncertainty: A review. *IIE Transactions*, 38(7):547–564, 2006.
31. P. J. M. van Laarhoven and E. H. L. Aarts. *Simulated annealing: theory and applications*. Kluwer Academic Publishers, Norwell, MA, 1987.
32. Laurence A. Wolsey. *Integer Programming*. Wiley-Interscience, 1998.
33. Laurence A. Wolsey and George L. Nemhauser. *Integer and Combinatorial Optimization*. Wiley-Interscience, 1999.
34. C. K. Wong and M. C. Easton. An efficient method for weighted sampling without replacement. *SIAM Journal on Computing*, 9(1):111–113, 1980.
35. Jiawei Zhang. Approximating the two-level facility location problem via a quasi-greedy approach. *Mathematical Programming, series A*, 108(1):159–176, 2006.

Part III

Supply Chain Operation

Event-Time Models for Supply Chain Scheduling

Ömer S. Benli

Department of Information Systems
California State University, Long Beach, CA 90840
obenli@csulb.edu

Summary This study presents a modeling paradigm for scheduling problems in supply chains. The constituents of a supply chain need to cooperate, rather than compete, in order to achieve maximum respective benefits. To analyze this, it is essential to have a concise but comprehensive formulation of this problem. This formulation, in addition to being computationally viable, must account for special characteristics of supply chains. It is argued that *lot streaming* provides an appropriate paradigm for scheduling problems in supply chains. It is shown that *event-time models* provide a general formulation for lot streaming problems. Furthermore, logic-based modeling framework of constraint programming makes it possible to handle special requirements of these models. This is illustrated by a small example. It is also shown that the fundamental results for single-job lot streaming problems can be systematically obtained as special cases of this generalized formulation. This demonstrates that event-time formulation accurately models all of the features of the problem; and that starting with a concise but comprehensive formulation, solutions for special cases of the problem can be obtained systematically. To that end, a number of additional special case formulations of single-job lot streaming problem is presented. An appendix presents a schema for lot streaming problem classifications.

1 Introduction

Today's supply chains differ from the integrated logistics systems of the past primarily because of the autonomous nature of its constituents. In the traditional approach to integrated logistics, the entire system is treated as a monolithic entity, whereas today's supply chains are usually composed of components that are autonomous entities with competing interests. These constituents of the supply chain, such as manufacturers, wholesalers, and retailers, will be better off if they operate in cooperation. Similar situation arises in *supply contracts*. [2] Cooperation via supply contracts results in a win-win outcome for all parties concerned. The same is true for supply chain scheduling, which is concerned with timing and amount of material handling moves throughout the supply chain. Supply chain scheduling has replaced the

W. Chaovalitwongse et al. (eds.), *Optimization and Logistics Challenges in the Enterprise*, Springer Optimization and Its Applications 30, DOI 10.1007/978-0-387-88617-6_9, © Springer Science+Business Media, LLC 2009

integrated production planning and scheduling systems of traditional logistics; and cooperation is essential in scheduling operations in supply chains.

At *"The Factory Scheduling Conference"* held at Carnegie Institute of Technology in May 1961, William Pounds argued that a production scheduling problem was not "... a visible one in many firms because other parts of the firm have absorbed much of the impact of poor scheduling" [15]. If the due dates were not routinely met, it was customary to give protracted due dates; if there were a bottleneck machine, the problem was solved by acquiring another machine. Although the changing nature of business competitiveness demands highly advanced production scheduling systems, sufficient emphasis is still not being given to scheduling in supply chains. In the indexes of two recent handbooks on supply chain management, 6 out of 765 pages in Graves and de Kok [7] and only one page out of 817 pages in Simchi-Levi et al. [17] directly refer to scheduling. de Kok and Fransoo [5] suggest the following explanation: *"Decisions with regard to the different components of planning of supply chain operations have traditionally been analyzed independently from one another by the researchers. Research addressing the scheduling problem, the (multi-echelon) inventory problem, and the aggregate capacity planning problem have hardly been interconnected while maintaining their own characteristics."*

Production planning problems customarily are posed as periodic review processes. On the other hand, detailed scheduling problems extend over relatively shorter planning horizons and require continuous time domain. Among the first to comment on this incongruity was Karmarkar [10]; comparing the different modeling perspectives in theory and practice, he stated that:

> [Analytical models] often treat capacity in terms of loading time buckets. However, in *practice*, it is much easier to think in terms of time lines and events and intervals. In some models, such as scheduling with Gantt charts, we use this kind of modeling, but we lack ways of dealing with decomposition or composition of these models. As a result, capacity and planning models are often formulated very differently. Perhaps this is one reason that time interval and release oriented methods like MRP and DRP are used in practice for planning even though they are completely unable to actually deal with resource allocation decisions.

Likewise, Bill Maxwell questioned the use of models that lump operations and events into large time buckets and suggested formulation of *"event-time"* models which are in essence "ordered sequence of time and data" [13].

2 Lot Streaming Paradigm

A major problem in production planning is how to handle sequencing requirements on resources, whereas models of traditional machine scheduling cannot handle lot sizing. *Lot streaming* may provide the necessary conceptual framework for integrating lot sizing and machine scheduling. Basically, lot streaming is moving some portion of a process batch ahead to begin a downstream operation. Classic machine scheduling theory envisions an operation as an elemental task to be performed. It is assumed that "[t]he processing times of successive operations of a particular job cannot be overlapped. A job can be in process on at most one operation at a time" [4]. This assumption is justified when jobs are monolithic entities. But for scheduling production lots, where each lot consists of a number of units, this assumption may be overly restrictive. The processing time of such a lot is composed of a (usually "detached") setup time and the sum of the processing times of each unit in the lot. For instance, when the machine is available, it is not reasonable to delay its setup until *all* the items arrive from the upstream machine.

Lot streaming, in this context, was introduced in papers by Baker [1] and Trietsch [18]. In a later joint work, they discuss the practical importance of this approach [19]. A number of manufacturing management innovations, such as *group technology* (leading to cell based manufacturing, resulting in shorter lead times and reduced work in progress inventories), *just-in-time systems* ("lot size of one"), and OPT/*synchronous manufacturing* (transfer *vs.* process batches) paved the way to lot streaming theory, which provides a rigorous analytical treatment of these issues. In the recent years lot streaming attracted considerable attention in machine scheduling research. Chang and Chiu [3] present a comprehensive review of lot streaming literature.

There does not seem to exist a unified approach to solving lot streaming problems. Probably this is due to the fact that a general model for lot streaming problems does not exist. The event-time modeling approach, introduced in the next section, attempts to provide such a paradigm.

3 Event-Time Models

In event-time modeling *events* are ordered sequences of material handling moves (interstage material transfers). There are two types of events:

exogenous events whose time of occurrence are given (as parameters of the problem), such as demand occurrences or order deadlines, and

endogenous events whose occurrence times are decision variables of the model, such as WIP movements.

Essentially, the model is formulated as a multi-item periodic review process with *variable period lengths*. Let

M be the index set of all items i, and

M_e be the index set of items i that have external (independent) demand in addition to possible internal (dependent) demand. The items $i \in M_e$ are called *end items*.

Also define

$P(i)$ as the predecessor set of item i, which is the index set of items that are required in the procurement of item i, and

$\mathbb{S}(i)$ as the successor set of item i, which is the index set of items that require
item i in their procurement.

Interstage transfers can occur at time points T_t, $t \in \mathbb{T}$, where $\mathbb{T} = \{1, \ldots, n\}$ is the index set of all such time points. These time points are decision variables of the problem except for the points corresponding to exogenous
events: the given deadlines or other external demand occurrences. Let a sub-
set of $\mathbb{T} \supset \mathbb{T}_e = \{\tau_1, \tau_2, \ldots, \tau_m\}$ denote a set of time points corresponding to
exogenous events.

A period, which has variable length, is defined as the time interval
$[T_{t-1}, T_t], t \in \mathbb{T}$, and is referred to as period t. The section of the process depicting the inventory and procurement operations corresponding to
item i is called *stage i*. In each period, stage i contains two *buffers*. During period t, the items that will be needed in the procurement of item i
are inventoried in the *input buffer* and the procured item i; that which
is not yet transferred to the input buffers of successor items is stored
in the *output buffer*. The corresponding material flow is depicted in
Figure 1.

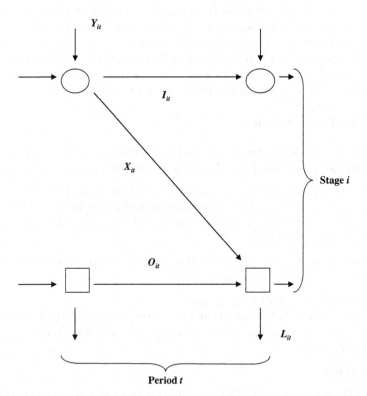

Fig. 1. Material flow through buffers

We can now define the following variables:

Y_{it} is the amount added to the input buffer of item i in period t, which consists of all predecessor items, in required proportions, needed for the procurement of item i. One unit of item i requires α_{ji} units of item j for all $j \in \mathbb{P}(i)$.

I_{it} is the input inventory level of item i at the end of period t (amount in the input buffer at T_t).

X_{it} is the amount of item i procured during $[T_{t-1}, T_t]$; that is, amount of procurement during period t in stage i.

O_{it} is the output inventory level of item i at the end of period t (amount in the output buffer at T_t).

L_{it} is the sum of external (independent) demand for item i at period t and the amount of item i made available for its successor items at time T_t, which is the size of the transfer batch at the end of period t in stage i: $L_{it} = D_{it} + \sum_{j \in \mathbb{S}(i)} \alpha_{ij} Y_{j,t+1}$.

Whether or not a transfer is to take place in stage i at the end of period t is indicated by a set of logical variables,

$Z_{it} = 1$, if there is a transfer batch from stage i to all stages $j \in \mathbb{S}(i)$ at the end of period t,

$\quad\; = 0$, otherwise.

There are two sets of *inventory balance equations*, for input and output buffers, respectively:

$$I_{i,t-1} + Y_{i,t} = I_{it} + X_{it}, \; \forall \, i, t,$$

$$O_{i,t} + X_{it} = O_{i,t+1} + L_{it}, \; \forall \, i, t,$$

$$\text{where } L_{it} = D_{it} + \sum_{j \in \mathbb{S}(i)} \alpha_{ij} Y_{j,t+1}, \; \forall \, i, t,$$

Recall that demand for end items occur only at time points $t \in \mathbb{T}_e$. D_{it} is the given external (independent) demand for item $i \in \mathbb{M}_e, t \in \mathbb{T}_e$, and $D_{it} = 0$ for all $i \notin \mathbb{M}_e, t \notin \mathbb{T}_e$.

There are resource constraints restricting the amount that can be procured during a period,

$$\sum_i \rho_{ki} X_{it} \leq r_k \left[T_t - T_{t-1} \right], \; \forall \, k \in \mathbb{K}, \; t \in \mathbb{T}$$

where,

$\mathbb{K}$ is the index set of resources used by items,

ρ_{ki} is the amount of resource $k \in \mathbb{K}$ required for procuring a unit of item i, and

r_k is the total availability of resource k per unit time.

Big$-M$ constraints, $L_{it} \leq MZ_{it}$, $\forall\, i \in \mathbb{M}, t \in \mathbb{T}$, can enforce the transfers and can take place only if they are indicated to do so.

Transfer times among stages are assumed to be negligible. If these times are significant, as it would be in goods movement in a supply chain, then the interstage transfers can be formulated as a separate "stage" with appropriate resource requirements.

The basic framework of logic-based modeling [8] makes event-time models computationally viable utilizing the specific features of constraint programming. In addition to its capability to handle big$-M$ constraints quite efficiently, constraint programming makes it possible to handle the exogenous events whose times are fixed and the endogenous events whose times of occurrence are decision variables in the same model.

ILOG's constraint programming software, OPL STUDIO [9], allows for *variable subscripts*, making it possible to handle the following *conditional constraint*. Let ν_ℓ denote the period at the end of which an exogenous event ℓ occurs. Recalling that $\tau_\ell \in \mathbb{T}_e = \{\tau_1, \tau_2, \ldots, \tau_m\}$ are the times at which the exogenous events occur, the *conditional constraint*

$$(\nu_\ell = t) \to (T_t = \tau_\ell), \quad t \in \mathbb{T} = \{1, \ldots, n\},$$

for each $\ell = 1, \ldots, m$, ensures that the exogenous event ℓ occurs at T_t. The following example illustrates how this constraint is incorporated in the program.

Example

Consider a planning horizon of 10 time units ("horizon"). Assume that there will be five time periods in the problem ("nbPeriods"). Suppose that there are following two exogenous events ("nbEvents"): external demands occurring at times three and eight. Then the following lines in the program make sure that two of the periods end exactly at time points three and eight. Timing of remaining three periods are determined based on other constraints and optimality criterion.

In terms of the notation and terminology used in model formulation, this example assumes that there will be five periods (nbPeriods, or $n = 5$) in the model. The end points of these periods correspond to *events* times, $T_1, T_2, \ldots, T_5$. They can assume any (integer) values within the planning horizon of 10 time units ("horizon"). Two of these events are exogenous (nbPeriods, $m = 2$) whose times are given as are fixed at $\tau_1 = 3\,and\,\tau_2 = 8$. The other three are endogenous events whose times are to be determined.

For example, a feasible ordering would be $\{T_1, T_2 = 3, T_3, T_4 = 8, T_5\}$; another would be $\{T_1 = 3, T_2, T_3, T_4, T_5 = 8\}$. These orderings and possible assignments are shown as Solution[22] and Solution[56] in the following illustration of the OPL code and its partial output.

```
int nbPeriods = 5;
int nbEvents = 2;
int horizon = 10;

range period 1..nbPeriods;
range exoEvent 1..nbEvents;

int givenTime[exoEvent] = [3,8];

var int time[period] in 1..horizon;
var int index[exoEvent] in 1..nbPeriods;

solve{

  forall(t in 2..nbPeriods)
     time[t-1] < time[t];

  forall(i in exoEvent)
  time[index[i]] = givenTime[i];
}
;
```

```
      Solution [22]                    Solution [56]

      time[1] = 1                      time[1] = 1
      time[2] = 3                      time[2] = 2
      time[3] = 4                      time[3] = 3
      time[4] = 8                      time[4] = 7
      time[5] = 9                      time[5] = 8

      index[1] = 2                     index[1] = 3
      index[2] = 4                     index[2] = 5
```

This section presented a generalized model for lot streaming problems. Furthermore, it is shown that constraint programming framework provides a computationally viable approach to these problems. The next section shows that the two fundamental results for single-job lot streaming problems can be systematically obtained as special cases of this generalized formulation. The purpose of that section is twofold. First, it is to validate, at least for a special class of lot streaming problems, that event-time formulation accurately models all of the features of the problem. Secondly, it maintains that, starting with a concise but comprehensive formulation, solutions for special cases of the problem can be obtained systematically. To that end, a number of additional special case formulations of single-job lot streaming problem is presented.

4 Single-Job Lot Streaming Problems

This section will show that two fundamental results for single-job lot streaming problems can be systematically obtained as special cases of an event-time modeling approach presented in the previous section. The special cases of single-job lot streaming problem are analyzed in [6, 19]. Consider the following operational situation. At each *stage* of production an *item* is produced. For example, the raw material enters stage 1 in which item 1 is produced. Item 1 goes into stage 2 for the production of item 2, and so on, until the finished good, item m, is produced in stage m. It has a very simple "series" product structure in which a single unit of item i is required for a unit production of item $i + 1$.

In each stage, there are s_i time points at which transfers in between stages can occur. Since, in general, these time points may not necessarily occur at the same instances in each stage, there are, at most, $n = \sum_{i=1}^{m} s_i$ time points at which the transfers can occur. Denote these time points by T_t, $t = 1, \ldots, n$ and let T_0 be the time at which the job starts on stage 1. The *periods* are defined by the time intervals $[T_{t-1}, T_t]$, $t = 1, \ldots, n$. Note that $T_t \leq T_{t+1}$, $t = 1, \ldots, n-1$, are decision variables in the model. Define the following variables:

X_{it} : Amount of item i produced during $[T_{t-1}, T_t]$, i.e., amount of production in period t in stage i.

Y_{it} : Amount of item i made available for the production of item i at time T_{t-1}, i.e., the size of the transfer batch at the end of period $t - 1$ in stage $i - 1$.

In stage i, the unprocessed material (item $i-1$) that has been transferred from stage $i-1$ is stored in an input buffer and the processed material (item i), if not transferred to stage $i+1$, is stored in an output buffer. In order to achieve the inventory balance, define two sets of inventory variables:

I_{it} : Input inventory level (amount in the input buffer at T_{t-1}) in stage i at the beginning of period t.

O_{it} : Output inventory level (amount in the output buffer at T_{t-1}) in stage i at the beginning of period t.

The earliest time period in stage i in which production can take place is period $t = i$ and the latest time period in which production can take place is $t = n - m + i$. For stage i to begin processing, a sublot must have been processed at $(i - 1)$ upstream stages. That is, at least, $(i - 1)$ transfers are required until the first sublot reaches stage i. Since each transfer takes place at the end of a stage, the earliest time period at which processing can start in stage i is period i. Similarly, at least $(m - i)$ transfers, and thus, time periods, are required for the last sublot in stage i to complete its processing in the downstream stages. Define the periods $t = i, \ldots, n - m + i$ as *active periods* for stage i. Thus, there are $n - m + 1$ active periods for each stage.

Finally, define a set of binary variables, one for each active period in every stage, in order to indicate whether or not a transfer takes place at stage i at the end of period t:

$Z_{it} = 1$, if there is a transfer batch from stage i to stage $i+1$ at the end of period t,

$= 0$, otherwise.

The transfers can take place only if they are indicated to do so,

$$Y_{it} \leq MZ_{it}, \ i = 1, \ldots, m, \ t = i, \ldots, n - m + i \tag{1}$$

where M is a very large number.

The total number of transfer batches in stage i is bounded by s_i,

$$\sum_{t=i}^{n-m+i} Z_{it} \leq s_i, \ i = 1, \ldots, m. \tag{2}$$

There are two sets of *inventory balance* equations, for input and output buffers, respectively,

$$I_{i,t-1} + Y_{i,t} = I_{it} + X_{it}, \ i = 1, \ldots, m, \ t = i, \ldots, n - m + i, \tag{3}$$
$$O_{i,t} + X_{it} = O_{i,t+1} + L_{i+1,t+1}, \ i = 1, \ldots, m, \ t = i, \ldots, n - m + i \tag{4}$$

There are *capacity* constraints restricting the amount that can be produced in a period,

$$\rho_i X_{it} \leq T_t - T_{t-1}, \ i = 1, \ldots, m, \ t = i, \ldots, n - m + i \tag{5}$$

with $T_0 = 0$.

Finally, the nonnegativity, integrality, and initial restrictions complete the model:

$$T_t \geq 0, \ t = 1, \ldots, n \tag{6}$$
$$X_{it}, \ Y_{it}, \ I_{it}, \ O_{it} \geq 0, \ i = 1, \ldots, m, \ t = i, \ldots, n - m + i \tag{7}$$
$$Y_{it} = 0 \text{ or } 1, \ i = 1, \ldots, m, \ t = i, \ldots, n - m + i. \tag{8}$$

and $T_0 = I_{i,i-1} = I_{i,h-m+i} = O_{i,i-1} = O_{i,h-m+i} = 0$, $\sum_{t=1}^{h-m+1} Y_{0,t} = \sum_{t=m}^{n} Y_{mt} = U$.

If the entire job, U units, is available at time T_0, then $Y_{0,0} = U$ and $\sum_{t=2}^{n-m+1} Y_{0,t-1} = 0$. For each stage and for each active period, production (X_{it}), transfer (Y_{it}), input buffer (I_{it}), and output buffer (O_{it}) variables are defined. Since the ending inventories are zero, two less inventory variables are needed for each stage. Thus, together with the n variables (T_t) defining the end points of the periods, there are altogether $4m(n-m+1) - 2m + n = 2m(2(n-m)-1) + n$ continuous variables and $m(n-m+1)$ binary variables ($Z_{i,t}$), where $n = \sum_{i=1}^{m} s_i$. There are two inventory balance and two capacity constraints for every active period and an additional m constraints restricting the maximum number of transfers in each stage, resulting in $4(n-m+1) + m = 4n - 3m + 4$ constraints.

4.1 Extensions of the Single-Job Model

The constraints described in the above general formulation describe the physics of the problem of lot streaming a single job in a flow shop. It should be noted that the setup times do not complicate the formulation, as long as the setup is *not* "attached" to the job and can be done while the job is not physically at the machine. Since there is a single job, it can be assumed that the setups are completed prior to the start of the first sublot in each machine.

In the case that there is a constraint on the total number of transfers, s, that can take place, rather than a separate restriction for each stage, it is sufficient to replace the set of constraints (2) by

$$\sum_{i=1}^{m} \sum_{t=i}^{n-m+i} Z_{it} \le s. \tag{9}$$

By defining a specific objective function, restricting values of some variables, and appending additional configurational constraints, we can obtain the specific formulations of different problem types. Some examples are given below.

Minimization of Make Span

Since there is a single job, this criterion is equivalent to minimizing the completion time of this job. The completion time of a job is determined by the time at which its last unit completes processing in the last machine. Thus, there is no need to have more than one sublot on the last machine, that is,

$$s_m = 1, \tag{10}$$

and thus $Z_{mt} = 0$, $t = m, \ldots, n-1$, and $Z_{mn} = 1$. The problem is

$$\min T_n,$$

subject to constraints (3) through (8) and (10).

Special cases of this problem are discussed in [6, 19].

Minimization of Mean Flow Time

In the traditional scheduling theory, the spirit of this criterion is to maximize "the rate at which jobs are completed." In the context of lot streaming with finite number of sublots, a reasonable interpretation would be to "weigh" each sublot with its size, i.e., the number of units in the sublot. These are not arbitrary weights that can be imposed in the problem data but intrinsic an property of the problem instance. Additional weights may be imposed as in the case of weighted mean flow time, only in the presence of multiple jobs.

With this interpretation, the minimization of mean flow time in a single-job lot streaming problem is minimization of total sublot completion time where each sublot is weighed by its size. Hence the problem becomes,

$$\min \sum_{t=m}^{h} Y_{mt} T_t,$$

subject to constraints (3) through (8).

Note that the objective function is quadratic. When the sublots are *consistent* (i.e., $L_{ij} = L_j$, $\forall\ i, j$; see Appendix). A quadratic programming formulation is given in [11]. In [16] specific results for the special cases of this problem are presented.

Minimization of Mean Unit Completion Time

This objective aims to maximize the rate at which the units complete their processing in the last stage. Practically, this means that there are as many transfer batches at the end of last stage as there are units in the lot. Since stage $(m-1)$ sends s_{m-1} sublots to stage m, we can think of stage m to process, uninterrupted, s_{m-1} sublots. Each unit, as it completes processing in the last stage, leaves the system. Each unit's contribution to the objective function value is its completion time on stage m. Consider the sublot that leaves stage m at time T_t of size Y_{mt}. This sublot must have started its processing on stage m at time $(T_t - \rho_m Y_{mt})$, and the first unit in this sublot must have been completed at time $(T_t - p_m Y_{mt} + \rho_m)$. In order to find the sum of completion times of all the units in this sublot, we have to compute the sum of Y_{mt} units in an *arithmetic progression*, in which each number differs from the previous number by a *common difference* of ρ_m and which has $T_t - \rho_m(Y_{mt} - 1)$ as its first term. This sum is equal to $Y_{mt}[T_t - \frac{1}{2}\rho_m(Y_{mt} - 1)]$. Since the earliest time at which a sublot can arrive to the last stage is T_m, the objective function becomes,

$$\min \sum_{t=m}^{n} Y_{mt}[T_t - (Y_{mt} - 1)\rho_m/2],$$

subject to constraints (3) through (8).

See [16] for specific results for special cases of this problem.

Minimization of Number of Transfer Batches
Subject to a Deadline

Suppose the job *must* be completed by a deadline $\bar{d}$. If the objective is to achieve this deadline using a minimum number of transfer batches, then set

$$T_n = \bar{d} \tag{11}$$

and optimize the objective function,

$$\min \sum_{i=1}^{m} \sum_{t=i}^{n-m+i} Z_{it},$$

subject to constraints (3) through (8) and (11).

Maximization of Throughput during a Planning Horizon

Suppose we have a planning horizon of $[0, \bar{d}]$. Given the maximum number of transfers between the stages, the problem is to process as many units as possible. Let U be the variable amount to be processed, then the objective function becomes,

$$\max U,$$

subject to constraints (3) through (8) and (11).

The next two subsections show that it is possible to obtain the well-known results available in the literature as a direct consequence of the special cases of the above general formulation. Notation used for the classification scheme for lot streaming problems is explained in the Appendix.

4.2 Single-Job $F2|s|C_{max}$ Problem

This problem is discussed in [1, 14, 18], and shown that the optimal sublot sizes are given by a geometric pattern. Using the notation introduced above, they have proved that the optimal sublot sizes are,

$$X_{i,t+i-1}^* = [(\pi^{t+i-2} - \pi^{t+i-1})/(1 - \pi^s)], \ i = 1, 2; \ t = 1, \dots, s,$$

where $\pi \equiv \rho_2/\rho_1$. Here, it will be shown that the same results can be obtained by restrictions on the general model.

The fractional values for sublot sizes are allowed; then, without loss of generality, for ease of notation, we can set $U \equiv 1$. Since the objective is make span minimization, $s_2 = 1$ and let $s_1 \equiv s$. Then $n = \sum_{i=1}^{2} s_i = s + 1$, and the *active periods* for stages 1 and 2 are, respectively, $\{1, 2, \dots, s\}$ and $\{2, 3, \dots, s + 1\}$. Hence, we can set $Z_{1t} = 1$, $t = 1, \dots, s$. The constraints (1) and (2), for $i = 1$, are redundant and therefore can be omitted. The resulting problem is a linear program, and therefore *the single job $F2|s, idlg|C_{\max}$ is polynomially solvable*.

The formulation can be further simplified. Since $\{Y_{1t} \geq 0, \ t = 1, \dots, s\}$ are no longer restricted, and a *transfer* can take place at the end of every active period, there is no need for the output buffer in stage 1. The amount produced in stage 1 during any (active) period can be transferred to (the input buffer of) stage 2 at the end of that period. Setting $O_{1t} = 0$, $t = 1, \dots, s$, in constraints (4) implies $X_{1t} = Y_{1t}$, $t = 1, \dots, s$, thus eliminating the Y_{1t} variables from the formulation.

The resulting inventory balance equations can be equivalently represented as,

$$\sum_{k=1}^{t} X_{1k} \geq \sum_{k=1}^{t} X_{2,k+1}, \; l = 1, \ldots, s \tag{12}$$

$$\sum_{t=2}^{s+1} X_{2t} = 1 \tag{13}$$

Moreover, since the objective function is minimizing T_{s+1}, the capacity constraints (5) simply require

$$T_{s+1} = \rho_1 X_{11} + \sum_{t=2}^{s} \left[\max_{1 \leq i \leq 2} \{ \rho_i X_{it} \} \right] + \rho_2 X_{2,s+1}$$

Thus the LP formulation of the single-job $F2|s, idlg|C_{\max}$ problem becomes,

$$\min \; T_{s+1} = \rho_1 X_{11} + \sum_{t=2}^{s} [\max_{1 \leq i \leq 2} \{ \rho_i X_{it} \}] + \rho_2 X_{2,s+1} \tag{14}$$

subject to:

$$\sum_{k=1}^{t} X_{1k} \geq \sum_{k=1}^{t} X_{2,k+1}, \; t = 1, \ldots, s \tag{15}$$

$$\sum_{t=2}^{s+1} X_{2t} = 1, \tag{16}$$

$X_{it} \geq 0, \; i = 1, 2; \; t = i, \ldots, s + i - 1.$

We need the following result.

Result: There exists an optimal solution $\{X_{it}^*, \; i = 1, 2; \; t = i, \ldots, s+i-1\}$ to the single-job $F2|s, idlg|C_{\max}$ problem such that

$$X_{1t}^* = X_{2,t+1}^*, \; t = 1, \ldots, s,$$

and,

$$\rho_1 X_{1t}^* = \rho_2 X_{2t}^*, \; t = 2, \ldots, s.$$

Proof: Clearly, this holds for $\rho_1 = \rho_2$. Therefore, assume $\rho_1 \neq \rho_2$. For convenience in notation let,

$$z \equiv T_{s+1},$$
$$x_t \equiv X_{1t}, \; t = 1, \ldots, s$$
$$x_{s+1} \equiv 0$$
$$y_1 \equiv 0$$
$$y_t \equiv X_{2t}, \; t = 2, \ldots, s + 1$$
$$a \equiv \rho_1$$
$$b \equiv \rho_2$$

Rewriting the problem in this notation,

$$\min z = \sum_{t=1}^{s+1} [\max\{ax_t, by_t\}] \tag{17}$$

subject to:

$$\sum_{k=1}^{t} x_k \geq \sum_{k=1}^{t} y_{k+1}, \ t = 1, \ldots, s \tag{18}$$

$$\sum_{t=1}^{s+1} y_t = 1, \tag{19}$$

with $x_t, y_t \geq 0$, $t = 1, \ldots, s+1$.

An equivalent representation of the problem is,

$$\min z = ax_1 + \sum_{t=2}^{s} w_t + by_{s+1}$$

subject to:

$$w_t - ax_t \geq 0, \ t = 2, \ldots, s, \tag{20}$$
$$w_t - by_t \geq 0, \ t = 2, \ldots, s, \tag{21}$$

and constraints (18), (19) and $x_t \geq 0$, $t = 1, \ldots, s$; $y_t \geq 0$, $t = 2, \ldots, s + 1$; $w_t \geq 0$, $t = 2, \ldots, s$.

Let the dual variables $\gamma_t, \delta, \alpha_t, \beta_t$, respectively, correspond to the primal constraints (18) through (21). The the dual problem is:

$$\max \delta$$

subject to:

$$\sum_{k=t}^{s} \gamma_k - a\alpha_t \geq 0, \ t = 1, \ldots, s, \tag{22}$$

$$-\sum_{k=t}^{s} \gamma_k + \delta - b\beta_{t+1} \geq 0, \ t = 2, \ldots, s, \tag{23}$$

$$\alpha_t + \beta_t \leq 1, \ t = 2, \ldots, s, \tag{24}$$

and $\alpha_t, \beta_t \geq 0$, $t = 2, \ldots, s$; $\gamma_t \geq 0$, $t = 1, \ldots, s$; and setting $\alpha_1 = 1, \beta_{s+1} = 1$.

The resulting complementary slackness conditions are:

$$x_t \left[\sum_{k=t}^{s} \gamma_k - a\alpha_t \right] = 0, \ t = 1, \ldots, s, \tag{25}$$

$$y_t \left[-\sum_{k=t-1}^{s} \gamma_k + \delta - b\beta_t \right] = 0, \ t = 2, \ldots, s+1, \tag{26}$$

$$w_t[1 - \alpha_t - \beta_t] = 0, \ t = 2, \ldots, s, \tag{27}$$

$$\alpha_t[w_t - ax_t] = 0, \ t = 2, \ldots, s, \tag{28}$$

$$\beta_t[w_t - by_t] = 0, \ t = 2, \ldots, s, \tag{29}$$

$$\gamma_t \left[\sum_{k=1}^{t} x_t - \sum_{k=1}^{t} y_{t+1} \right] = 0, \ t = 1, \ldots, s. \tag{30}$$

To show that there exists an optimal solution to the primal such that

$$x_t > 0, \ t = 1, \ldots, s, \tag{31}$$

$$y_t > 0, \ t = 2, \ldots, s+1, \tag{32}$$

$$x_t = y_{t+1}, \ t = 1, \ldots, s, \tag{33}$$

$$ax_t = by_t, \ t = 2, \ldots, s, \tag{34}$$

it is sufficient to show the existence of a dual feasible solution that satisfies the complementary slackness conditions.

Because of (31) and (32), the conditions (25) and (26), respectively, require,

$$\sum_{k=t}^{s} \gamma_k = a\alpha_t, \ t = 1, \ldots, s, \tag{35}$$

$$\sum_{k=t}^{s} \gamma_t = \delta - b\beta_{t+1}, \ t = 1, \ldots, s. \tag{36}$$

Finally, because of (31), (32), and (34), condition (3) requires

$$\alpha_t + \beta_t = 1. \tag{37}$$

Recalling that $\alpha_t, \beta_{s+1} = 1$ and using Equations (35) and (36),

$$a\alpha_t = \delta - b\beta_{t+1}, \ t = 1, \ldots, s.$$

Then, using Equation (37),

$$-a\alpha_t + b\beta_{t+1} = b - \delta, \ t = 1, \ldots, s - 1,$$
$$-a\alpha_s = b - \delta.$$

Let $\pi = b/a$,

$$\alpha_t = \delta/a - \pi + \pi\alpha_{t+1}, \ t = 1, \ldots, s - 1,$$
$$\alpha_s = \delta/a - \pi.$$

Defining $\varphi = \delta/a - \pi$,

$$\alpha_s = \varphi,$$
$$\alpha_t = \varphi + \pi\alpha_{t+1}, \ t = s - 1, \ldots, 1.$$

Solving the above equations of φ,

$$\alpha_t = \varphi \left[\sum_{j=0}^{s-t} \pi^j \right], \ t = 1, \ldots, s.$$

Recalling that $\alpha_1 = 1$,

$$1 = \varphi \left[\sum_{j=0}^{s-1} \pi^j \right] = \varphi \left[\frac{1 - \pi^s}{1 - \pi} \right].$$

Then,

$$\varphi = \frac{1 - \pi}{1 - \pi^s} > 0.$$

Thus,

$$0 < \alpha_t = \frac{1 - \pi^{s-t+1}}{1 - \pi^s} \leq 1, \ t = 1, \ldots, s,$$

and

$$0 < \beta_t = \frac{\pi^{s-t+1} - \pi^s}{1 - \pi^s} \leq 1 \ t = 2, \ldots, s+1,$$

and finally, using Equation (35), $\gamma_t \geq 0$, $t = 1, \ldots, s$.

The above result also shows that there will be no intermittent idling in an optimal solution to single-job $F2|s, idlg|C_{\max}$ problem; thus it simply can be stated as single job $F2|s|C_{\max}$. Furthermore, there exists an optimal solution with consistent sublots.

Now, to show that

$$X^*_{i,t+i-1} = [(\pi^{t+i-2} - \pi^{t+i-1})/(1 - \pi^s)], \ i = 1, 2; \ t = 1, \ldots, s,$$

where $\pi \equiv \rho_2/\rho_1$, solves the single job $F2|s|C_{\max}$ optimally, note that $X_{1t} = \pi X_{2t} = \pi X_{1,t-1} = \pi^{t-1} X_{11}$, $t = 1, \ldots, s$. But since $\sum_{t=1}^{s} X_{1t} = 1$,

$$1 = X_{11} \sum_{t=0}^{s-1} \pi^t = [(1 - \pi^s)/(1 - \pi)] X_{11}$$

That is, $X_{11} = [(1 - \pi)/(1 - \pi^s)]$. Thus,

$$X_{1t} = \pi^{t-1}[(1 - \pi)/(1 - \pi^s)], \ t = 1, \ldots, s,$$

$$= [(\pi^{t-1} - \pi^t)/(1 - \pi^s)], t = 1, \ldots, s.$$

And since, $X_{2,t+1} = X_{1t}$,

$$X_{2t} = [(\pi^{t-2} - \pi^{t-1})/(1 - \pi^s)], \ t = 2, \ldots, s+1.$$

4.3 Single Job $F|s_i|C_{\max}$

The single-job $F|s_i|C_{\max}$ problem, where idling is not permitted, can be solved in polynomial time by applying the result for a two-stage problem to adjacent machine pairs [19]. In the following, the same result is derived as a special case of the general formulation.

Suppose no intermittent idling is allowed in between sublots in any stage. This means that there is continuous production in a stage once the production starts. Let the period at which the production starts in a stage $i+1$ is k_{i+1} where,

$$k_{i+1} = \min_{i \leq t \leq h-m+i} \{t|Y_{it} = 1\} + 1,$$

and the production is completed at the end of period l_{i+1}, where,

$$l_{i+1} = \max_{i \leq t \leq h-m+i} \{t|Y_{it} = 1\} + 1$$

Thus the continuous production takes place in stage $(i+1)$ throughout the periods $\{k_{i+1}, \ldots, l_{i+1}\}$. The make span is minimized only if T_{l_i} is as small as possible in every stage $i = 1, \ldots, m$.

There will be exactly s_i transfer batches in between stages i and $i+1$. Recalling the definitions above, the first transfer batch occurs at the end of period $(k_{i+1} - 1)$, so that the production can start at period k_{i+1} in stage $(i+1)$.

For notational convenience, reindex the periods at which the transfers take place from stage i to $i+1$, as $i(\tau)$, $\tau = 1, \ldots, s_i$, where $i(1) = k_{i+1} - 1$. This results in fixing the values of the indicator variables,

$$Y_{it} = 1, \ t = i(\tau), \ \tau = 1, \ldots, s_i,$$
$$= 0, \text{otherwise}.$$

Furthermore, the sizes of the transfer batches are given by

$$\rho_{i+1}L_{i,i(\tau)} = \rho_i L_{i,i(\tau+1)}, \ \tau = 1, \ldots, s_i - 1.$$

That is,

$$L_{i,i(\tau)} = \pi_i^{\tau-1} L_{i,i(1)}, \ \tau = 1, \ldots, s_i - 1.$$

where $\pi_i \equiv \rho_{i+1}/\rho_i$. But, since $\sum_{\tau=1}^{s_i} L_{i,i(\tau)} = 1$,

$$1 = L_{i,i(1)} \sum_{\tau=1}^{s_i} \pi_i^{\tau-1}$$
$$= L_{i,i(1)} \sum_{\tau=0}^{s_i-1} \pi_i^{\tau}$$
$$= [(1 - \pi_i^{s_i})/(1 - \pi_i)]L_{i,i(1)}$$

or

$$L_{i,i(1)} = [(1 - \pi_i)/(1 - \pi_i^{s_i})]$$
$$L_{i,i(\tau)} = \pi_i^{\tau-1}[(1 - \pi_i)/(1 - \pi_i^{s_i})]$$
$$= [(\pi_i^{\tau-1} - \pi_i^{\tau})/(1 - \pi_i^{s_i})], \ \tau = 1, \ldots, s_i,$$

which is the desired result.

5 Conclusions

The problems that we have dealt with in this study are scheduling problems in supply chains. A modeling paradigm is proposed for scheduling problems in supply chains. The primary contention has been that the constituents of a supply chain need to cooperate, rather than compete, in order to achieve overall, as well as individual, maximum benefits. In order to analyze this, it is essential to have a concise but comprehensive formulation. The formulation approach proposed, event-time models, in addition to being computationally viable, account for exogenous events, such as demand occurrences and other deadlines, as well as the endogenous events that are decision variables in the model.

After providing the motivation for this study in Section 1, Section 2 argued that lot streaming is a suitable paradigm for supply chain scheduling. Section 3 introduced event-time modeling as a general modeling approach for lot streaming problems. Furthermore, logic-based modeling framework of constraint programming makes it possible to handle exogenous and endogenous models in the same model. This is demonstrated by an example using OPL (optimization programming language), which can handle variable subscripts in conditional constraints. In Section 4, it is shown that the event-time modeling approach can account for all features of a single-job lot streaming problem that are available for this problem in the literature. This section also presents a number of modeling extensions for single job problem.

A The Notation

There is no widely accepted classification for lot streaming problems. Being basically scheduling problems, lot streaming problems can be classified by using, with possible extensions, the established scheme of [12]. In this scheme a problem type is specified in terms of a three-field classification $\alpha|\beta|\gamma$, where α specifies the machine environment, β indicates a number of job characteristics, and γ refers to the optimality criterion. Using the definitions given in [14, 19], the following addendum to the list of job characteristics can be proposed for lot streaming problems.

In [12], the second field is defined as $\beta \subset \{\beta_1, \ldots, \beta_4\}$, where the elements $\beta_1, \beta_2, \beta_3$, and β_4 are concerned, respectively, with preemption, precedence, release dates, and unit processing requirements.

Further streaming characteristics can be incorporated as $\{\beta_5, \ldots, \beta_8\}$ where $\beta_5 \in \{s_i, s, s = \kappa, \circ\}$. (The symbol $\circ$ denotes an empty symbol which is omitted in the statement of problem types.)

$\beta_5 = s_i$: *Streaming* is allowed, i.e., any job may be split into sublots so that its operations can be overlapped and its progress accelerated; furthermore the maximum number of transfer batches ("sublots"), s_i, allowed from machine i to machine $i+1$ is variable and can be different for all $i = 1, \ldots, m$.

$\beta_5 = s$: *Streaming* is allowed, but the maximum number of transfer batches ("sublots") allowed from machine i to machine $i+1$ is variable but the same for all $i = 1, \ldots, m$.

$\beta_5 = s = \kappa$ where κ is any positive integer: The number of sublots is a constant and specified as part of the problem.

$\beta_5 = \circ$: No streaming is allowed, i.e., $\kappa = 1$.

$\beta_6 \in \{idlg, \circ\}$:

$\beta_6 = idlg$: Intermittent idling of machines are allowed in between processing of the sublots.

$\beta_6 = \circ$: No idling is allowed, i.e., each machine, once started, must process the entire job continuously, without idling.

$\beta_7 \in \{dscr, \circ\}$:

$\beta_7 = dscr$: Each sublot ("transfer batch") must contain discrete number of units.

$\beta_7 = \circ$: No integer requirement on the size of the sublots.

$\beta_8 \in \{equl, cnst, \circ\}$:

$\beta_8 = equl$: *Equal sublots*, each sublot ("transfer batch") size must be equal, that is, work is allocated equally among all sublots, on all machines.

$\beta_8 = cnst$: *Consistent sublots*, sublot ("transfer batch") size must be the same at each machine, that is, the allocation of work to sublots is the same on all machines.

$\beta_8 = \circ$: *Variable sublots*, no restriction, other than that they sum up to lotsize, on the size of the sublots, that is, transfer batches between machines i and $(i+1)$ may differ from the transfer batches between machines $(i+1)$ and $(i+2)$.

In the classic scheduling theory, *single-job* problems do not exist in the sense that they are trivial. Other than very few exceptions, the number of jobs practically play no significant role in defining problem *types*. But in lot streaming, the number of jobs is not only important but most of the current research deals with *single-job* problems. In order not to further complicate the above schema, an additional field should not be introduced but the statement as to whether the problem type is *single job* or *multiple jobs* can be mentioned along with the three-field classification.

The *dominance* relation among the problems discussed in [19] corresponds to *relaxation* of the dominating problem resulting in the dominated problem. If problem Q is a relaxed version of problem P, denote the relationship as $P \prec Q$. Thus the objective function value of P can never be better than that of Q. The "job characteristics" $\beta_6, \beta_7, \beta_8$ imply the following dominance relations:

$$\boxed{\text{equl}} \prec \boxed{\text{cnst}} \prec \boxed{\circ}$$

$$\boxed{\text{dscr}} \prec \boxed{\circ}$$

$$\boxed{\circ} \prec \boxed{\text{idlg}}$$

On the other hand, reducibility among the lot streaming problems, in the complexity sense, can be stated for job characteristic β_5. It can be easily verified that if $P \longrightarrow Q$ implies that the decision version of P reduces to the decision version of Q, then:

$$\boxed{s_i} \longrightarrow \boxed{s} \longrightarrow \boxed{s = \kappa} \longrightarrow \boxed{\circ}$$

References

1. K. R. Baker. Lot streaming to reduce cycle time in a flow shop. Working Paper 203, Amos Tuck School, Dartmouth College, 1988.
2. G. P. Cachon. Supply chain coordination with contracts. In S. C. Graves and A. G. de Kok, editors, *Supply Chain Management: Design, Coordination and Operation*, volume 11 of *Handbooks in Operations Research and Management Science*, chapter 6, pages 229–340. North-Holland, 2003.
3. J. H. Chang and H. N. Chiu. Comprehensive review of lot streaming. *Int. J. Prodn Res.*, 43:1515–1536, 2005.
4. R. W. Conway, William L. Maxwell, and Louis W. Miller. *Theory of Scheduling*. Addison-Wesley, 1967.
5. T. G. de Kok and Jan C. Fransoo. Planning supply chain operations. In S. C. Graves and A. G. de Kok, editors, *Supply Chain Management: Design, Coordination and Operation*, volume 11 of *Handbooks in Operations Research and Management Science*, chapter 12, pages 596–676. North-Holland, 2003.
6. C. A. Glass, J. N. D. Gupta, and C. N. Potts. Lot streaming in three-stage production processes. *European Journal of Operational Research*, 75:378–394, 1994.
7. S. C. Graves and A. G. de Kok. *Supply Chain Management: Design, Coordination and Operation*. North-Holland, 2003.
8. J. Hooker. *Logic-Based Methods for Optimization*. Wiley, 2000.
9. ILOG. ILOG OPL Studio 3.7.1, 2005. http://www.ilog.fr/, (5 June 2007).
10. U. S. Karmarkar. New directions in modelling production and distribution problems. Paper presented at the 15th International Symposium on Mathematical Programming, August 1994, Ann Arbor, 1994.

11. D. H. Kropp and T. L. Smunt. Optimal and heuristic models for lot splitting in a flow shop. *Decision Sciences*, 21:691–709, 1990.

12. E. L. Lawler, J. K. Lenstra, A. H. G. Rinnooy Kan, and D. B. Shmoys. Sequencing and scheduling: Algorithms and complexity. In S. C. Graves, A. H. G. Rinnooy Kan, and P. H. Zipkin, editors, *Logistics of Production and Inventory*, volume 4 of *Handbooks in Operations Research and Management Science*, chapter 9, pages 445–522. North-Holland, 1993.

13. W. L. Maxwell. Comments made at the NSF Workshop on Manufacturing Logistics, Lehigh University, 1997.

14. C. N. Potts and K. R. Baker. Flow shop scheduling with lot streaming. *Operations Research Letters*, 8:297–303, 1989.

15. W. F. Pounds. The scheduling environment, 1961. Paper presented at The Factory Scheduling Conference, Carnegie Institute of Technology, May 10–12, 1961.

16. A. Sen, Engin Topaloğlu, and Ömer Benli. Optimal streaming of a single job in a two stage flow shop. *European Journal of Operational Research*, 110:42–62, 1998.

17. David Simchi-Levi, S. David Wu, and Zuo-Jun Shen. *Handbook of Quantitative Supply Chain Analysis: Modeling in the E-Business Era.* Springer, 2004.

18. D. Trietsch. Optimal transfer lots for batch manufacturing: A basic case and extensions. Technical Report NPS-54-87-010, Naval Postgraduate School, Monterey, CA, 1987.

19. D. Trietsch and K. R. Baker. Basic techniques for lot streaming. *Operations Research*, 41:1065–1076, 1993.

A Dynamic and Data-Driven
Approach to the News Vendor Problem
Under Cyclical Demand

Gokhan Metan[1] and Aurélie Thiele[2]

[1] American Airlines, Fort Worth, TX 78155
gom204@lehigh.edu
[2] Department of Industrial and Systems Engineering
Lehigh University, Bethlehem, PA 18015
aut204@lehigh.edu

Summary We consider the problem of managing the inventory of perishable goods when the demand is cyclical with unknown distribution. While traditional inventory management under uncertainty assumes the precise knowledge of the underlying probabilities, such information is difficult to obtain for time-varying processes in many real-life applications, as the evolution over time of the deterministic seasonal trend interferes with the analysis of the stochastic part of the demand. To address this issue, we propose a *dynamic and data-driven approach* that builds directly upon the historical observations and addresses seasonality by *creating and recombining clusters* of past demand points. This allows the decision maker to place his order at each time period based only on the most relevant data. The algorithm we present requires the estimation of only one parameter, the demand periodicity; furthermore, system performance is protected against estimation errors through a cluster aggregation subroutine, which recombines clusters as needed. We present extensive numerical experiments to illustrate the approach. The key contribution of the chapter is to address a logistics challenge faced by many practitioners, namely, the lack of distributional information for nonstationary demand, by integrating historical data directly into the decision-making module through an algorithm devised specifically for cyclical processes.

1 Introduction

The news vendor problem, in which a manager seeking to maximize his profit must decide how many perishable goods to order in the presence of demand uncertainty, has been the focus of extensive research efforts since the 1950s (see, e.g., Arrow et al. [1], [2] for early work, and Porteus [17] for a review). The difficulty in estimating future demand distributions has been pointed out by Scarf [18], who computes the optimal order quantity when only the first two moments are known. These results have been extended by Gallego and

W. Chaovalitwongse et al. (eds.), *Optimization and Logistics Challenges* 277
in the Enterprise, Springer Optimization and Its Applications 30,
DOI 10.1007/978-0-387-88617-6_10, © Springer Science+Business Media, LLC 2009

Moon [4] to a wide range of single-period inventory problems. More recently, Bertsimas and Thiele [3] have proposed an approach that is entirely data-driven, in the sense that it incorporates historical demand directly into the decision-making process without requiring the estimation of the underlying distribution, and has the added benefit of capturing risk aversion through a single scalar parameter rather than a utility function. While these two groups of researchers approach the lack of distributional knowledge from very different perspectives (based on the demand mean and variance for the former, historical data for the latter), the authors all assume stationarity of the demand. In practice, however, nonstationary demand processes are much more difficult to estimate than stationary ones, as the presence of a seasonal pattern adds a layer of complexity to the estimation procedure.

Hence, the need arises for an approach to the news vendor problem that will address the lack of distributional information for nonstationary demand and exhibit strong empirical performance. As a first step in that direction, Metan and Thiele [13] propose an adaptive algorithm for the number of most recent data points to be incorporated in the decision-making process at each time period, for general nonstationary demand processes. When demand is cyclical, however, using data points that do not correspond to the current cycle phase burdens the decision maker with irrelevant information; this motivates the development of an algorithm specifically tailored to the case of seasonal effects. While an obvious methodology would consist of estimating demand periodicity and computing the next order based only on the past data points that were observed during the (estimated) next phase of the system, a flaw in this technique is its extreme dependence on a high-quality estimate of the periodicity. In this chapter, we present a dynamic and data-driven approach that builds directly upon the historical observations and addresses seasonality by creating and recombining clusters of past demand points. Such a methodology is well-suited for applications where customer behavior exhibits cyclical trends (e.g., boom-bust periods in manufacturing, daily newspaper sales with some columns or sections published weekly), but is also affected by exogenous drivers whose influence is difficult to quantify accurately (e.g., new technological advances, specific news items). Errors in estimating the demand process can result in costly inventory buildup, as the capacity glut that followed the burst of the dot-com bubble in 2001 demonstrated, or missed profit opportunities due to unmet demand, for instance in the real estate market in the early 1990s.

Our work is at the interface of two broad research areas: (i) inventory management under nonstationary demand, and (ii) data clustering. Since our core contribution falls within the field of inventory management and the clustering technique we use is motivated by the specific, cyclical demand process at hand, the main part of the literature review below focuses on the first research stream with only a brief mention of the second area. This apparent neglect of the clustering literature is also due to the fact that most clustering techniques have been developed to group multiattribute customers in marketing applications, and hence are unapplicable here. The vast majority of researchers in inventory management under nonstationary demand investigate the structure of the optimal policy, following Karlin's pioneering work [8]. The demand at each time period is assumed to be an independent random variable with *known distribution,* which might change over time. The optimal strategy is then derived in terms of an infinite sequence of order-up-to levels for back-ordered demand as well as lost sales. In a follow-up study, Karlin [9] computes optimal order-up-to levels for seasonal stochastic demand. The results are similar to those obtained in Karlin [8], with the exception that the optimal strategy is a *finite* sequence of order-up-to levels, where the number of levels is equal to the demand periodicity. Veinott [19] investigates the multiproduct inventory policies for nonstationary demand processes, and derives optimality conditions for independent and dependent demands over time. Morton [15] presents upper and lower bounds for the nonstationary infinite-horizon inventory problem; these bounds are based on the information acquired during the previous time periods. Calculating the bounds after period k requires solving a k-period inventory problem, which becomes computationally unattractive as the number of time periods increases.

Zipkin [20] studies the infinite-horizon inventory problem under nonstationary seasonal demand, and provides an alternative and relatively simpler approach to derive the optimal solution when the average logistics cost is minimized. Lovejoy [11] describes simple operational policies that result in optimal or near-optimal solutions for nonstationary inventory models; in particular, he proposes an inventory policy based on critical fractiles when the exact demand parameters are unknown but estimated dynamically. In a more recent study, Morton and Pentico [16] investigate the finite-horizon nonstationary stochastic inventory problem, derive stronger lower and upper bounds and compare the performance of four heuristic techniques. Kapuscinski and Tayur [7] analyze the optimal policy for a single product, finite-horizon inventory problem with seasonal demand. Tractability issues arise when one attempts to solve the problem analytically, which requires approximating the optimal solution through a simulation-based technique. Metters [14] highlights the difficulty in finding an exact optimal solution for multiproduct inventory problems under seasonal demand, and presents heuristics based on myopic decision rules. The best heuristic is reported as having a 2.5% gap with the optimal solution on average for a class of simple problems. Most recently, Ketzenberg et al. [10] study multi-item production/inventory problems with seasonal demand. They develop a heuristic procedure under a lost-sales assumption for a constrained production environment.

To the best of our knowledge, this chapter is the first to use a clustering procedure to address demand seasonality in inventory management. *Clustering* can be defined as grouping a set of instances into a given number of subsets, called clusters, so that instances in the same cluster share similar characteristics. Most research efforts in clustering focus on developing efficient clustering algorithms and techniques, in particular in the context of marketing applications and customer data, which are beyond the scope of this chapter; instead, we refer the reader to Jain et al. [6] for an in-depth discussion of these methods. We only mention here the method of k-means, which was originally developed by MacQueen [12] and is one of the simplest clustering approaches available in the literature. It partitions the data into k clusters by minimizing the sum of squared differences of objects to the centroid or mean of the cluster. In other words, the objective is to minimize total intracluster variability. The objective can be formulated in mathematical terms as: $\min \sum_{i=1}^{k} \sum_{j \in S_i} |x_j - \mu_i|^2$, where there are k clusters S_i, $i = 1, 2, ..., k$ and μ_i is the mean point of all the instances x_j in S_i. This method presents similarities with the clustering approach we propose; in particular, we sort the demand and assign the data points to clusters in a manner that also attempts to minimize intracluster variability. A difference is that, while the K-means algorithm proceeds iteratively by updating the centroids of each cluster but keeps the same number of clusters throughout, we update the number of clusters by merging subsets as needed.

Our contributions to the literature on inventory management under stochastic nonstationary demand are threefold:

1. We propose a *data-driven approach* to the news vendor problem under seasonal demand that does not require any distributional assumption on the demand process.
2. We address cyclical patterns in the historical demand by applying ideas from the field of *data clustering* to aggregate demand points into subgroups.
3. We present a *dynamic cluster-recombination procedure* to mitigate the risk of data misclassification.

The remainder of this chapter is structured as follows. We review the classical news vendor problem and highlight the issues raised by nonstationary demand in Section 2. We propose a dynamic, data-driven algorithm to utilize the available information efficiently in Section 3. We conduct extensive numerical experiments in Section 4. Finally, Section 5 contains concluding remarks.

2 Problem Setup

2.1 Stationary Demand

The news vendor problem has been extensively studied in the literature under a wide range of assumptions (see, e.g., Porteus [17].) We review here properties of the optimal order when the demand D is a continuous nonnegative random variable with probability density function (pdf) f and cumulative density function (cdf) F, and there is no setup cost for placing an order. The news vendor decides how many items to order before knowing the exact value of the demand, with the goal of maximizing his expected profit. Since the goods are perishable, no inventory can be carried over to the next time period. We use the following notation:

c: the unit ordering cost,

p: the unit selling price,

s: the salvage value,

Q: the order quantity,

D: the demand level,

c_u: undershoot cost $(c_u = p - c)$,

c_o: overshoot cost $(c_o = c - s)$,

α: critical ratio $(\alpha = \frac{c_u}{c_u + c_o})$,

$\phi(.)$: pdf of standard normal distribution,

$\Phi(.)$: cdf of standard normal distribution.

The classic news vendor problem is then formulated as:

$$\max_{Q \geq 0} (p - c) Q - (c - s) E \left[\max(0, Q - D)\right], \tag{1}$$

Theorem 1 (Optimal order in classic model (Porteus [17])).
(i) The optimal solution to Problem (1) is:

$$Q^* = F^{-1}(\alpha) \tag{2}$$

(ii) When the demand is normally distributed with mean μ and standard deviation σ, the optimal order is:

$$Q^* = \mu + \sigma \Phi^{-1}(\alpha). \tag{3}$$

Remark: If $c_u < c_o$, we have $\alpha > 1/2$ and the decision maker seeks to protect himself from not having enough items in inventory. In that case, the first term, μ, in Equation (3) is called the *cycle stock* and the second term, $\sigma \Phi^{-1}(\alpha)$, is called the *safety stock*. ($\alpha > 1/2$ yields $\Phi^{-1}(\alpha) > 0$.) The purpose of the cycle stock is to meet the expected demand, while the safety stock provides a buffer against demand uncertainty. The values of both the cycle stock and the safety stocks in this model depend heavily on the decision maker's assumptions on the underlying demand distribution. In practice, however, such precise knowledge is difficult to obtain, because we lack historical observations to compute meaningful estimates. This data scarcity may be due to the introduction of a new product (no historical data is available) or to the nonstationarity of the demand (historical data are then in large part irrelevant, since past demands obey different distributions than the demand at the current time period). Under such circumstances, attempts at estimating demand distributions are fraught with errors, and lead to the implementation of order quantities that are not optimal for the true demand. This motivates the *direct use* of the empirical data to guide the manager, so that the decision-making process builds upon the precise amount of information available without requiring additional assumptions.

In what follows, we focus on nonstationary demand as the main reason for scarce data. We assume that we do not have any information about the distribution of the underlying demand process, but we do have a set of historical observations at our disposal. We introduce the following notation, in addition to the parameters presented in the classical model:

S: the set of data points obtained from previous observations,
N: the number of historical observations in S,
d_i: the ith demand observation in set S, with $i = 1, 2, \ldots, N$,
$d_{<i>}$: the ith smallest demand in set S, with $i = 1, 2, \ldots, N$.

The data-driven counterpart of the classic news vendor problem (1) becomes:

$$\max_{Q \geq 0} (p - c) Q - \frac{(c - s)}{N} \sum_{i=1}^{N} \max(0, Q - d_i), \tag{4}$$

where the news vendor now maximizes his sample average revenue.

Theorem 2 (Optimal order in data-driven model (Bertsimas and Thiele [3])). *The optimal order Q^* for Problem (4) is given by:*

$$Q^* = d_{<j>}, \quad with \ j = \lceil \alpha N \rceil. \tag{5}$$

Remark: In the data-driven model, the optimal order quantity is equal to the α-quantile of the data sample, with α the critical ratio. When N becomes large and the demand obeys a stationary demand process, the optimal order given by Equation (2) converges towards the value given by Equation (5). In the stationary case, it is desirable to include all the data points available in the decision-making process to achieve more accurate demand predictions. Section 2.2 highlights several issues that arise when demand is nonstationary.

2.2 Issues for Nonstationary Demand

In practice, demand for many perishable items evolves over time; for instance, sales of the daily *New York Times* might depend on the op-ed columnist publishing his or her column on that day, or of the specific sections (Business, Real Estate) included in that issue of the newspaper, with additional uncertainty due to the events in the news. The nonstationarity of the demand process adds a layer of complexity to inventory management, because it makes fitting all the data points to a unique distribution meaningless: the current demand will not obey such a distribution, and any decision-making process based on stationary demand will result in inaccurate orders and poor system performance.

We briefly illustrate this point on an example, which allows us to introduce the clustering approach. We consider the (admittedly oversimplistic) case where a deterministic, sinusoidal demand, $D_t = a + b \sin(\frac{2\pi t}{T})$, has been misclassified as a stationary stochastic process. The optimal solution when demand is deterministic is to order an amount equal to the demand. The decision maker believing instead that the demand is stationary will estimate mean and standard deviation based on N data points using the following formulas:

$$\widehat{\mu} = a + \frac{b}{N} \sum_{t=1}^{N} \sin\left(\frac{2\pi t}{T}\right),$$

$$\widehat{\sigma}^2 = \frac{b}{N-1} \sum_{t=1}^{N} \left[\sin\left(\frac{2\pi t}{T}\right) - \frac{1}{N} \sum_{s=1}^{N} \sin\left(\frac{2\pi s}{T}\right)\right]^2.$$

For instance, if $N = T$ and T is a multiple of 4, we can prove using straightforward trigonometric equations that $\widehat{\mu} = a$ and $\widehat{\sigma}^2 = (b^2/2) \cdot \frac{T}{T-1}$. The error in the amount ordered at time t is then given by: $b \mid \sin(\frac{2\pi}{T}t) - \frac{\Phi^{-1}(\alpha)}{\sqrt{2}}\mid$, with a maximum at $b\left(1 + \frac{|\Phi^{-1}(\alpha)|}{\sqrt{2}}\right)$. For instance, with $a = 40$, $b = 10$, $T = 20$, $\alpha = 0.75$ and $t = T/4$, the decision maker will order 45 units while the true demand is 50, resulting in a relative error of 10%. Figure 1 shows the demand values on the unit circle, as well as their respective frequency on the right. This provides some preliminary insights into the clustering mechanism presented in Section 3. Here, there are 20 data points and 11 clusters (9 with 2 data points and 2 with 1 data point), and all data points in any one cluster share the same value; in more general cases, the values of the demand points inside a cluster will be close but not necessarily equal. As discussed in Section 3 below, the proper classification of the data points into the correct cluster is critical to

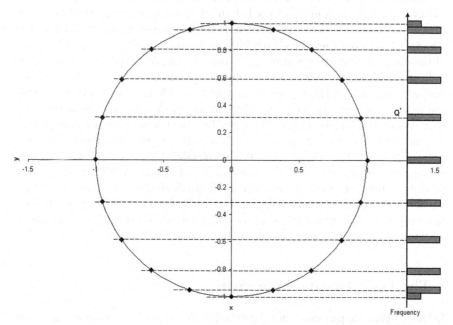

Fig. 1. Deterministic sinusoidal demand with a period of 20 time units

good algorithm performance. The frequency plot exhibits a behavior similar to the sampled realization of a discrete uniform distribution on the interval $[a - b, a + b]$. If we include all the data points in the decision-making process, we treat the observations as if they were generated from a stationary distribution, and select as our optimal order the data point corresponding to the α-quantile of the sample set, that is, $Q^* = d_{<\lceil \alpha N \rceil >}$ in Figure 1. The impact of misinterpreting seasonal deterministic demand as a stochastic process is proportional to the difference between (the projections on the y-axis of) Q^* and the actual data point.

Identifying seasonality is only the first hurdle in the decision-making process; many decision makers in practice do have an inkling whether demand for their products is cyclical. However, periodicity is in general difficult to estimate accurately because stochastic effects interfere with the observation of highs and lows in the demand process. A simple algorithm that estimates periodicity and computes the optimal order based on the data points corresponding to the next phase will be highly sensitive to estimation errors; furthermore, it will not capture adequately complex trends such as phases of different lengths, e.g., consecutive time periods obeying the same distribution during the core of the season but changing rapidly at the season's beginning and end. For instance electricity consumption in the Northeast is on average high in August for air-conditioning purposes and high from December to February for heating purposes, which makes winter three times as long as summer from that perspective.

Hence, the need arises, in environments featuring nonstationary demand, to develop novel techniques to predict the demand levels and take full advantage of past observations, while acknowledging nonstationarity. Metan and Thiele [13] propose such a methodology, and investigate the impact of the size of the data set when demand obeys a time-varying distribution. Their results indicate that the number of historical observations plays a key role in system performance. The authors present an updating rule that dynamically adjusts the number of most recent data points to use in the computations; however, they do not specifically address the case of cyclical demand, where the pattern of nonstationarity is more structured than in the general non-stationary case. The purpose of this chapter is to devise an efficient algorithm that captures seasonality through the use of a clustering mechanism. While clustering has received much attention in marketing, in particular in the field of multiattribute data analysis, there has been to the best of our knowledge no attempt to incorporate such a tool in inventory management. We describe the approach in detail in Section 3, and perform extensive computational tests in Section 4.

3 The Data-Driven Approach

In this section, we propose a novel methodology to determine the optimal order for the news vendor problem under cyclical demand. Our main contribution is to develop a clustering algorithm that utilizes the historical data efficiently to make high-quality predictions, by isolating the past data points that provide relevant information for the next time period. The main motivation behind this algorithm is to differentiate between the deterministic *seasonal effect* and the *stochastic variability* so that we can set cycle stock and safety stock levels accurately. Before presenting the algorithm in detail, we need to introduce the following terminology.

Definition 1 (k-neighborhood).
*(i) We say that the data point d_t is a k-neighborhood maximum point for some $k \geq 1$ integer if and only if d_t is greater than or equal to $d_{t-k}, d_{t-k+1}, \ldots, d_{t+k}$.
(ii) The set of demands at time periods $t - k$, $t - k + 1$, $\ldots$, $t + k$ is called the k-neighborhood of demand at time t.*

We define the following parameters:

k: neighborhood level,
$\widehat{T}$: estimated (integer) value of the periodicity,
M_k: list of time epochs corresponding to k-neighborhood maxima,
t_l: l^{th} time epoch in set M_k, $l = 1, 2, \ldots, |M_k|$,
φ: phase of the next decision point, where $0 < \varphi < \widehat{T}$,
C_j: list of data points assigned to cluster j, $j = 1, \ldots, \widehat{T}$, ranked in increasing order,
m: number of data points assigned to each cluster.

In the notation above, the term "cluster" refers to the clusters *before* recombination. The master algorithm consists of five main steps, as described in Algorithm 1.

Algorithm 1 (Master algorithm for cyclical demand)
Step 1. Estimate the periodicity $\widehat{T}$ of the demand process (Algorithm 2).
Step 2. Sort the historical demand data in set S in ascending order to form the list $\widehat{S}$.
Step 3. Create $\widehat{T}$ data clusters using $\widehat{S}$ (Algorithm 3).
Step 4. Estimate the phase φ of the demand at the next time period.
*Step 5. **Repeat** (until the end of the planning horizon)*
*   Detect which cluster C_j the next data point is expected to fall into ($j = 1, 2, \ldots, \widehat{T}$).*
*   Aggregate C_j with adjacent clusters if needed, to form temporary cluster C'_j (Algorithm 4).*
*   Select the next order based only on the data points in cluster C'_j: $Q^* = d_{<\lceil \alpha |C'_j| \rceil>}$.*
*   Assign the new demand observation to appropriate cluster.*
*   Update the phase for the next decision point ($\varphi \longleftarrow (\varphi + 1) \bmod \widehat{T}$)*
*   **End repeat.***

We describe each step of Algorithm 1 in further detail below.

Step 1: Period estimation (Algorithm 2)

Because the estimated periodicity determines the number of clusters created before aggregation, Algorithm 2 plays an important role in the overall performance of the approach. If the decision maker overestimates the periodicity ($\widehat{T}$ too large), he will allocate data points drawn from *similar* distributions to *different* groups, whereas if he underestimates the periodicity ($\widehat{T}$ too small), he will aggregate into *one* cluster data points that actually do *not* obey similar distributions. (Note that aggregating clusters will mitigate the consequences of periodicity overestimation, but not underestimation. Hence, it is recommended to tune Algorithm 2 with a focus on preventing underestimation, as explained below.) Furthermore, the value of the estimated periodicity is used to *update the phase of the system* φ for the next time period. The phase determines our current position on the demand cycle and, as such, plays a crucial role to select the data cluster appropriately.

We first state Algorithm 2 in mathematical terms, then provide its interpretation at a higher level. We remind the reader that $k \geq 1$ is the neighborhood level, M_k is the set of data points that are k-neighborhood maxima, t_l with $l = 1, \ldots, |M_k|$ is the time at which the l-th such maximum occurs, and S is the set of previous observations.

Algorithm 2 (Period estimation) *Set* $i = k + 1$, $l = 1$, $M_k = \emptyset$.
Repeat
  if $d_i \geq d_{i+j}$ *for all* $j \in \{-k, -k+1, \ldots, k-1, k\}$, {
  $M_k \longleftarrow M_k \cup \{t_l = i\}$, $l \longleftarrow l + 1$,
  },
  $i \longleftarrow i + 1$,
End repeat *when* $i = |S| - k + 1$.
Return $\widehat{T} = r\left(\frac{t_{|M_k|} - t_1}{|M_k| - 1}\right)$ *where* $r(\cdot)$ *is the rounding operator to the nearest integer.*

The quality of the estimation made by Algorithm 2 depends on the system parameter k. This value defines how many immediate neighbors of a data point are taken into account in the period estimation process (specifically, $2k + 1$). Algorithm 2 identifies the k-neighborhood maximum points in the data set S to isolate the peaks in demand, and sets the estimated period length to the average time difference between two successive maximum points, rounded to the nearest integer. The introduction of the concept of k-neighborhood is done to mitigate the effect of "false maxima," which arise when a large deviation of the stochastic component of the demand from its mean turns a data point into a local maximum, although the deterministic, cyclical component is not at its peak. Figure 2 depicts the period estimation procedure on a data set of size 100 generated from a normally distributed function with mean $\mu_t = 200 + 50\sin(\frac{2\pi t}{20})$ and standard deviation $\sigma = 7$. The intervals between the 10-neighborhood maxima are also shown in Figure 2, here, 20, 20, 19 and 21. The estimated period is then: $\widehat{T} = \frac{20+20+19+21}{4} = 20$, which is exactly the true periodicity of the unknown demand function.

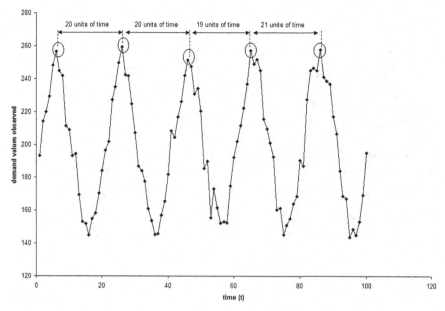

Fig. 2. Estimation of periodicity $\hat{T}$ when $D \sim N(200 + 50\sin(\frac{2\pi t}{20}), 7)$ and $k = 10$

The circled data points are the k-neighborhood maxima with $k = 10$. Note that if we had simply rounded to the nearest integer the average distance between the local maxima, i.e., the 1-neighborhood maxima, we would have obtained the much lower and wildly inaccurate value of 6, due to the numerous volatility-driven tiny peaks in Figure 2. As a rule of thumb, a value of k close to the half-period $\frac{T}{2}$ appears to yield accurate estimations. Since T is not known at the beginning of the algorithm, it is often useful to update k once an estimate of the period has been found, and run the algorithm again until the estimate $\hat{T}$ stabilizes (or decreases, since we want to protect the system against *underestimating* the periodicity). For instance, once we have found the erroneous value of $\hat{T} = 6$, iterating the procedure with $k = 3$ will remove a substantial number of the small peaks that were interfering with the estimation process.

Steps 2 and 3: Data sorting and cluster creation (Algorithm 3)
After Algorithm 2 completes the period estimation process, Algorithm 1 continues with Steps 2 and 3: first, the historical data is sorted and stored in the list $\hat{S}$, which is then used by Algorithm 3 to create the initial clusters. Algorithm 3 creates as many clusters as the periodicity of the seasonal demand function, i.e., one cluster for each phase of the seasonality, and assigns $m = \left\lfloor \frac{N}{\hat{T}} \right\rfloor$ data points to each cluster, with the $N - m\hat{T}$ oldest data points being discarded. Alternatively, the last cluster could receive more points than

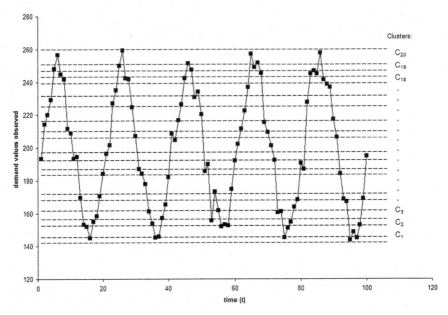

Fig. 3. Clustering the historical data for $D \sim N(200 + 50\sin(\frac{2\pi t}{20}), 7)$

the others; for the sake of clarity, we present here the version where all the clusters have the same number of points.

Algorithm 3 (Initial clustering) *Set* $m = \left\lfloor \frac{N}{\widehat{T}} \right\rfloor$, $j = 1$.
repeat
    Assign observations $d_{<m(j-1)+1>}, \ldots, d_{<m(j-1)+m>}$ *to cluster* C_j,
    $j \longleftarrow j + 1$,
end repeat when $j = \widehat{T} + 1$.

Figure 3 shows the repartition of 100 data points into clusters for a cyclical demand with an estimated periodicity of $\widehat{T} = 20$. Since, $m = \frac{N}{\widehat{T}} = \frac{100}{20} = 5$, the algorithm creates 20 clusters with 5 data points in each. The data points that lie within the boundaries of two consecutive horizontal lines in Figure 3 all belong to the same cluster.

Remark 1: By default Algorithm 3 creates $\widehat{T}$ clusters regardless of $\widehat{T}$ being an even or odd number. In some special cases, such as $\widehat{T}$ being an even number, there is a potential modification in Algorithm 3 for the initial number of clusters. For example, assuming a periodicity of 20 and peaks at 0 and 20, the demand periods 5 and 15 exhibit similar demand quantities. In this case, these data points can be considered to be assigned in the same data cluster. Subsequently, the required number of data clusters reduces to half of the estimated periodicity (i.e., $\widehat{T}/2$). This might increase the quality of the clusters due to increased number of observations within each cluster. However, there is an associated risk with it. Basically, the algorithm does not know the true periodicity, T, of the demand a priori, but it estimates it with $\hat{T}$. Although the estimation procedure of $\hat{T}$ is quite precise, there is no guarantee that it will be exact. In this sense, there is a probability of estimating $\hat{T}$ as an even number even though the true periodicity, T, is odd. In this case, having $\hat{T}/2$ clusters increases the error rate of assigning demand observations coming from different phases to the same cluster. Hence, although it seems like creating $\hat{T}/2$ would also work within the approach, especially for even values of periodicities, it brings additional risks alongside.

Remark 2: The performance of the proposed algorithm depends primarily on: (i) the algorithm's ability to estimate the period correctly, and (ii) the algorithm's ability to classify data points accurately. Our experiments do show that Algorithm 2 is quite capable of estimating the periodicity correctly; furthermore, a number of techniques, such as two-sided trimming, can be used to strengthen the quality of estimations based on sample averages (Huber [5]). But even when the demand periodicity is estimated accurately, observations might be *misclassified* during the clustering process, as demand volatility can be misinterpreted as deterministic cyclical trends. (A point is said to be misclassified when it is out-of-sequence after the sorting procedure, that is, the order of the data points has been affected by *volatility* rather than *seasonality*. Misclassification is an issue because each cluster created by Algorithm 3 should only contain demands corresponding to a specific phase of the cyclical process.) This is illustrated in Figure 4, with the average sinusoidal

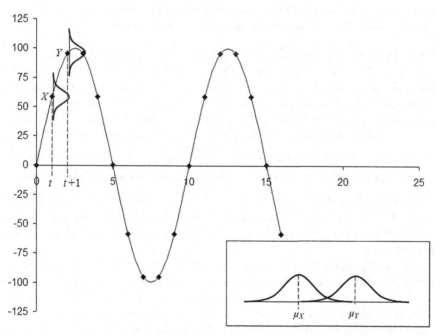

Fig. 4. Misclassification probability of data points for $D \sim N\left(a + b\sin(\frac{2\pi t}{T}), \sigma\right)$

demand being plotted as well as small sinusoids at sample points X and Y to illustrate stochasticity. Lemma 1 gives an upper bound on the probability of misclassifying a single data point when the demand at each time period obeys a normal distribution.

Lemma 1. *Assume that the demand process obeys a normally distributed distribution with a sinusoidal mean: $D \sim N(a + b\sin(\frac{2\pi t}{T}), \sigma)$, and the demand at each time period is independent of the demand at other time periods. Then the probability of misclassifying a single observation is bounded from above by $\Phi\left(-b\left[1 - \cos\left(\frac{2\pi}{T}\right)\right]/(\sigma\sqrt{2})\right)$.*

Proof: Let X, Y be the random demand at time periods t and $t+1$, respectively (see Figure 4). Then we have:

$$X \sim N\left(a + b\sin\left(\frac{2\pi t}{T}\right), \sigma\right), Y \sim N\left(a + b\sin\left(\frac{2\pi(t+1)}{T}\right), \sigma\right)$$

We assume that $\frac{2\pi t}{T}$ and $\frac{2\pi(t+1)}{T}$ fall into the first quadrant. (The proof in the other cases is similar and yields the same bound.) Then the probability of misclassification can be written as:

$$Pr(Y \leq X) = Pr(Y - X \leq 0) = \Phi\left(\frac{-b[\sin(2\pi(t+1)/T) - \sin(2\pi t/T)]}{\sigma\sqrt{2}}\right)$$

using that $Y - X \sim N\left(b\left[\sin(\frac{2\pi(t+1)}{T}) - \sin(\frac{2\pi t}{T})\right], \sigma\sqrt{2}\right)$. We conclude by noting that the probability of misclassification is greatest when $\sin(2\pi(t + 1)/T) = \sin(\pi/2) = 1$, and using that $\sin(\frac{\pi}{2} - a) = \cos(a)$ for all a.

Remark: $\Phi\left(-b\left[1 - \cos\left(\frac{2\pi}{T}\right)\right]/(\sigma\sqrt{2})\right)$ is at most 0.5, which occurs when $\frac{b}{\sigma} \to 0$ or $T \to \infty$: in both cases, the seasonality effect is negligible, either because the amplitude of the seasonality is much smaller than the standard deviation of the demand process or because the periodicity is so large that it is virtually impossible to detect. For instance, consider the demand function $D \sim N\left(200 + 100\sin(\frac{2\pi t}{10}), 7\right)$. Then the probability of misclassification for a single observation is at most 3% ($\Phi\left(\frac{-100[1-\cos(\pi/5)]}{7\sqrt{2}}\right) = 0.027$.) If, instead, we assume that $D \sim N\left(200 + 5\sin(\frac{2\pi t}{100}), 7\right)$, then the probability of misclassification is at most 50 % ($\Phi\left(\frac{-5[1-\cos(\pi/50)]}{7\sqrt{2}}\right) = 0.5$). This is due to the large value of T, which makes it difficult to differentiate between seasonality and stochasticity; hence, the decision maker will in general assume that demand is stationary.

While Algorithm 3 creates as many clusters as the estimated period length, the probability of misclassification increases when $\frac{b}{\sigma}$ decreases or T increases, resulting in a negative impact on the performance of the proposed approach. Therefore, the next step of the algorithm focuses on utilizing the clusters more efficiently, in order to achieve better predictions. We rely on the concept of *c-level cluster aggregation*, which represents the process of temporarily combining $2c + 1$ adjacent clusters (c below and c above the cluster under consideration).

Upon completion of the initial clustering procedure, the control returns to Algorithm 1, in order to detect the phase of the next decision point. The phase is simply computed as: $\varphi = (t + 1) \bmod \widehat{T}$. After this step, the cluster corresponding to that phase is activated and Algorithm 4 aggregates that cluster with neighboring ones. We present below several methods that can be implemented by the decision maker to aggregate clusters, with the goal of reducing the risk of misclassifying data. These methods should in particular be used if the amplitude of the cycle is small compared with the standard deviation, if the periodicity is large, and more generally if there are numerous small peaks in the demand curve plotted over time, as these bursts in volatility are the primary indicators of misclassified data. (The algorithm can identify this situation by computing the average periodicity $\widehat{T}$ using 1-, 2-, ..., k-neighborhood maxima; in the presence of such peaks, $\widehat{T}$ increases sharply when the neighborhood level increases.) Note that it is also possible to create any prespecified number of clusters (rather than creating $\widehat{T}$ and subsequently aggregating them); however, the algorithm performance improves when the manager has an estimate of $\widehat{T}$ to help guide the decision-making process, as we want to avoid mixing data points from very different distributions.

Algorithm 4 (Cluster aggregation)

Method 1. (Static homogeneous aggregation) Select $c \geq 0$ integer at the beginning of the decision-making process. After completion of Algorithm 3 (cluster creation), set $j = 1$.

> **Repeat**
> Create the new cluster $C'_j \leftarrow \{C_{c(j-1)+1}, \ldots, C_{cj}\}$,
> $j \leftarrow j + 1$.
> **End repeat** when $j = \left\lceil \frac{\widehat{T}}{c} \right\rceil$.
> Create the new cluster $C'_{\lceil \widehat{T}/c \rceil} \leftarrow \{C_{c\lfloor \widehat{T}/c \rfloor}, \ldots, C_{\widehat{T}}\}$ with the remaining clusters.

Method 2. (Static heterogeneous aggregation) For $i = 1, \ldots, \widehat{T}$, let K_i the list of time periods t expected to share the same phase i, i.e., $K_i = \{i, \widehat{T}+i, 2\widehat{T}+i, \ldots\}$. For all j, let $\sigma(j)$ be the index of the dominant phase in C_j, i.e., most data points in C_j have time periods in $K_{\sigma(j)}$, and let $\alpha \in (0, 1)$. Set $j = 1$.

> **Repeat**
> If at least $100\,\alpha\%$ of the data points in C_{j+1} come from $K_{\sigma(j)}$, then {
> $C_{j+1} \leftarrow C_j \cup C_{j+1}$
> };
> $j \leftarrow j + 1$.
> $C'_j \leftarrow C_j$.
> **End repeat** when $j = \widehat{T} - 1$.

Method 3. (Dynamic homogeneous aggregation) Compute the optimal order and realized profits for N values of the cluster aggregation level c, implementing *the order obtained for the active value of c and switching to a better-performing value c' if the orders obtained for c' yield higher profits $100\,\beta\%$ of the time with $\beta \in (1/2, 1)$. Cluster aggregation for any aggregation level of c is performed as described in Method 1, static homegeneous aggregation. As opposed to Method 1, the aggregation level c is now dynamic in which the decision maker switches to a better-performing c' value that results in more adequate cluster aggregation.*

Method 4. (Dynamic heterogeneous aggregation) At each time period, start with the clusters created in Algorithm 3 and implement Method 2 for the cluster C_j corresponding to the next phase only, with the modification that you also consider merging with C_{j-1}, as well as C_{j+1}.

Methods 2 and 4 are particularly well-suited for cases where several consecutive time periods obey similar distributions; in that case, the algorithm will merge all the data collected in these time periods.

The new, aggregate cluster is then set as active data set and the optimal order quantity is computed according to Equation (5). Note that the number of data points N used in Equation (5) should now be the number of observations in the active cluster only. Upon observing the actual demand, the decision maker adds this new data point to the appropriate cluster (i.e., the active cluster that is used to make the prediction for the current phase) as a new observation. Then the phase of the system is updated to the next time period and the process repeats itself (Step 5 in Algorithm 1).

Figure 5 illustrates the cluster aggregation process under Method 4. In this figure, the periodicity of the demand process is 21 time units and therefore

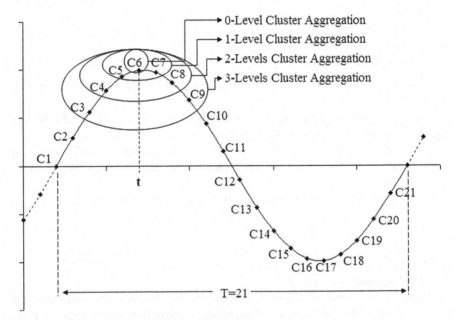

Fig. 5. Cluster aggregation

21 clusters are constructed by Algorithm 3. The assignment of clusters to specific time epochs over a demand cycle is also shown in Figure 5. Let t be the next time period. If cluster aggregation is not used, the demand is predicted by using only one cluster, $C6$. If 1-level cluster aggregation is used, clusters $C5, C6, C7$ are temporarily combined and the prediction is made from this aggregate data set. Similarly, 2-level cluster aggregation requires combining $C4, C5, C6, C7$, and $C8$ at time t. This aggregation scheme becomes particularly important when the periodicity in the demand process is large (i.e., cyclical effects are small) compared to demand stochasticity.

The algorithms described in this chapter are especially designed but not limited to sinusoidal demand patterns. They can be further enhanced to be implemented in such settings where the demand exhibits linear trends, asymmetric seasonalities, seasonalities with trends, etc. Asymmetric seasonalities are originated from the difference between the upswing and downswing periods' lengths. For instance, consider the sales of hot drinks in a moderate climate with long winters and short summers. The demand pattern of such a commodity might look like Figure 6 in which the winter seasons last longer than summer seasons (i.e., $T_1 > T_2$). Note that the sinusoidal demand pattern in Figure 6 is only given for descriptive purposes and the demand is not required to be a sinusoidal. In this case, the algorithms can be modified to estimate two different half periods, namely $\hat{T}_1$ and $\hat{T}_2$, and the clustering of data points can be performed on two separate sets of observations characterized by the location of observations being upswing or downswing. In other words, $\hat{T}_1$

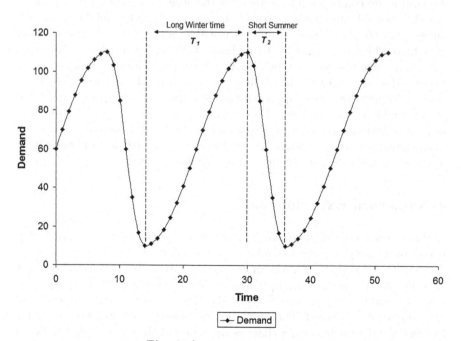

Fig. 6. Asymmetric seasonality

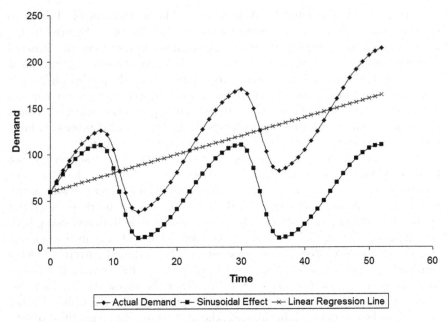

Fig. 7. Seasonal demand with linear trend

clusters are constructed for classifying the observations in winter periods and $\hat{T}_2$ clusters are constructed for classifying the observations in summer periods. Another possible application area of the proposed models might be seasonalities with trends (Figure 7). One reasonable modification on the methods to achieve this goal might be to perform statistical regression on the data set to estimate the linearity of demand. A decomposition technique can be developed to differentiate the cyclical behavior from the linearity. Then, the proposed clustering-based approach can be implemented to characterize the cyclical behavior of the demand process. Note that this kind of an approach would work nicer especially when the linearity and the seasonality effects act independently from each other. In other words, the amplitude and periodicity of the seasonality do not get affected by the linear trend.

4 Numerical Experiments

In this section, we perform extensive numerical experiments to test the performance of the proposed algorithm for the following family of demand processes, $D_t \sim \mathcal{N}(a + b\,sin(2\pi t/T), \sigma)$, under various parameter settings. In all the experiments, we use a warm-up period of 500 time units, which in turn provides 500 data points to initialize the algorithm. The statistics are collected after the warm-up period. The simulation run length is set to 2000 time units. Because the parameter a only changes the center of the sinusoidal, it does not

Table 1. Performance for experiment group 1 ($b = 200$)

T	k	CAL	Best	Algorithm	Exp. Smth.	Mov. Avg.
10	5	0	13.42	14.52	195.76	309.99
16	5	0	13.42	13.54	126.32	315.39
18	5	0	13.42	14.40	112.60	316.26
20	5	0	13.42	13.51	101.83	316.89
25	5	0	13.42	13.69	82.27	317.92
30	5	0	13.42	14.37	69.03	318.15
50	5	0	13.42	14.40	43.74	318.08
100	40	0	13.42	13.93	26.97	317.12
160	40	0	13.42	13.89	22.73	318.36

affect the quality of the results and is set to 500 throughout. Furthermore, the value of the parameter σ must be analyzed in conjunction with the time period T to quantify seasonality versus stochasticity; hence, we set σ to 7 and vary T instead. We perform three groups of experiments, characterized by the amplitude in seasonality: (1) high ($b = 200$, Table 1), (2) moderate ($b = 100$, Table 2), and (3) low ($b = 5$, Table 3). The case of small amplitude compared to the standard deviation is particularly important as the risk of misclassifying data points is highest; therefore, we present more results in that more difficult case. We shall see that the data-driven algorithm exhibits excellent performance in all three groups of experiments. Within each group we vary parameters such as the true period T, the neighborhood level k and the cluster aggregation level CAL in the manner indicated in Tables 1-3. We consider four performance measures:

- The average regret, called *Best*, based on the difference between the realized demand values and the classic news vendor predictions, which are computed using *perfect information* on the demand distribution,
- The average regret achieved by our algorithm, called *Algorithm*,

Table 2. Performance for experiment group 2 ($b = 100$).

T	k	CAL	Best	Algorithm	Exp. Smth.	Mov. Avg.
10	5	0	13.42	14.52	99.88	156.22
12	5	0	13.42	13.51	84.25	157.87
13	5	0	13.42	13.59	78.69	159.20
17	5	0	13.42	13.69	61.29	159.36
25	5	0	13.42	13.68	43.74	159.27
30	5	0	13.42	14.22	37.55	159.64
40	5	0	13.42	13.68	30.72	158.54
50	5	0	13.42	14.09	27.33	159.53
100	40	0	13.42	14.01	21.20	158.92

Table 3. Performances for experiment group 3 ($b = 5$)

T	k	CAL	Best	Algorithm	Exp. Smth.	Mov. Avg.
10	4	0	13.42	17.49	15.85	15.41
10	4	1	13.42	16.96	15.85	15.41
10	4	2	13.42	16.35	15.85	15.41
10	4	3	13.42	15.21	15.85	15.41
10	4	4	13.42	15.85	15.85	15.41
15	8	0	13.42	17.34	15.84	15.53
15	8	1	13.42	17.52	15.84	15.53
15	8	2	13.42	16.68	15.84	15.53
15	8	3	13.42	16.15	15.84	15.53
15	8	4	13.42	15.64	15.84	15.53
15	8	5	13.42	15.22	15.84	15.53
19	10	0	13.42	17.16	15.75	15.43
19	10	5	13.42	16.85	15.75	15.43
19	10	6	13.42	16.05	15.75	15.43
24	12	0	13.42	14.48	15.66	15.48
24	12	1	13.42	14.28	15.66	15.48
31	15	0	13.42	17.43	15.63	15.52
31	15	5	13.42	16.43	15.63	15.52
31	15	7	13.42	16.08	15.63	15.52
50	26	0	13.42	17.05	15.14	15.60
50	26	7	13.42	16.35	15.14	15.60
50	26	12	13.42	15.66	15.14	15.60
99	50	0	13.42	16.28	14.66	15.19
99	50	20	13.42	15.73	14.66	15.19

- The regret values based on the Moving Average prediction technique for the demand,
- The regret values based on the Exponential Smoothing prediction technique for the demand.

Note that *Best* gives us the minimum regret value that can ever be achieved under complete information so *Algorithm* will never fall below *Best*. We use the following values for the news vendor's cost parameters: $p = 10$, $c = 7$, $s = 5$.

Experiment group 1

For high amplitude, the performance of the proposed method is found to be significantly better than the performance of the other forecasting techniques (Figure 8). Moreover, *Algorithm* almost equals *Best* for all values of T. The performance of exponential smoothing improves when T increases, but even for $T = 160$ the average regret achieved from exponential smoothing is twice as much as the regret observed for the proposed algorithm. Figure 9 shows the actual demand values along with the predictions of the proposed algorithm and exponential smoothing (for clarity purposes, we have omitted the results for moving average). We plot the data for the simulation runs from the demand

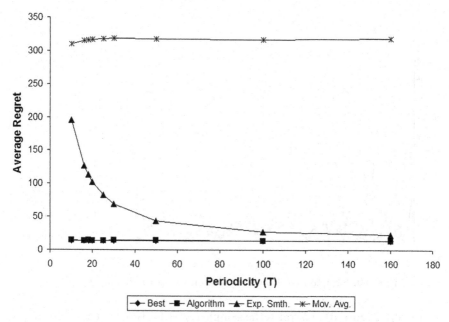

Fig. 8. Performance for high-amplitude seasonality

function $D \sim N\left(500 + 200\sin(\frac{2\pi t}{20}), 7\right)$. From Figure 9, we see that the predictions of the algorithm are very close to the actual demand values; on the other hand, exponential smoothing follows the demand function with a time lag and results in significant losses.

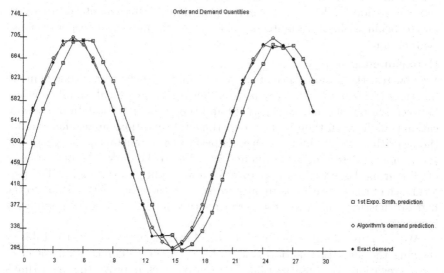

Fig. 9. Demand and order observations for $D \sim N\left(500 + 200\sin(\frac{2\pi t}{20}), 7\right)$

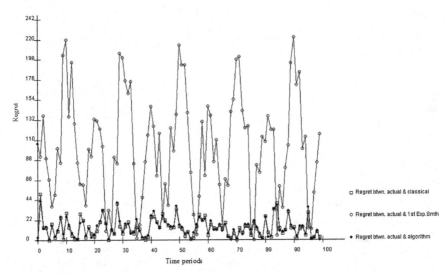

Fig. 10. Sample regret on a simulation run (experiment group 1)

Finally, Figure 10 illustrates the sample regret for three of the performance measures. (For clarity purposes, we have omitted moving average due to the different scales.) Since the large amplitude allows the algorithm to correctly identify and capture seasonality, the methodology performs much better a more traditional method.

Experiment group 2

For the experiments with moderate amplitude value, $b = 100$, we observe the same behavior as in the case of high amplitude (see Table 2). We set the cluster aggregation level to 0 because the probability of misclassification is very low to begin with, so recombining clusters would only lead to performance degradation.

Experiment group 3

We now test the performance of the algorithm for the low amplitude demand functions ($b = 5$). Note that $b < \sigma$ here; therefore, it is difficult to differentiate between seasonality and stochasticity and the probability of data misclassification is high. Even though these experimental conditions are challenging for the algorithm, results indicate that its performance remains quite acceptable (performance deterioration is about 20%). Figure 11 shows the performance of all four methods with respect to T for low-amplitude seasonality. The performance of the algorithm is slightly worse than the other methods due to the high stochasticity; however, the performance gap decreases as the periodicity increases.

We also investigate the benefits of cluster aggregation, with Table 3 showing the algorithm performance achieved under different levels of cluster aggregation. We observe that aggregation does improve the algorithm's performance, which was expected given our previous discussions about the

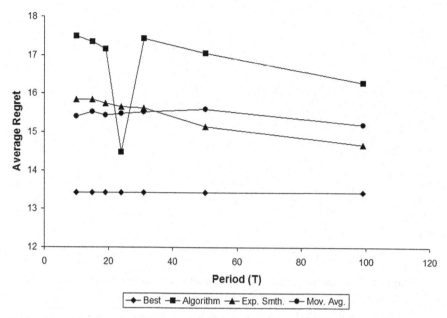

Fig. 11. Performance for low-amplitude seasonality with $CAL = 0$

clusters and the probability of misclustering. Figure 13 shows the performance improvement of the algorithm due to cluster aggregation for the demand process $D \sim N\left(500 + 5\sin(\frac{2\pi t}{10}), 7\right)$. An improvement of approximately 15% is achieved by using $CAL = 4$ compared to the case of no cluster aggregation at all. This is a remarkable improvement compared to the 20% in overall performance deterioration due to the low amplitude in seasonality when there is no cluster aggregation. As a rule of thumb, $CAL \approx T/3$ gives the best results when the periodicity T is small ($T \leq 20$), with CAL decreasing towards $T/5$ for higher values of T (e.g., $T = 99$). Finally, Figure 12 shows the behavior of the demand realizations and the algorithm's predictions for the demand process $D \sim N\left(500 + 5\sin(\frac{2\pi t}{10}), 7\right)$. Since the effect of the seasonality is insignificant, demand realizations appear to be independent and identically distributed. Therefore, the decision maker should assume that the demand process is an independent and identically-distributed (i.i.d.) random variable.

Figure 14 illustrates the behavior of the sample regret under three demand prediction models (moving average is omitted again for scale reasons) on a simulation run. We observe that, due to the small ratio of amplitude to standard deviation, the decision maker cannot easily distinguish seasonality from stochasticity.

As part of our experiments, we also investigate the clusters' mean and standard deviation values. Table 4 gives the mean and standard deviation values of the demand clusters after the warm-up period and end of the simulation run. The statistics are collected for the demand function

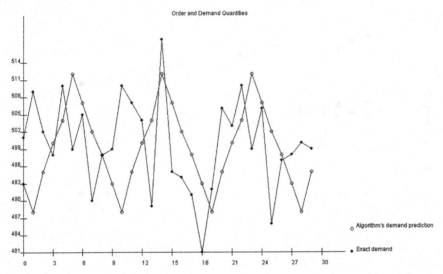

Fig. 12. Demand and order observations for $D \sim N\left(500 + 5\sin(\frac{2\pi t}{10}), 7\right)$

$D \sim N\left(500 + 100\sin(\frac{2\pi t}{10}), 7\right)$; hence, we have 10 clusters. Table 4 also gives the theoretical cluster mean and standard deviation values in the last three columns. The abbreviations *W.*, *E. S.*, and *T.* stand for *Warm-up*, *End of Simulation*, and *Theoretical*, respectively. The numbers of data points in each

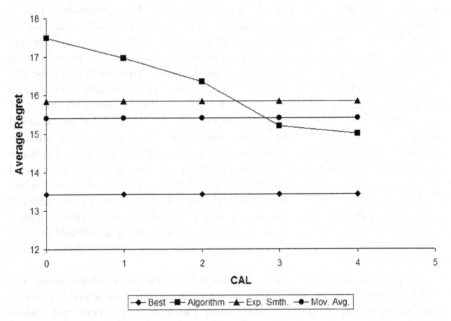

Fig. 13. Effect of *CAL* on performance for demand process $D \sim N\left(500 + 5\sin(\frac{2\pi t}{10}), 7\right)$

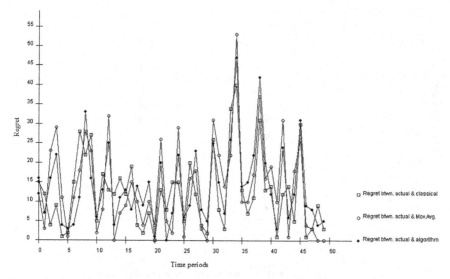

Fig. 14. Sample regret on a simulation run (experiment group 3)

cluster after the warm-up and at the end of the simulation, respectively, are 48 and 248. After the warm-up, mean and standard deviations of the clusters differ from their theoretical counterparts. This difference is attributable to the misclassification of observations in the cluster construction phase. But when we look at the end-of-simulation values, we see that the gap between the theoretical and the actual values has narrowed. This is because we add new demand observations to clusters during the simulation, so (provided misclassification is not too frequent) each cluster converges with its correct population in time. In other words, clusters that are ill-structured at the very beginning tend to "heal" autonomously over time. Figures 15 and 16 show the convergence of the mean and standard deviation values.

Table 4. Evolution of cluster means and standard deviations for $D \sim N\left(500 + 100\sin(\frac{2\pi t}{10}), 7\right)$

Cluster	W. Mean	W. Std.	E. S. Mean	E. S. Std.	T. Mean	T. Std.
1	398.78	4.08	403.67	6.98	404.89	7
2	410.17	4.32	405.55	6.78	404.89	7
3	436.65	3.42	440.79	6.36	441.22	7
4	447.48	4.15	442.43	7.01	441.22	7
5	494.92	5.05	499.14	6.73	500.00	7
6	506.52	4.34	500.95	6.42	500.00	7
7	553.55	4.56	558.00	7.08	558.78	7
8	564.40	4.22	560.08	6.99	558.78	7
9	589.91	4.09	594.35	7.22	595.11	7
10	600.40	3.43	595.37	7.11	595.11	7

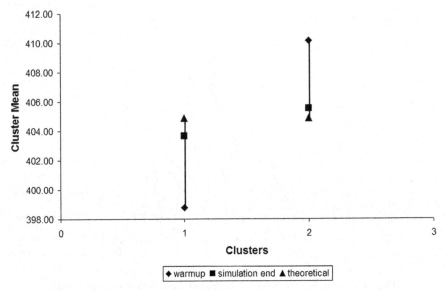

Fig. 15. Convergence of cluster 1 and 2 mean values for $D \sim N\left(500 + 100\sin\left(\frac{2\pi t}{10}\right), 7\right)$

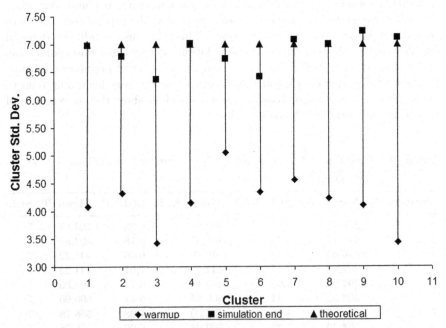

Fig. 16. Convergence of cluster standard deviation values for $D \sim N\left(500 + 100\sin\left(\frac{2\pi t}{10}\right), 7\right)$

5 Conclusions

We have proposed a dynamic, data-driven algorithm to the news vendor problem under cyclical demand. The algorithm, which relies on the clustering of the historical data, requires the estimation of only one parameter—the demand periodicity—is robust to estimation errors, and exhibits excellent empirical performance. Future research directions include incorporating the data-driven framework to finite-horizon inventory management problems, as well as extending the clustering approach to multiproduct environments with correlated demand.

Acknowledgments

The authors would like to thank two anonymous reviewers for their helpful suggestions. This work was supported in part by NSF Grant DMI-0540143.

References

1. Arrow K, Harris T, Marschak J (1951) Optimal Inventory Policy. Econometrica 19:250-272.
2. Arrow K, Karlin S, Scarf H (1958) Studies in the Mathematical Theory of Inventory and Production, Stanford University Press Palo Alto CA.
3. Bertsimas D, Thiele A (2004) A Data-Driven Approach to Newsvendor Problems. Technical report, MIT Cambridge, MA.
4. Gallego G, Moon I (1993) The Distribution-Free Newsboy Problem: Review and Extensions. J Oper Res Soc 44:825-834.
5. Huber P (2003) Robust Statistics. Wiley Series in Probability and Statistics, Wiley New York, NY.
6. Jain A, Murty M, Flynn P (1999) Data Clustering: A Review. ACM Comput Surv 31:264-323.
7. Kapuscinski R, Tayur S (1998) A Capacitated Production-Inventory Model with Periodic Demand. Oper Res 46:899-911.
8. Karlin, S (1960) Dynamic Inventory Policy with Varying Stochastic Demands. Man Sci 6:231-258.
9. Karlin S (1960) Optimal Policy for Dynamic Inventory Process with Stochastic Demands Subject to Seasonal Variations. J Soc Industr Appli Math 8:611-629.
10. Ketzenberg M, Metters R, Semple J (2006) A Heuristic for Multi-item Production with Seasonal Demand. IIE Trans 38:201-211.
11. Lovejoy W (1990) Myopic Policies for Some Inventory Models with Uncertain Demand Distributions. Man Sci 36:724-738.
12. MacQueen J (1967) Some Methods for Classification and Analysis of Multivariate Observations. Proceedings of 5th Berkeley Symposium on Mathematical Statistics and Probability, Berkeley, University of California Press 31:264-323.
13. Metan G, Thiele A (2006) An Adaptive Algorithm for the Optimal Sample Size in the Non-Stationary Data-Driven Newsvendor Problem. In: Extending the Horizon: Advances in Computing, Optimization, and Decision Technologies, pp.77-96. Springer, New York.

14. Metters R (1998) Producing Multiple Products with Stochastic Seasonal Demand and Capacity Limits. J Oper Res Soc 49:263-272.
15. Morton T (1978) The Nonstationary Infinite Horizon Inventory Problem. Man Sci 24:1474-1482.
16. Morton T, Pentico D (1995) The Finite Horizon Nonstationary Stochastic Inventory Problem: Near-Myopic Bounds, Heuristics, Testing. Man Sci 41:334-343.
17. Porteus E (2002) Stochastic Inventory Theory, Stanford University Press Palo Alto CA.
18. Scarf H (1958) A Min-Max Solution to an Inventory Problem. In: Studies in the Mathematical Theory of Inventory and Production, Stanford University Press Palo Alto CA.
19. Veinott A (1965) Optimal Policy for a Multi-Product, Dynamic, Nonstationary Inventory Problem. Man Sci 12:206-222.
20. Zipkin, P (1989) Critical Number Policies for Inventory Models with Periodic Data. Man Sci 35:71-80.

Logic-based Multiobjective Optimization for Restoration Planning

Jing Gong[1], Earl E. Lee[2], John E. Mitchell[3], and William A. Wallace[4]

[1] Department of Decision Sciences and Engineering Systems
Rensselaer Polytechnic Institute, Troy, NY 12180
gongj@rpi.edu
[2] Department of Civil and Environmental Engineering
University of Delaware, Newark, DE 19702
elee@udel.edu
[3] Department of Mathematical Sciences
Rensselaer Polytechnic Institute, Troy, NY 12180
mitchj@rpi.edu
[4] Department of Decision Sciences and Engineering Systems Rensselaer
Polytechnic Institute, Troy, NY12180
wallaw@rpi.edu

Summary After a disruption in an interconnected set of systems, it is necessary to restore service. This requires the determination of the tasks that need to be undertaken to restore service, and then scheduling those tasks using the available resources. This chapter discusses combining mathematical programming and constraint programming into multiple objective restoration planning in order to schedule the tasks that need to be performed. There are three classic objectives involved in scheduling problems: the cost, the tardiness, and the make span. Efficient solutions for the multiple objective function problem are determined using convex combinations of the classic objectives. For each combination, a mixed integer program is solved using a Benders decomposition approach. The master problem assigns tasks to work groups, and then subproblems schedule the tasks assigned to each work group. Hooker has proposed using integer programming to solve the master problem and constraint programming to solve the subproblems when using one of the classic objective functions. We show that this approach can be successfully generalized to the multiple objective problem. The speed at which a useful set of points on the efficient frontier can be determined should allow the integration of the determination of the tasks to be performed with the evaluation of the various costs of performing those tasks.

1 Problem Description

Our previous work [1], [2] introduced the interdependent layered network model (ILN). This model was a network flow-based model of civil infrastructure systems incorporating their interdependencies (or interconnectedness).

W. Chaovalitwongse et al. (eds.), *Optimization and Logistics Challenges in the Enterprise*, Springer Optimization and Its Applications 30, DOI 10.1007/978-0-387-88617-6_11, © Springer Science+Business Media, LLC 2009

The work identified five types of interdependency and mathematical representations of each were developed. This model of the system of systems could demonstrate the cascading effects of a disruption; allow for collaborative restoration planning across the set of systems; and could show vulnerability in a system due to its reliance on other systems.

In [1], a scenario was developed to exercise the model. Using data provided by the respective system managers, a realistic representation of the power, communications and subway systems of a large portion of Manhattan was developed. A disruption with effects similar to the September 11, 2001 attacks on the World Trade Center was proposed. The disruption was entered into the ILN with the output showing the service outages which resulted directly from the disruption and the impact due to the cascading effect of the disruption due to the interconnectedness of the systems.

The next step was the development of a restoration plan which met the constraints placed upon planners by the various management agencies involved. In general, the restoration plan consisted of the running of temporary lines along the streets of Manhattan to restore power and phone service until permanent repairs could be made. New constraints included the capacity of the temporary lines, limits on streets where lines may or may not be run along or crossed, etc. Consider the street sections in the area of interest as arcs. The output of the ILN was a subset of those arcs which met the constraints of the planners and restored the services. With a plan developed, another module of the ILN developed a schedule for the set of tasks. This scheduling module, like the ILN was a mixed-integer problem.

This chapter builds upon that work. In this case, a hybrid mixed integer and constraint programming modeling framework is presented for scheduling. Based on the optimal restoration plan developed in the ILN, this hybrid model determines how to accomplish the plan, i.e., the assignment and sequence of activities.

In the example provided in this chapter, all resources are considered unary. Workers and equipment are bundled into work groups and have sufficient skills to accomplish any of the tasks in the set. Each task only requires one work group. The only differences between the work groups are their cost and time required to complete each task. So the decision problem is how to assign repair tasks to these groups and schedule tasks for each work group in an optimal fashion. Future work will include shared resources. For example, supervisors might be a shared resource. Each task would have its own unique requirement for the number of supervisors at each job site, with a fixed number of supervisors available during each time interval.

Requirements from a planner or a manager could be: spend as little money as possible; complete all tasks by their due dates; or finish all tasks as soon as possible. So there are three measures that need to be minimized: the cost, the tardiness of each task and the make span of all tasks. They are formulated as objectives in our model. Our approach to address this multiobjective issue is to minimize the weighted sum of the three objectives. Decision making for the optimal solution can be viewed as a procedure to trade-off among them.

2 Literature Review

Constraint programming (CP) developed as a computer science technology which employs developments in artificial intelligence and computer programming languages [3]. It provides the capability of defining the structure of the search space and specifying the algorithm for exploring the search space, which make it possible to solve some particular problems efficiently, e.g., some combinatorial optimization problems [4]. Constraint programming is viewed as more like a method of solving the problem, not just a modeling language, like AMPL, GAMS and so on, although languages supporting constraint programming have a strengthened expressiveness compared with those traditional mathematical programming languages. More and more researchers discuss incorporation of this technology into the operations research field.

Recently, research interest in combining constraint programming with mathematical programming arose. Kim and Hooker [5] applied a hybrid solution method which combines constraint programming and linear programming to fixed charge network flow problems. They solved the problem by combining constraint propagation and a projected relaxation and got a significant computational performance improvement compared with a commercial mixed integer programming code. Hooker [6] compared those two technologies and pointed out that they have complementary strengths in solving integer and mixed integer problems, although one originated from mathematics and the other from engineering. Hooker [7] proposed a search-infer-relax framework for integrating solution methods. Searching is a procedure of enumerating all values in the domain. Inference derives implicit constraints to reduce the domain. Relaxation solves a relaxed problem for a bound on the original problem. Branch-and-bound in mathematical programming, specifically, integer programming and mixed integer programming, is a particular strategy of searching the whole space. It can be viewed as a sophisticated enumeration, which makes it possible to combine two solution methods in this integration framework.

Another integration scheme is applying constraint programming into classic Benders decomposition for some problems whose subproblem is easily solved by constraint programming technology. Hooker [8] extended the idea of the classic Benders decomposition to a logic-based Benders. The logic-based Benders cuts are obtained by solving the inference dual of the subproblem. The solution of the inference dual can prove optimality when variables of the master problem take certain values. The difference between the logic-based Benders cut and classic Benders cut is that no standard form exists for the logic-based cut. The subproblem could be a linear program (LP), mixed integer program (MIP), or CP, and cuts are generated by logical inference on the solution of the subproblem. However, generally the master problem is an MIP; therefore, cuts are formulated as linear constraints. The classic Benders decomposition is strengthened by introducing the logic-based Benders cut.

Scheduling is a decision-making process of allocating limited resources to tasks over time with the goal of optimizing a certain objective [9]. Scheduling problems can be solved by dynamic programming or integer programming models. Most scheduling problems do not have a polynomial time algorithm and are NP-hard problems. Pinedo [9] presented a complexity hierarchy of scheduling problems. The IP model for scheduling problems contains much symmetry, which makes the model hard to be solved by branch-and-bound. Lustig and Puget [3] pointed out that constraint programming is often better than integer programming in application to sequencing and scheduling.

Jain and Grossmann [10] applied the logic-based Benders cut into a planning and scheduling problem that involves cost minimization. Harjunkoski and Grossmann [11] extended the decomposition strategy for a multistage planing and scheduling problem. Maravelias and Grossmann [12] applied the decomposition to a scheduling of batch plants formulated as the state task network. Hooker [13, 14] developed the logic Benders cuts for three different objectives (minimum cost, minimum make span and minimum total tardiness) in general planning and scheduling problems. He modeled them as three different problems, solving each of them one at a time. However, the requirements from a real-world application might be that several goals need to be achieved simultaneously. Multiple objective optimization might be the case. This chapter discusses how to implement logic-based Benders cuts for multiple-objective optimization in planning and scheduling problems. It can be viewed as an extension to Hooker's work.

As mentioned before, a logic-based Benders cut is based on the inference of the subproblem, so there is no standard formulation for the cut. It is problem specific. Different objectives require different logical inference for cuts, so complicated objectives might result in difficulty in cut generation. This chapter proposes a logic-based Benders cut approach for a planning and scheduling problem with combined multiple objectives.

3 Integrated Solution of the Problem

As stated in Section 1, an assignment and scheduling problem must be solved for a set of tasks which comprise the restoration plan. The optimal solution assigns the tasks to a work group and then arranges the tasks into a schedule. Each task has a due date and the objective function will include a term to minimize the amount of time each task exceeds its due date by (referred to later as tardiness). The cost and time to complete a task depends on the work group to which it is assigned. The objective function will include terms to minimize the cost of completing all the tasks, the total tardiness, and the time to complete the last task (the make span). Each of these three terms of the objective function will be weighted to reflect the priorities of the decision maker. The problem is fundamentally an assignment and scheduling problem. It falls into a category of problems that have proved to be difficult to solve [10, 13]. An integrated algorithm based on a MILP master problem and several CP subproblems is presented in this chapter. The idea behind it is that the master problem assigns tasks to groups; for each work group, a subproblem schedules the tasks assigned to it. When subproblems prove the optimal solution from the master problem is also feasible (for each work group, a schedule which achieves the same optimality as the master problem can be found with satisfaction of all constraints), the algorithm will stop and the solution from the master problem is optimal to the original problem. Otherwise, relevant cutting planes are added to the master problem and the above procedure iterates. Basically, the original problem is decomposed to several smaller problems as long as an assignment is set. Because parallel groups will not interact with each other, each subproblem can be solved individually, which causes a dramatic improvement in computational performance.

3.1 Master Problem

The master problem determines the assignment of tasks to work groups. The formulation requires the following notation:

i : a task,
m : a work work group,
I : the set of tasks,
M : the set of work groups,
I_m : the set of tasks assigned to work group m,
x_{im} : $= 1$ if task i is assigned to work group m,
t_i : starting time of task i,
s_i : tardiness of task i, always ≥ 0, since we only consider penalty,
y : make span, i.e., completion time of the last task,
R_m : the amount of resource bundled to work group m,
c_{im} : the cost of work group m completing task i,
p_{im} : time of work group m requiring to complete task i,
d_i : the due time of task i,
q_i : the work group that task i is assigned to,
r_{im} : the amount of resource work group m requires to complete task i.

The master problem is formulated as a mixed integer program.

$$\min_{x,s,y,t} \alpha \sum_{i \in I} \sum_{m \in M} c_{im} x_{im} + \beta \sum_{i \in I} s_i + \gamma y \qquad (1)$$

$$\text{subject to} \qquad t_i + \sum_{m \in M} p_{im} x_{im} - s_i \leq d_i \quad \forall i \in I \qquad (2)$$

$$t_i + \sum_{m \in M} p_{im} x_{im} \leq y \quad \forall i \in I \tag{3}$$

$$\sum_{m \in M} x_{im} = 1 \quad \forall i \in I \tag{4}$$

$$\sum_{i \in I} x_{im} p_{im} \leq y \quad \forall m \in M \tag{5}$$

$$\texttt{integer cuts} \tag{6}$$

$$x_{im} = 0 \quad \text{if } r_{im} > R_m \tag{7}$$

$$x_{im} \in \{0, 1\} \ \forall i \in I \text{ and } \forall m \in M \tag{8}$$

$$s_i \geq 0 \quad \forall i \in I \tag{9}$$

$$y \geq 0 \tag{10}$$

The master problem includes all three measures: total cost, tardiness and make span. Weights α, β, γ are used to trade off among those three objectives. The objective function tries to minimize the weighted sum of them. Equation (2) ensures that task i is completed before the due time, otherwise tardiness s_i is incurred. Equation (3) ensures the make span is no smaller than the completion time of every task. Equation (4) is an assignment constraint: each task can only be assigned to one group. Equation (5) ensures that total performing time of each group should be less than or equal to make span. Equation (6) are integer cuts from all subproblems. These integer cuts impose more and more strict restrictions on variable x_{im}, s_i, y as the algorithm proceeds, and eventually drive those variables to the optimal solution of the original problem. Equation (7) ensures each task is only assigned to a work group which has enough resources to complete it.

Note that the master problem does not have sequencing constraints, which means for each task no restrictions exist on starting time, so starting times in the solution of master problem might not be feasible. Since starting times imply a schedule of tasks, the master problem cannot yield feasible schedules because of infeasible starting times. Tardiness s_i and make span y are associated with schedules, infeasible schedules result in infeasible s_i and y. Therefore, the master problem yields the optimal assignment x_{im}, optimal tardiness s_i and optimal make span y, but their feasibility cannot be guaranteed. That requires the subproblems to play a role in checking feasibility of the optimal solution from the master problem.

3.2 Subproblem (CP model)

The master problem cannot guarantee what it yields is feasible to the original problem, so it only solves the problem partially. The goal of the subproblem is to examine whether this partial solution can be extended to a full solution for the original problem, i.e., to feasible schedules. When the subproblem finds the schedule which can achieve the same minimum tardiness and make span as the master problem, the original problem is solved. The cost is not involved in the subproblem because it is only related to the assignment result and will not change. The subproblem only checks the result of the master problem and it will not change the assignment result.

The subproblem requires as input the assignment results from the master problem. For each group, the set of tasks composing the group is determined by the master problem. That is, the assignment result x_{im} from the master problem can derive the value of q_i indicating to which group task i is assigned. For each group m, a set of tasks assigned to it, $I_m := \{i \in I : q_i = m\}$, can be determined. Then a subproblem m is modeled as:

$$\min \ \beta \sum_{i \in I_m} s'_i + \gamma y' \tag{11}$$

$$\text{subject to} \ \ i.end \leq y^* + y' \quad \forall i \in I_m \tag{12}$$

$$i.start \leq d_i + s'_i - p_{iq_i} \quad \forall i \in I_m \tag{13}$$

$$i.duration = p_{iq_i} \quad \forall i \in I_m \tag{14}$$

$$i_1 \ \text{precedes} \ i_2 \quad \forall (i_1, i_2) \in \text{precedence pairs}, i_1 \neq i_2 \in I_m \tag{15}$$

$$i \ \text{require} \ q_i \quad \forall i \in I_m \tag{16}$$

$$s'_t \geq 0 \ \ \forall i \in I_m \tag{17}$$

$$y' \geq 0 \tag{18}$$

The model is a pure CP model and seeks a schedule which minimizes the weighted sum of total tardiness and make span. Variable task i is a special object in OPL [17] and has some attributes such as starting time, ending time, duration and so on. Variable s'_i is the new tardiness. Compared with s_i from the master problem, it is the exact tardiness since the subproblem imposes effective constraints on it. Parameter y^* is the optimal make span of the master problem. Variable y' is the slack in the make span of the master problem, as defined in Equation (12). Equation (13) ensures every task is completed before its due time. Equation (14) specifies the processing time of task i. Equation (15) is a precedence constraint. A work group m is a unary resource. Equation (16) is a unary resource constraint and makes sure no two tasks requiring it are scheduled at the same time.

3.3 Algorithm

Let x^{k^*}, s^{k^*}, y^{k^*} be the optimal solution of the master problem at iteration k. For each group m, a subproblem is formulated. So the number of subproblems is the number of groups and the original problem is decomposed into several small size problems. They determine our three objectives: cost, tardiness and make span, respectively. Let $s'^{(m,k)^*}$, $y'^{(m,k)^*}$ be the optimal solution of subproblem m at iteration k. Different schedules in subproblems will not change the optimal cost, because the cost is only related to the assignment. Thus, all we are concerned with at iteration k is whether or not the minimum tardiness and make span from the master problem can also be achieved in the subproblem. That is, whether the sum of the differences between the master problem and the subproblem, i.e., difference in tardiness and difference in make span, equals zero. As mentioned before, constraints on those two terms in the subproblem are always tighter than those in the master problem; consequently, those two values in the subproblem are larger than those in the master problem. Let's introduce a new term defined as follows:

$$D^{(m,k)} = \sum_{i \in I_m^k} (s')_i^{(m,k)*} - \sum_{i \in I_m^k} s_i^{k*} + (y')^{(m,k)*} \tag{19}$$

where $I_m^k = \{i \in I : x_{im}^{k*} = 1\}$.

The *subproblem difference* $D^{(m,k)}$ is a measure of the difference in tardiness and make span between the master problem solution and the subproblem solution. When $\sum_{m \in M} D^{(m,k)} = 0$, at iteration k the minimum tardiness and make span from the master problem can also be achieved by optimal schedules from the subproblems, so the optimal solution for the original problem, i.e, an optimal assignment and schedules, is obtained. Otherwise, the current optimal solutions from master problem x_{im}^{k*}, s_i^{k*}, y^{k*} needs to be cut off. Different values of those two measurements determine the cutting plane. There are four cases:

1. Subproblem m is infeasible, indicating that the assignment from the master problem is incorrect, so the cutting plane at iteration k is to cut off the current assignment:

$$\sum_{i \in I_m^k} x_{im} \leq \sum_{i \in I_m^k} x_{im}^{k*} - 1 \tag{20}$$

2. $D^{(m,k)} > 0$ indicates that tardiness s_i^{k*} and make span y^{k*} from the master problem might not be large enough, so the cutting plane at iteration k either cuts off the current assignment or increases tardiness and make span:

$$\sum_{i \in I_m^k} x_{im} \leq \sum_{i \in I_m^k} x_{im}^{k*} - 1$$

or

$$\sum_{i \in I_m^k} \beta s_i + \gamma y \geq \sum_{i \in I_m^k} \beta(s')_i^{(m,k)*} + \gamma \left(y^{k*} + (y')^{(m,k)*} \right) \tag{21}$$

3. $D^{(m,k)} \leq 0$ indicates that current solution from the master problem is feasible for subproblem m.

4. If $\sum_{m \in M} D^{(m,k)} = 0$ holds for all subproblems, then the current solutions are optimal. The optimal assignment is from the master problem and optimal schedule is from the subproblems.

There are two fundamental constraints comprising the above cuts. Constraint (20) cuts off the current assignment by restricting the sum of assignment variables x_{im}, $i \in I_m^k$ to be less than the current value. This cut only involves the assignment variables currently assigned to a group, so it is tight enough to make the feasible region shrink efficiently. Constraint (21) increases the tardiness and make span by imposing a lower bound, i.e., the optimal weighted sum of them from the current subproblem, which implies the weighted sum of tardiness and make span should be at least as large as $\sum_{i \in I_m^k} \beta s'^{(m,k)^*}_i + \gamma(y^{k^*} + y'^{(m,k)^*})$. In other words, constraint (21) employs a nice bound to drive variable s_i, $i \in I_m^k$ and y towards optimality.

The optimal solution of the master problem can be infeasible in subproblem m in one of two ways: either (1) the assignment is wrong and not all of the jobs currently assigned to m can be assigned to group m, or (2) it is possible to assign all these jobs to group m, but then either the tardiness or the make span will be larger than the values currently calculated in the master problem. So the cut is either to cut off the assignment or to increase the make span and tardiness values in the master problem. Hooker's cut [13] tried to build up the relationship between the objective value and the assignment variables. When the problem has one of the classic single objectives of cost or make span or tardiness, it is possible to place bounds on how the objective value is changed as the values of the assignment variables change. Thus, Hooker is able to derive more specialized cuts for these objectives. However, when the objective function is a convex combination of the classic objectives, it is hard to derive such a relationship. So we use different logic to derive the cut and drive the solution to optimality. Nonetheless, as will be seen in the section on computational results, these general cuts are still powerful and enable the fast solution of realistic problems.

We use disjunctive constraints to represent this logic for case 2. The binary variable $z^{(m,k)}$ is involved in the formulation of disjunctive constraints. Whether constraint (20) and constraint (21) are effective or not is dependent on values of this binary variable. So case (2) can be rewritten as follows:

$$\sum_{i \in I_m^k} x_{im} \leq \mathsf{M} z^{(m,k)} + \sum_{i \in I_m^k} x_{im}^{k^*} - 1$$

$$\sum_{i \in I_m^k} \beta s_i + \gamma y \geq z^{(m,k)} \left[\sum_{i \in I_m^k} \beta(s')_t^{(m,k)^*} + \gamma \left(y^{k^*} + (y')^{(m,k)^*} \right) \right]$$

M is equal to the total number of tasks, which is an upper bound which is large enough for the sum of some assignment variables, since all assignment variables are binary. When $z^{(m,k)}$ takes a value of 0, constraint (20) will take effect. Otherwise, the cut becomes a redundant upper bound for x_{im}. Likewise, both variable s_i and y have a lower bound 0. When $z^{(m,k)}$ takes a value of 1, constraint (21) will take effect. When it takes a value of 0, the cut will turn out to be a redundant lower bound for s_i and y. The optimal value of $z^{(m,k)}$ is determined by the algorithm; hence the algorithm determines whether it is better to enforce (20) or (21).

We are currently investigating methods to strengthen this cut, which should make the algorithm even faster. Hooker [15] (page 83) notes that "the success of a Benders method often rests on finding strong Benders cuts that rule out as many infeasible solutions as possible." For example, Hu et al. [16] found that strenghtening cuts was essential in a Benders type of approach for linear programming problems with equilibrium constraints.

Figure 1 depicts the whole solving procedure. The original problem is divided into two parts: the assignment problem and the scheduling problem. The former is the master problem and includes all assignment constraints. As the algorithm proceeds, more and more integer cuts are added to it. The latter is the subproblem and includes all sequencing constraints. The algorithm begins with solving a master problem, solutions of the master problem specify each subproblem. If a subproblem is infeasible, the cut which cuts off the current assignment will be added to the master problem, and the algorithm goes back to the beginning. If the subproblem is feasible, its optimal value will be compared with the corresponding part of master problem value. The algorithm ends up with equality of those two values. If they are not equivalent, disjunctive cuts will be added to the master problem and the algorithm goes back to the beginning. Briefly, subproblems at each iteration check if

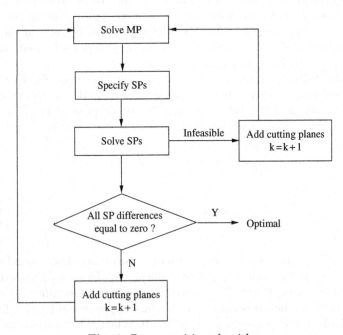

Fig. 1. Decomposition algorithm

the master problem solution can be extended to feasible schedules. If so, the optimal solution for the original problem is found. Otherwise, cutting planes are added to the master problem.

4 Computational Results

All CP subproblems were solved by ILOG Solver and ILOG Scheduler [18]. All MIP master problems were solved by ILOG CPLEX 8.0 [19]. The decomposition algorithm was implemented using the script language in ILOG OPL Studio 3.7 [17]. There are three work groups and 30 tasks in our problem. The problem is of a similar size to those solved by Hooker [14]. At present, the constraint programming solver for the subproblem has difficulty with instances that are much larger than this.

4.1 Objective Function Scaling

As defined in Equation (1), the goal of our model is to find the optimal solution for the combined objectives (the cost, the tardiness and the make span). They are not comparable because of the different units they use, so it is hard to reflect the priorities of the decision maker by weights. Our approach to address this issue is to convert different units of measures into the same unit. The conversion can be viewed as a process of finding the relationship of different units, i.e., how much does it cost when there is a one-day delay? How much does it cost for every working day? We try to measure make span and tardiness by money. Therefore, the objective can be rewritten as follows:

$$\alpha \sum_{i \in I} \sum_{m \in M} c_{im} x_{im} + \beta \eta_1 \sum_{i \in I} s_i + \gamma \eta_2 y \tag{22}$$

where η_1 is the ratio of the cost to the tardiness, η_2 is the ratio of the cost to the make span.

Generally, ratios η_1 and η_2 need to be set by model users. Sometimes, their values can be derived mathematically. Take η_2 as a example, set α to be 100, β to be 0 and γ to be 0, solve the problem for the minimum cost ($cost^*$), and then set α to be 0, β to be 0 and γ to be 100, solve the problem for the minimum make span ($makespan^*$). Then η_2 can be set as follows:

$$\eta_2 = \frac{cost^*}{makespan^*}$$

The argument for this method is that given the same amount of resources and same constraints two goals can be achieved individually, so the ratio of their optimal values can be viewed as the ratio of the cost to the make span for those resources and constraints. This method only applies to the problem with nonzero optimal value for each objective; otherwise, an infinitely large ratio could be derived.

For the problem discussed in this chapter, we get three extreme points by considering only one objective at a time. They are (442,240,60), (503,0,60) and (502,240,20), corresponding to three sets of (α, β, γ) equal to (100,0,0), (0,100,0) and (0,0,100), respectively. So, in order to complete all restoration tasks, we need to pay at least 442 dollars, which is the lower bound of the cost. If we do not care about how much completing the task will cost or how long completing all tasks will take, then we can complete all tasks by its due date. If we do not care about the cost or if every task is done by the due date, then we can finish all the tasks in 20 days:

$$\text{cost}^* = 442$$
$$\text{tardiness}^* = 0$$
$$\text{make span}^* = 20$$

The value of η_2 is obtained by dividing $cost^*$ by $makespan^*$, giving 22. While the above method does not apply to η_1 since $tardiness^*$ is 0, it is set to be 1 by user's experience.

4.2 Computational Performance

Table 1 shows computational performance for different values of weight. The # cuts gives the number of cuts generated to solve the problem. *Iterations* lists the number of times the master problem was solved. *Solving time* is the CPU

Table 1. Computational performance of algorithm for different combinations of α, β, γ

No.	α	β	γ	# Cuts	Iterations	Solving time (seconds)
1	0	75	25	28	22	1.834
2	5	70	25	28	21	1.812
3	10	65	25	27	20	1.545
4	15	60	25	27	20	1.672
5	20	55	25	21	17	1.187
6	25	50	25	18	15	0.766
7	30	45	25	14	12	0.547
8	35	40	25	14	12	0.673
9	40	35	25	10	9	0.500
10	45	30	25	9	8	0.391
11	50	25	25	7	6	0.172
12	55	20	25	5	5	0.156
13	60	15	25	2	3	0.079
14	65	10	25	2	3	0.078
15	70	5	25	2	3	0.078
16	75	0	25	0	1	<0.001

time for solving the model. It was calculated by summing up solving times of the MIP model and three CP models. Different combinations of the three weights α, β, γ lead to different complexities of problems. As the proportion of α decreases, deviation of the master problem solution from the optimal solution to the original problem becomes larger. Thus more iterations are involved in the algorithm.

4.3 Optimal Solutions

Tables 2–10 display optimal solutions for different priorities on three objectives: cost, tardiness and make span. Almost all of the problems could be solved in less than four seconds. The exceptions are when the make span has low priority, in particular γ is equal to 5 or smaller. We found that the problem could not be solved in two hours for many of these priority combinations with $\gamma = 0$ so these results have been omitted. The sum of the three weights is 100. Each weight is held constant at three different levels, i.e., 25, 50 and 75, and the other two weights varied by multiples of five. It can be seen that most of the time optimal solutions do not change as weights change. This uneven distribution of optimal solutions was noted by Das and Dennis [20]. They pointed out that an evenly distributed set of weights fails to produce an evenly distributed set of points from all parts of the Pareto set. The result set presents frontier values for different weights. The algorithm returns a Pareto optimal point as long as all the weights are strictly positive.

Let us take Table 2 as an example. In this case, the weight of make span, γ, is held constant at 25 and five optimal solution sets are obtained by varying

Table 2. γ held constant at 25

No.	α	β	γ	Cost (dollars)	Tardiness (days)	Make span (days)	Solving time (seconds)
1	0	75	25	497	3	20	1.834
2	5	70	25	495	3	20	1.812
3	10	65	25	495	3	20	1.545
4	15	60	25	495	3	20	1.672
5	20	55	25	495	3	20	1.187
6	25	50	25	495	3	20	0.766
7	30	45	25	495	3	20	0.547
8	35	40	25	495	3	20	0.673
9	40	35	25	494	4	20	0.500
10	45	30	25	494	4	20	0.391
11	50	25	25	483	3	21	0.172
12	55	20	25	483	3	21	0.156
13	60	15	25	483	3	21	0.079
14	65	10	25	483	3	21	0.078
15	70	5	25	483	3	21	0.078
16	75	0	25	483	240	21	<0.001

weights of cost and tardiness, α and β. It can be seen that when make span is 25% of the total objective, the three efficient (nondominated) solutions are $(495, 3, 20)$, $(494, 4, 20)$, and $(483, 3, 21)$. Those three solutions compose the frontier value set of optimal solutions at $\gamma = 25$ level. In this way, frontiers for different levels of the weights can be presented. The set of such frontiers provides the decision maker a nice picture of the correspondence between weights and optimal solutions.

Tables 3 and 4 show the efficient solutions are $(495,3,20)$ and $(494,4,20)$ when γ is held constant at 50 or 75. Table 5 shows the efficient solutions are $(495,3,20)$, $(494,4,20)$, $(488,0,21)$ and $(480,0,22)$ when α is held constant at 25. Table 6 shows the efficient solutions are $(494,4,20)$, $(483,3,21)$, $(472,3,23)$ and $(470,1,24)$ when α is held constant at 50. Table 7 shows the efficient solutions are $(483,3,21)$, $(471,7,23)$, $(461,7,26)$ and $(452,14,30)$ when α is held constant at 75. Table 8 shows the efficient solutions are $(495,3,20)$, $(494,4,20)$, $(483,3,21)$, $(472,3,23)$ and $(452,14,30)$ when β is held constant at 25. Table 9 shows the efficient solutions are $(495,3,20)$, $(494,4,20)$, $(483,3,21)$, $(480,0,22)$ and $(470,1,24)$ when β is held constant at 50. Table 10 shows the efficient

Table 3. γ held constant at 50

No.	α	β	γ	Cost (dollars)	Tardiness (days)	Makespan (days)	Solving time (seconds)
1	0	50	50	497	3	20	1.264
2	5	45	50	495	3	20	1.391
3	10	40	50	495	3	20	1.359
4	15	35	50	495	3	20	0.906
5	20	30	50	495	3	20	0.485
6	25	25	50	494	4	20	0.328
7	30	20	50	494	4	20	0.110
8	35	15	50	494	4	20	0.156
9	40	10	50	494	4	20	0.094
10	45	5	50	494	4	20	0.063
11	50	0	50	494	240	20	0.016

Table 4. γ held constant at 75

No.	α	β	γ	Cost (dollars)	Tardiness (days)	Make span (days)	Solving time (seconds)
1	0	25	75	495	3	20	1.001
2	5	20	75	495	3	20	1.609
3	10	15	75	495	3	20	0.469
4	15	10	75	494	4	20	0.172
5	20	5	75	494	4	20	0.094
6	25	0	75	494	240	20	0.031

Table 5. α held constant at 25

No.	α	β	γ	Cost (dollars)	Tardiness (days)	Make span (days)	Solving time (seconds)
1	25	0	75	494	240	20	0.016
2	25	5	70	494	4	20	0.047
3	25	10	65	494	4	20	0.095
4	25	15	60	494	4	20	0.203
5	25	20	55	494	4	20	0.235
6	25	25	50	494	4	20	0.280
7	25	30	45	495	3	20	0.485
8	25	35	40	495	3	20	0.469
9	25	40	35	495	3	20	0.625
10	25	45	30	495	3	20	0.718
11	25	50	25	495	3	20	0.828
12	25	55	20	495	3	20	1.110
13	25	60	15	488	0	21	3.751
14	25	65	10	488	0	21	3.769
15	25	70	5	480	0	22	14.203

Table 6. α held constant at 50

No.	α	β	γ	Cost (dollars)	Tardiness (days)	Make span (days)	Solving time (seconds)
1	50	0	50	494	240	20	0.016
2	50	5	45	494	4	20	0.046
3	50	10	40	494	4	20	0.031
4	50	15	35	494	4	20	0.109
5	50	20	30	494	4	20	0.141
6	50	25	25	483	3	21	0.204
7	50	30	20	483	3	21	0.109
8	50	35	15	483	3	21	0.780
9	50	40	10	472	3	23	3.655
10	50	45	5	470	1	24	2923

Table 7. α held constant at 75

No.	α	β	γ	Cost (dollars)	Tardiness (days)	Make span (days)	Solving time (seconds)
1	75	0	25	483	240	21	0.031
2	75	5	20	483	3	21	0.140
3	75	10	15	471	7	23	0.062
4	75	15	10	461	7	26	3.765
5	75	20	5	452	14	30	500

Table 8. β held constant at 25

No.	α	β	γ	Cost (dollars)	Tardiness (days)	Make span (days)	Solving time (seconds)
1	0	25	75	495	3	20	1.077
2	5	25	70	495	3	20	1.344
3	10	25	65	495	3	20	0.657
4	15	25	60	495	3	20	0.734
5	20	25	55	495	3	20	0.658
6	25	25	50	494	4	20	0.265
7	30	25	45	494	4	20	0.282
8	35	25	40	494	4	20	0.250
9	40	25	35	494	4	20	0.172
10	45	25	30	494	4	20	0.187
11	50	25	25	483	3	21	0.266
12	55	25	20	483	3	21	0.110
13	60	25	15	472	3	23	1.703
14	65	25	10	472	3	23	4.453
15	70	25	5	452	14	30	1998

Table 9. β held constant at 50

No.	α	β	γ	Cost (dollars)	Tardiness (days)	Make span (days)	Solving time (seconds)
1	0	50	50	497	3	20	1.297
2	5	50	45	495	3	20	1.375
3	10	50	40	495	3	20	1.313
4	15	50	35	495	3	20	1.718
5	20	50	30	495	3	20	0.875
6	25	50	25	494	4	20	0.797
7	30	50	20	494	4	20	1.342
8	35	50	15	483	3	21	1.156
9	40	50	10	480	0	22	6.705
10	45	50	5	470	1	24	724

Table 10. β held constant at 75

No.	α	β	γ	Cost (dollars)	Tardiness (days)	Make span (days)	Solving time (seconds)
1	0	75	25	496	3	20	1.844
2	5	75	20	495	3	20	2.250
3	10	75	15	495	3	20	2.220
4	15	75	10	488	0	21	2.751
5	20	75	5	480	0	22	3.111

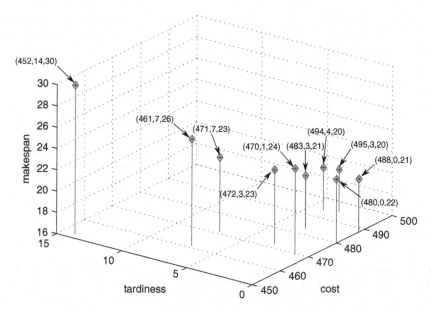

Fig. 2. Efficient points for the problem

solutions are (495,3,20), (488,0,21), (480,0,22) and (470,1,24) when β is held constant at 75.

Altogether, 11 efficient solutions were found and 10 of these are displayed in Figure 2. The additional solution is the efficient solution (442,240,60) found when minimizing the single objective of cost; this is not plotted because it would change the scaling of the picture too dramatically. The solutions found by minimizing the single objectives make span and tardiness were not efficient. As can be seen from the tables, these solutions are found very quickly, making it practical to determine them all. It appears that either an integrated MIP approach or a Benders decomposition approach where the subproblems were solved using integer programming techniques would require far more time. Because of the speed of the algorithm, all these efficient solutions could be presented to a decision maker, who could then choose an appropriate schedule of tasks to work groups in order to restore the interdependent layered network.

4.4 The Computational Benefit of a Hybrid Solver

Our computational results demonstrate that the hybrid decomposition approach can solve these scheduling problems effectively. In this subsection, we investigate pure integer programming and pure constraint programming formulations. The pure formulations were very similar to those in Jain and Grossmann [10]. Results are contained in Table 11 for a representative selection of problem instances from Table 1 with 30 tasks. The pure constraint programming approach was unable to solve these instances due to memory

Table 11. Computational performance comparison (solving time: seconds)

Tasks	α	β	γ	IP model	CP model	Decomposition
10	25	50	25	0.59	0.68	0.051
	50	25	25	0.55	0.79	0.030
	70	5	25	0.48	0.38	0.070
20	25	50	25	3144.01	*	0.633
	50	25	25	51.93	*	0.410
	70	5	25	16.39	149.5	0.090
30	25	50	25	>7200	*	0.766
	50	25	25	>7200	*	0.172
	70	5	25	1434.43	*	0.078

* Ran out of memory

requirements, so results for similar instances with fewer tasks are also included. The results clearly demonstrate the benefit of the hybrid approach, with run times several orders of magnitude better with the hybrid approach than with either of the other approaches.

5 Conclusions and Future Work

This chapter presents a general framework of a logic-based Benders cut for objective functions that combine the cost, the tardiness and the make span. Disjunctive cuts are generated based on logical inference. Our computational results show that the algorithmic framework allows rapid solution of these problems, enabling the determination of representative points on the efficient frontier set of optimal solutions for a multiple objective optimization problem.

Future work will address the following issues:

1. **Take shared resources into account**. As described earlier, the resources considered in the problems are bundled into work groups, i.e., unary resources. When shared resources are considered, each subproblem cannot be solved individually. The shared resources must be considered across all the subproblems. This will impose a challenge on the decomposition algorithm and cutting planes.

2. **Integrate determination of restoration plan with assignment and scheduling**. The example given separates the determination of the restoration plan from its cost and schedule. Solving the assignment and scheduling problem is the second step in restoration. Separation of the procedure into two steps could result in a case where the optimal restoration plan found in the first step is hard to implement in the second step for some reason, for example, limited budget, limited resources, and so on. Integration of the two steps into a single process might yield more efficient

restoration plans. The speed of solution of the scheduling problem should make this integration possible.

6 Acknowledgments

This research is supported by NSF grant CMS 0301661, Decision Technologies for Managing Critical Infrastructure Interdependencies

References

1. E.E. Lee, J.E. Mitchell, W.A. Wallace. Restoration of services in interdependent infrastructure systems: A network flows approach. *IEEE Transactions on Systems, Man, and Cybernetics, Part C: Applications and Reviews*, 37(6):1303–1317, 2007.
2. E.E. Lee. Assessing vulnerability and managing disruptions to interdependent infrastructure systems: A network flows approach. Ph.D. Thesis, Rensselaer Polytechnic Institute, 2006.
3. I.J. Lustig, J.F. Puget. Program does not equal program: Constraint programming and its relationship to mathematical programming. *Interfaces*, 31:29–53, 2001.
4. P.V. Hentenryck, L. Perron, J.F. Puget. Search and strategies in OPL. *ACM Transactions on Computational Logic*, 1:282–315, 2000.
5. H.J. Kim, J.N. Hooker. Solving fixed-charge network flow problems with a hybrid optimization and constraint programming approach. *Annals of Operations Research*, 115:95–124, 2002.
6. J.N. Hooker. Logic, optimization and constraint programming. *INFORMS Journal on Computing*, 14:295–321, 2002.
7. J.N. Hooker. A search-infer-and-relax framework for integrating solution methods. In Roman Barták and Michela Milano, editors, *Integration of AI and OR Techniques in Constraint Programming for Combinatorial Optimization Problems (CPAIOR)*, pages 243–257. Springer, 2005.
8. J.N. Hooker. *Logic-based Methods for Optimization: Combining Optimization and Constraint Satisfaction.* John Wiley, 2000.
9. M. Pinedo. *Scheduling: Theory, Algorithms and Systems.* Prentice Hall, 2002.
10. V. Jain, I.E. Grossmann. Algorithms for hybrid MILP/CP models for a class of optimization problems. *INFORMS Journal on Computing*, 13:258–276, 2001.
11. I. Harjunkoski, I. E. Grossmann. Decomposition techniques for multistage scheduling problems using mixed-integer and constraint programming methods. *Computers and Chemical Engineering*, 26:1533–1552, 2002.
12. C.T. Maravelias, I.E. Grossmann. A hybrid MILP/CP decomposition approach for the continuous time scheduling of multipurpose batch plants. *Computers and Chemical Engineering*, 28:1921–1949, 2004.
13. J.N. Hooker. Planning and scheduling by logic-based benders decomposition. *Operations Research*, 55(3):588–602, 2007.
14. J.N. Hooker. A hybrid method for planning and scheduling. *Constraints*, 10:385–401, 2005.

15. J.N. Hooker. *Integrated Methods for Optimization*. Springer, 2007.
16. J. Hu, J.E. Mitchell, J.S. Pang, K.P. Bennett, G. Kunapuli. On the global solution of linear programs with linear complementarity constraints. *SIAM Journal on Optimization*, 19(1):445–471, 2008.
17. ILOG Inc. *ILOG OPL Studio 3.7.1 Language Manual*. ILOG Inc. Mountain View, 2002.
18. ILOG Inc. *ILOG OPL Studio 3.7.1 User's Manual*. ILOG Inc. Mountain View, 2002.
19. ILOG Inc. *ILOG CPLEX 8.0 User's Manual*. ILOG Inc. Mountain View, 2002.
20. I. Das, J. Dennis. A closer look at drawbacks of minimizing weighted sums of objectives for Pareto set generation in multicriteria optimization problems. *Structural Optimization*, 14:63–69, 1997.

Part IV

Networking and Transportation

The Aircraft Maintenance Routing Problem

Zhe Liang[1] and Wanpracha Art Chaovalitwongse[2]

[1] Department of Industrial and Systems Engineering
Rutgers University, Piscataway, NJ 08854
liangzhe@eden.rutgers.edu

[2] Department of Industrial and Systems Engineering
Rutgers University, Piscataway, NJ 08854
wchaoval@rci.rutgers.edu

Summary The airline network is one of the world's most sophisticated, yet very complex, networks. Airline planning and scheduling operations have posed many great logistics challenges to operations researchers. Optimizing flight schedules, maximizing aircraft utilization, and minimizing aircraft maintenance costs can drastically improve the airlines' resource management, competitive position and profitability. However, optimizing today's airline complex networks is not an easy task. There are four major optimization problems in the airline industry including flight scheduling problem, fleet assignment problem, crew pairing problem, and aircraft maintenance routing problem. These problems have been widely studied over the past few decades. Yet, they remain unsolved due to the size and complexity. In this chapter, we provide a review of advances in optimization applied to these logistics problems in the airline industry as well as give a thorough discussion on the aircraft maintenance routing problem. Several mathematical formulations and solution methods for the aircraft maintenance routing problem will also be presented. Later, we conclude the current research and discuss possible future research of this problem.

1 Introduction

The airline network is among the most sophisticated and complicated networks in the world. In Europe, a typical major international airline operates 1,400 flights per day to over 150 cities in 76 countries. In the US, major domestic carriers operate 5,000 flights per day and offer over 4,000,000 fares to serve over 10,000 markets. New York City is the top domestic market. Airlines offer their products for sale more than one year in advance. On a typical day, a major carrier changes about 100,000 fares. The schedules change twice each week. Total operating expense is about 147,413 million dollars. Airlines plan their US domestic flights in a hub and spoke model typically with three-seven hubs. Given airlines' large-scale operations with very high degrees of freedom, airline planning and scheduling pose great logistics and optimization challenges

W. Chaovalitwongse et al. (eds.), *Optimization and Logistics Challenges in the Enterprise*, Springer Optimization and Its Applications 30, DOI 10.1007/978-0-387-88617-6_12, © Springer Science+Business Media, LLC 2009

to airline industry. On average, a major airline uses about 350 aircrafts of over 10 types and employs about 3,400 cockpits, 14,000 cabin, and 8,300 ground crews. Usually airlines have the same schedule for all weekdays but somewhat different schedules on weekends. Due to the size and complexity, the airline planning and scheduling operations are usually divided in several stages and managed sequentially. In each stage, the planning and scheduling operation is usually modeled in a time line network, in which three flow streams (airplanes, passengers, and crews) will be managed. The carrier overall performance is focused on three key areas: short-term scheduling, network planning, and fleet optimization. Generally, the logistics of airline planning and scheduling consisted of four major optimization problems: flight scheduling, fleet assignment, crew scheduling, and aircraft maintenance routing.

1.1 Flight Scheduling

Usually, airline planners build flight schedules based on historical data, traffic forecasts, airline network analysis, and profitability analysis [10, 20]. Traditionally, flight schedules are built several months in advance. Resulting from the market analysis like the choice of hub-and-spoke or distributed network, the demand forecast is normally based on origin-destination pairs. The schedule is often built by the airline marketing department and once it is published it will last for a number of months. However, it is almost impossible to match capacity with demand on a daily basis, especially given the recent booking trends and unforeseeable circumstances. If planners could match schedule capacity with market demand, they could reallocate common flight deck equipment and maximize profits. When a schedule is close in (35 to 45 days out), there might be some slight changes required in the schedule. Recently, revenue management system has been employed by several major airlines in attempt to produce an accurate forecast for a close-in schedule. As a result, airlines may decide which direct flights should be offered in the new schedule and the best departure times will be optimized based on the new operational constraints like fleet sizes.

1.2 Fleet Assignment

An airline usually has a variety of fleets. Considering factors such as passenger demands (both point-to-point and continuing services), revenues, operation costs, the fleet assignment problem is to assign predetermined aircraft types to a given airline timetable in order to maximize profit [14, 23]. In other words, the fleet assignment is to assign a fleet to each flight of the schedule so as to maximize the total profit. A fleet type prescribed by the manufacturer is a particular class of aircrafts which has a given seating capacity and fuel consumption. Therefore, in network optimization, airplane types can be viewed as commodities. Given a flight schedule and available fleet of aircraft, the objective of the fleet assignment problem is to maximize operating profit subject to the following physical and operational constraints. The cover constraint is to ensure that each flight in the schedule must be assigned exactly one aircraft type. The plane count constraint is to ensure that the total number of aircraft assigned cannot exceed the number available in the fleet. The balance constraint is to assure that aircrafts cannot appear or disappear from the network. In practice, there may be other additional constraints like time windows, integration routing (code-share), or yield management. In practice, airlines usually impose the condition that all planes should have the same physically aging mechanisms, such as wear, tear and fatigue. With a uniform aircraft usage and structural condition, every aircraft is likely to have the same maintenance demands. The maintenance constraint to ensure that each plane in the fleet should fly all flight legs given to a given plane can be incorporated in the model. In addition, the same plane is normally restricted to being used midday and late night since the same crew can not be used to fly different types of planes.

1.3 Crew Pairing and Rostering

The planning process of crew scheduling usually consists of two steps: crew pairing and crew rostering. A crew pairing is a crew trip spanning one or more work days separated by periods of rest. Each crew member has a base. Airlines prefer that the crew spends most of its night at its homebase; otherwise the airlines would have to provide them with accommodations. In reality, the costs of overnights are 7-8% of crew costs and the max number of days a crew is on the road is usually 4 to 5 days. The crew pairing problem is to determine the best set of crew pairings to cover the flights [2]. Normally, this problem is solved by using a lot of cheap pairings and solving a shortest route problem. In a closely related problem, each cockpit crew is qualified to fly a set of closely related fleet types, known as a fleet family. The flight assignment to the corresponding fleet family can be modeled as a crew pairing problem as well. The crew rostering problem is to construct personalized monthly schedules (rosters) for crew members by assigning them pairings and rest periods [12].

1.4 Aircraft Maintenance Routing

After the fleet assignment is arranged, airlines have to manage the aircraft rotation such that each aircraft gets enough maintenance opportunities. A maintenance routing problem is to determine how an individual aircraft is managed in the rotation of comprehensive maintenance, repair and overhaul system [6]. The configuration management and work-flow capabilities of aviation maintenance, repair and overhaul have presented their own unique complexities associated with maintenance-related flight delays, regulatory fines, inefficient maintenance and surplus or obsolete inventory. This is mainly due to different types of maintenance checks required by Federal Aviation

Administration (FAA). The different types of maintenance checks require different amounts of work to be done. Generally, FAA requires A, B, C and D checks [11]. Type A checks inspect all the major systems and are performed frequently (every 65 flight hours). Type B checks entail a thorough visual inspection plus lubrication of all moving parts, and are performed every 300 to 600 flight hours. Types C and D checks require the airlines to take the aircraft out of service for up to a month at a time and are performed about once every 1 to 4 years. Also, different airlines are required to operate with slightly different maintenance regulations. It is a common practice in the industry to allow at most 35 to 40 hours of flying before the aircraft undergoes a process called *transit check*. The transit check involves a visual inspection and a visualization to identify if the aircraft carries what is called a *minimum equipment list*. The maintenance routing problem is to schedule the most frequent maintenances (i.e., type A), whereas the less frequent maintenances (i.e., types B, C and D) can be incorporated into the fleet assignment problem. The combination of the rotation and the fleeting problems is possible because the numbers of maintenance hours do not vary so much across different types of aircrafts.

1.5 Organization of the Chapter

In this chapter, we will focus on the aircraft maintenance routing problem. We will provide the classification of various methodologies for the aircraft maintenance routing problem. These methodologies can be first divided into three groups: network flow-based formulation, string-based formulation and heuristics algorithms. The rest of this chapter is organized as follows. In Section 2, we discuss some basics of the aircraft maintenance routing problem including operational rules, cost structure, and network structure. In Section 3, mathematical models as well as solution techniques for the aircraft maintenance routing are presented. In Section 4, many extensions of the aircraft maintenance routing problem will be discussed. Integrated aircraft maintenance routing problem with other airline planning problems like the fleet assignment and crew pairing problems will be presented. Finally in Section 5, concluding remarks and discussions of the future research are provided.

2 Aircraft Maintenance Routing: Basics

After the fleet assignment is managed, airlines then route and schedule their aircrafts to perform the assigned legs at minimal cost and maximal profit. In fact, the costs associated with operating aircraft and crews form an important component of total distribution costs. Consequently, a small percentage of savings in these expenses could result in substantial savings over a number of years. In the last two decades, aircraft maintenance routing problem has been extensively studied in the Operations Research literature [3, 6, 9, 13, 16, 17,

22, 24, 25]. Furthermore, it should be noted that recent advances in routing and scheduling research make it possible to integrate two or more planning stages together to gain potential savings. For example, an integration of a fleet assignment model with the aircraft maintenance routing problem was proposed in [1, 5]. An integration of the crew pairing problem with aircraft maintenance routing problem was studied in [7, 8, 18, 19]. The importance of achieving these potential savings has become increasingly prominent due to escalating fuel costs, higher capital costs (e.g., purchasing new aircrafts, growing salaries for crews).

The basic aircraft maintenance routing problem can be stated as follows. Given a set of flight segments to be served by a fleet of aircrafts in which each aircraft must be routed to a maintenance station within a given number of days, the objective is to minimize the total maintenance costs without violating maintenance checks for each aircraft subject to the resource constraints of the maintenance stations. Generally, the solution will provide a rotation of maintenance schedules for each aircraft. The rotation specifies the sequence of flight segments to be flown and identifies the times at which the maintenance at the specific locations is to be carried out.

2.1 Operational Rules

There are several operational rules which are widely used in the airline industry. For example, most airlines required an *unlocked cyclic rotation* schedule to be provided. If the flights covered by an aircraft are separated by the flights covered by another aircraft, then we call this situation a *broken rotation*. If two sequences of flight segments in a broken rotation do not share any common location, we say it is a *locked rotation*. During the maintenance routing process, most airlines try to avoid the locked rotation. In the ideal case, airlines design to have only one rotation for each fleet, i.e., all aircrafts in the same fleet should follow the same sequence of flights.

In general, the resource constraints considered in aircraft maintenance routing problem contain two parts. The first part is the availability of the maintenance slots and the second part is the man power availability. However, in most cases, these two constraints are handled simultaneously.

2.2 Cost Structure

The objective of the aircraft maintenance routing problem is to maximize the profit and minimize the cost. The profit can be measured by a *through value* of routing flights through airports. The through value can be measured by the number of passengers who stay on the same aircraft between two consecutive flight segments (in one-stop flight service) in the rotation. For example, consider the four flight segments that connect five airports located in Los Angeles (LAX), San Diego (SAN), Dallas (DFW), Boston (BOS) and New York City (JFK). The details of these flight segments are as follows:

$LAX \rightarrow DFW$, $BOS \rightarrow DFW$, $DFW \rightarrow SAN$ and $DFW \rightarrow JFK$. Assume that the first two flights arrive in Dallas at noon and the last two flights depart from Dallas at early afternoon. It is obvious that the rotation should connect $LAX \rightarrow DFW$ with $DFW \rightarrow JFK$, and $BOS \rightarrow DFW$ with $DFW \rightarrow SAN$. This is because Los Angeles is closed to San Diego geographically; as a result, there is less market demand between these two cities. It is similar for New York City and Boston.

The cost of a rotation can be viewed in several ways, e.g., the robustness measurement of the rotation or the penalty value of the undesired connection in the rotation. In general, the cost can be decomposed into two parts, shown as follows:

$$c_{rotation} = \sum_{connection \, \in \, rotation} c_{connection} + consecutive \; penalty \quad (1)$$

The first part is contributed by each connection in the rotation. For example, a penalty cost could be placed if the connection time is too short or too long. The second part is the accumulative penalty based on the total flying hours or some other operational factors.

2.3 Network Structure

There are two widely used network structures in the research of aircraft maintenance routing problem, namely, *time-space network* and *connection network*. These two network structures provide platforms for studying most airline operations research problems. In the following sections, we will discuss these two network structures in detail.

Time-Space Network

The *time-space network* first appeared in the fleet assignment model [14]. In a time-space network, a time line is associated with a station, consisting of series of event nodes which occur sequentially with respect to a specific time at this station. Each event node represents an event of flight arrival or departure at the station. In order to allow connection between different flight segments, a turn-time is added on the actual flight arrival time for calculating time of the arrival node. In the time-space network, there are three types of arcs: ground, flight and overnight (or sometimes call warp-around) arcs. Ground arcs represent one or more aircrafts staying at the same station for a given period of time. With all the ground arcs in the network, it is possible to preserve the aircraft balance and allow all possible connections. Flight arcs represent flight segments. Overnight arcs connect the last event of the planning period with the first event of the planning period at a specific station. Overnight arcs ensure the continuity of the aircraft routing from the current planning period to the next one. For most American domestic airlines, the

planning period is one day. The time lines of different stations are connected
by incoming and outgoing flight arcs.

In Figure 1, we show eight flights between three stations. The time progresses horizontally from left to right, and each node represents a flight
arrival or departure at a particular station. The arc types are shown in the
legend.

A set of preprocessing methods to reduce the size of the time-space network was presented in [14]. As we notice in Figure 1, when a flight departs
from Station A at 12:30, it does not matter when the last arrival is, as long
as there is no departure between the last arrival and current departure. Consequently, it is feasible to place the last arrival at the next departure time
or to place the current departure just after the last arrival. Therefore, the
feasibility and optimality of the problem remains when we combine the last
arrival node with the current departure node. The combination of arrival
and departure nodes is called *node aggregation*. In Figure 2a, we show a
time-space line in a station. In Figure 2b, we show the graph after node
aggregation.

The second preprocessing method is called *island isolation*. This method
can be widely used in hub-spoke networks. In some spoke stations, there are
no aircraft on the ground during some certain period. Therefore, by removing
the ground arcs, we can still maintain the optimality of the current solution.
After deleting these unused ground arcs, we generate some "islands" on the
time-space line. Because no ground arcs appear before or after the island,
the in degree and out degree of the island remain the same. In Figure 2c,
we show the network resulted by removing the zero-valued ground arcs.
Also, we can see that in Figure 2c, flight A must be followed by flight B.
Therefore, the island isolation also eliminates the problem size by reducing one
variable.

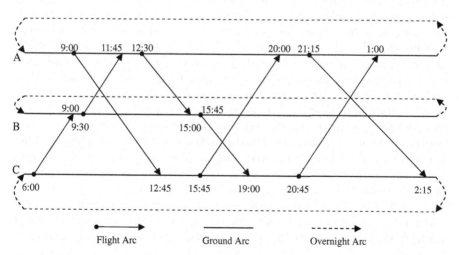

Fig. 1. Time-space network with eight flights and three stations

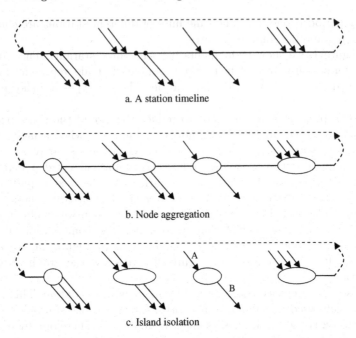

a. A station timeline

b. Node aggregation

c. Island isolation

Fig. 2. Preprocessing of the time-space network

Connection Network

In the connection network, a node represents an arrival or departure event. There are three types of arcs: flight arcs, connection arcs and originating/terminating arcs. The flight arcs are similar to the ones in time-space network, whereas the connection arcs represent possible connections between two flights. The originating/terminating arcs represent the aircraft status at the beginning or end of the day. In Figure 3, we show an example of a connection network that represents the same topology as in Figure 1. Here each station is represented by two vertical lines: one is $Station_{in}$ and the other is $Station_{out}$. $Station_{in}$ connects all the arrival flights to that station, and $Station_{out}$ connects all the departure flights from that station. The arcs between different stations represent the flights to be in service. The dash lines between each pair of $Station_{in}$ and $Station_{out}$ represent all possible connections. The vertical bold lines represent the initial and terminating conditions at the stations.

It should be noted that sometimes the flight arcs are simplified as flight nodes in the connection network. Therefore, the nodes represent the flight segment and arcs represent the possible linkages between flight segments.

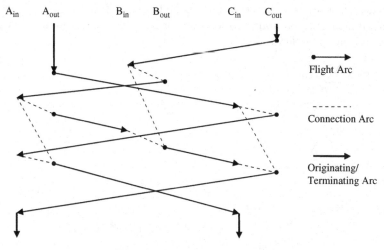

A_{in} A_{out} B_{in} B_{out} C_{in} C_{out}

Flight Arc

Connection Arc

Originating/
Terminating Arc

Fig. 3. Connection network

3 Aircraft Maintenance Routing: Models and Algorithms

3.1 Flight-Based Formulation Models

The time-space network for a given fleet is a *Eulerian* digraph, i.e., each node has its in degree equal to its out degree and it is connected. This criterion is ensured by the schedule design and fleet assignment problem. The studies in [6] solve the aircraft maintenance routing problem by forming a *Euler tour* in the time-space network, in which all the *service violation paths* are eliminated. A Euler tour is a cycle that includes all the arcs exactly once. A service violation path is a path with a length longer than the specified service period. By excluding the service violation path in the Euler tour, the maintenance constraints are satisfied. The objective is to maximize the benefit derived from making specific connections, which is referred to as a *through value*. The problem is solved using the Lagrangian relaxation and subgradient optimization.

Formally, we let the time-space network $G = (N, A)$ be defined with N the set of nodes and A the set of arcs. Define the t^- and t^+ as the time just before and after the time t. For any arc $a \in A$, define the head node of a as $h(a)$ and the tail node of a as $t(a)$. Define x_{ij} as the decision variable for the connection from $i \in A$ to $j \in A$ such that $h(i) = t(j)$. Therefore, $x_{ij} = 1$ if arc i connects with j, and $x_{ij} = 0$ otherwise. Define r_{ij} as the through value of the connection. Also, define S^k as the set of maintenance feasible paths and $\bar{S}^k$ as the set of all *minimal violation paths* for maintenance type $k \in K$, where K is the set of maintenance types. Here, a path is a minimal violation path if all of its subpaths are feasible. Let A_s denote the set of arcs of s ($s \in \bar{S}^k$) and

$A'_s = A_s-$ last arc of s. Let $f(i)$ denote the follower of $i \in A'_s$. We present the mathematical programming formulation for the aircraft maintenance routing problem in [6] as follows:

$$\max \sum_{ij \in R} r_{ij} x_{ij} \qquad (2)$$

$$\text{s.t.} \sum_{j:h(i)=t(j),i \neq j} x_{ij} = 1 \qquad \forall\, i \in A \qquad (3)$$

$$\sum_{i:h(i)=t(j),i \neq j} x_{ij} = 1 \qquad \forall\, j \in A \qquad (4)$$

$$\sum_{i \in A',j \in A \backslash A',\ h(i)=t(j)} x_{ij} \geq 1 \qquad \forall\, A' \subset A,\, 2 \leq |A'| \leq |A| - 2 \qquad (5)$$

$$\sum_{i \in A'_s,j \in A \backslash f(i),\ h(i)=t(j)} x_{ij} \geq 1 \qquad \forall\, s \in \bar{S}^k,\, k \in K \qquad (6)$$

$$x_{ij} \in \{0,1\} \qquad \forall\, i,j \in A. \qquad (7)$$

In the objective function in Eq. (2), we are trying to maximize the total through value. Constraints in Eqs. (3)-(4) enforce the flow balance constraints of the Euler tour. The constraints in Eq. (5) eliminate all subtours when forming the Euler tour. The constraints in Eq. (6) eliminate all the service violation paths from the Euler tour.

It was also pointed out in [6] that the above formulation can be viewed as an *asymmetric traveling salesman problem*. In particular, if the problem is projected in a simplified connection network, in which nodes represent flight segments and arcs represent possible connections, the objective is to find a *Hamiltonian cycle* with maximum through value. It is well known that a Hamiltonian cycle problem is equivalent to an asymmetric traveling salesman problem. In fact, any Euler tour in the time-space network corresponds to a Hamiltonian cycle in the simplified connection network.

The model presented above is capable of capturing multiple maintenance considerations. However, since there are an exponential number of possible $A' \subset A$ sets and A_s for $s \in \bar{S}^k, k \in K$ sets, it is not easy to enumerate all the subtour constraints and service violation constraints and solve the model directly. A Lagrangian relaxation procedure is used to attack this problem in the real implementation. In particular, the constraints in Eqs. (5)-(6) are dualized and a Lagrangian relaxation formulation is obtained with only constraints in Eqs. (3)-(4), (7). Computational results show for most of the test cases, near optimal solutions are obtained.

In [9], an MIP rotation planning model was proposed to solve the aircraft rotation planning problem in a European airline. It was pointed out that hub-and-spoke network structure is not employed in European airlines; therefore, the maintenance issues are handled in a different way. For European airlines, they only capture the flow balance constraints of the flight segments in the rotation planning. The objective function is to minimize the total delay risk, which is contributed by two parts: one from individual flight connections and another from the complete path formed by flight segments. Therefore, the path variables are also introduced in this model. However, it is noted that the path in the formulation only changes the cost but does not affect the feasibility of the problem as in the previous models. A Lagrangian relaxation method is used to solve the problem.

In [16], a network flow models is proposed to solve the aircraft maintenance routing problem for a single type maintenance. A D day time-space network is built (in Figure 4), where D is the maximum number of days allowed between the two consecutive maintenances for an aircraft. However, different from the traditional time-space network, the new network does not contain overnight arcs. Instead, a set of maintenance arcs is constructed, which start at the end of each day at a station and end at the beginning of the same station time line (go backwards). The capacity of a maintenance arc is the number of maintenances allowed at the particular maintenance station per day. Therefore, by enforcing the flow balance constraints, an aircraft must go through the maintenance arcs. As a result, the maintenance requirements are satisfied.

Mathematically, define D as the number of the maximal days allowed between maintenances. The set of maintenance arcs in a day is denoted as M. Define a set of flight segments in a day as F. Therefore, to represent a flight arc or a maintenance arc in a multiple day network, we have to use flight index f (maintenance index m) associated with a day index d. Define the number of aircrafts available as L. Define the number of the maintenances allowed per day at maintenance station b as Q_b, where $b \in B$ and B is a set

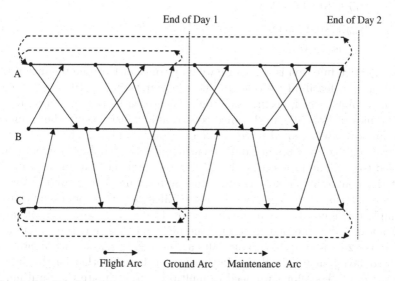

Fig. 4. Modified time-space network for aircraft maintenance routing problem

of maintenance stations. Define the set of maintenance arcs associated with a station $b \in B$ as M_b. The indicator α_{fdn}^+ has value 1 if flight f on day d starts at node n and 0 otherwise. The indicator α_{fdn}^- has value 1 if flight f on day d ends at node and 0 otherwise. Similarly, we have β_{mdn}^+ and β_{mdn}^- for maintenance. The decision variable z_{fd} has value 1 if flight f is flown on day d and 0 otherwise. The decision variable z_{md} has positive integer value 1 if aircraft maintenance m are done on day d and 0 otherwise. The ground arc variable w_n^- represents the number of the aircrafts on the ground before node n, and ground arc variable w_n^+ represents the number of the aircrafts on the ground after node n. Define G as the size of the fleet. The MIP formulation for this model is given by

$$\min \quad 0 \tag{8}$$

$$\text{s.t.} \sum_{d \in D} z_{fd} = 1 \qquad \forall\, f \in F \tag{9}$$

$$\sum_{f \in F, d \in D} \alpha_{fdn}^+ z_{fd} + \sum_{m \in M, d \in D} \beta_{mdn}^+ z_{md} + w_n^+$$
$$= \sum_{f \in F, d \in D} \alpha_{fdn}^- z_{fd} + \sum_{m \in M, d \in D} \beta_{mdn}^- z_{md} + w_n^- \quad \forall\, n \in N \tag{10}$$

$$\sum_{m \in M_b, d \in D} z_{md} \le Q_b \qquad \forall\, b \in B \tag{11}$$

$$\sum_{m \in M_b, d \in D} d z_{md} \le G \tag{12}$$

$$z_{fd} \in \{0, 1\} \qquad \forall\, f \in F \tag{13}$$

$$z_{md} \in \{0, 1, ..., Q_b\} \qquad \forall\, m \in M_b,\, b \in B \tag{14}$$

$$w_n^-, w_n^+ \ge 0 \qquad \forall\, n \in N. \tag{15}$$

The objective function is 0 because we only consider the maintenance routing problem as a feasibility problem. The constraints in Eq. (9) are the covering constraints, which ensure each flight is covered by only one aircraft. The constraints in Eq. (10) are balance constraints, which ensure the number of inbound aircrafts is equal to the number of outbound aircrafts at each node. The capacity constraints in Eq. (11) ensure that the number of maintenances needed to be done in a station is not greater than the station capacity. The fleet size constraint in Eq. (12) ensures the total number of aircrafts used is not greater than the fleet size. To better illustrate this model, consider an example in Figure 4. If a maintenance arc in day 1 is used once, it implies that an aircraft is needed to service the flight on that day. If a maintenance arc in day 2 is used once, two aircrafts are needed to service the flights in the last two days (one for the flights in the current day, the other for the flights in the last day). Therefore, we need to multiply d to estimate the total number of aircrafts needed. The constraints in Eqs. (13)-(14) are binary constraints, and capacity and integrality constraints for maintenance arcs. The ground arcs can be defined as nonnegative continuous constraints in Eq. (15) through Eqs. (9), (13) and (14).

It is noted that the above formulation does not provide detailed routing sequence. Instead, the model provides a set of arcs used in the optimal rotation known as a Euler tour. Therefore, one can always use polynomial time algorithm to find out the sequence of the arcs [4]. Two model enhancements were also presented in [16] to cater the cost/profit of the maintenance routing problem by constructing some functional arcs on the time-space network. The computational results show the model performs very well for large size aircraft maintenance routing problem.

A set of algorithmic solutions are presented in [13] and [25]. They considered a similar configuration as in [24], where one-day trips are fixed. A polynomial time algorithm is presented for finding a 3-day maintenance Euler tour if there exists one in a flight network. It is also proved that in the general case, a D day maintenance Euler tour problem is NP-complete for $D \geq 4$. However, if there is only one maintenance station, a polynomial time algorithm is given by solving a bipartite matching problem. Since authors only focus on finding a feasible maintenance solution, there is no cost factor involved in any of the algorithms. However, no computational results are reported in both papers.

3.2 String-Based Models

The first string-based model was proposed in [1] to solve the aircraft maintenance routing problem for one maintenance type. A *string* was defined as a sequence of connected flights that satisfy the following conditions. The sequence must begin and end at maintenance stations with a maintenance check at the end. The sequence must satisfy the flow balance and must be maintenance feasible. It is noted that the time between any two consecutive flights in the string cannot be less than the *minimum turn time*. This time is used for aircraft to change gates, clean up and so on. The string may begin and end at different maintenance stations. The model formulates the maintenance routing problem as a set partitioning problem with side constraints.

Formally, we define S as the set of feasible route strings. The cost of string s is denoted as c_s. Binary indicated parameter $\gamma_{fs} = 1$ if route string s covers flight f; 0 otherwise. Define decision variable $u_s = 1$ if route string s is included in the solution; 0 otherwise. N is the set of nodes representing points in space and time at which route strings begin or end. Also define ground arcs as w_n^- and w_n^+. Define S_O as a set of route strings spanning time O, which is an arbitrary time known as the *countline*. δ_s denotes the number of times string s crosses the countline O. N_O denotes the set of nodes with corresponding ground arcs g_n^+ spanning the countline. The formulation is as follows:

$$\min \sum_{s \in S} c_s u_s \tag{16}$$

$$\text{s.t.} \sum_{s \in S} \gamma_{fs} u_s = 1 \qquad \forall\, f \in F \tag{17}$$

$$\sum_{h(s)=n} u_s + w_n^- = \sum_{t(s)=n} u_s + w_n^+ \qquad \forall\, n \in N \tag{18}$$

$$\sum_{s \in S_O} \delta_s u_s + \sum_{n \in N_O} w_n^+ \le G \tag{19}$$

$$u_s \in \{0, 1\} \qquad \forall\, s \in S \tag{20}$$

$$w_n^+, w_n^- \ge 0 \qquad \forall\, n \in N. \tag{21}$$

The objective function in Eq. (16) minimizes the cost of the chosen route strings. The constraints in Eq. (17) are covering constraints, which state that each flight must be included in exactly one chosen route string. The constraints in Eq. (18) are balance constraints. The constraint in Eq. (19) makes sure that the total number of aircrafts at time O does not exceed the fleet size. Consequently, the number of aircrafts does not exceed the fleet size at any time because of the constraints in Eq. (18). The constraints in Eq. (20) ensure that the string decision variables are binary. Thus the integrality of the ground arc variables can be relaxed as denoted in Eq. (21). This problem is solved using a branch-and-price algorithm using branch-on-follow-on technique [21].

In [3] and [17], the aircraft maintenance routing problem is solved as an *asymmetric traveling salesman problem with replenishment arcs* (RATSP). They add a set of replenishment arcs in the original time-space network presented in [6]. The replenishment arcs represent the possible maintenance opportunities. It should be noted that by constructing the replenishment arcs, only one type of maintenance is considered in the model. They also introduce another set of dummy arcs which represent maintenance feasible strings. Here, the maintenance feasible string is different with the string in [1] since it does not contain a maintenance at the end of the string. The basic formulation they used is also similar with the model in [6]. However, instead of having individual variables for each flight arc, they only allow variables for replenishment arcs and dummy path arcs in the solution. Formally, this can be done by replacing all the x variables in [6] by new y variables for dummy arcs using the following equation:

$$x_{ij} = \sum_{ij \in A_s, s \in S} y_s \tag{22}$$

Here, S is a set of maintenance feasible paths. Decision variable $y_s = 1$ if path s is selected in the solution and $y_s = 0$ otherwise. The maintenance constraints can be written in the following way:

$$\sum_{s \in S_n} y_s \le 1 \qquad \forall\, n \in N \tag{23}$$

The above constraints ensure at most one dummy arc connects with a vertex in the network. Consequently, the other arc connecting to that vertex must be a maintenance arc. By doing so, an exponential number of dummy arc variables are introduced into the model. On the other hand, they eliminate an exponential number of constraints for maintenance feasibility check since there are only a limited number of vertices in the network. In [3], the RATSP was solved by a branch-and-price-and-cut algorithm. In [17], the RATSP was solved by three different types of heuristics. One is a simulated annealing algorithm and the other two are based on the Lagrangian relaxation of different side constraints. The result shows that the algorithm based on simulated annealing performs well overall.

In [24], a model for handling weekly schedule and two types of maintenance were presented. A set of one-day trips is constructed first, then a time-space network is built on these trips. It is noted that a flight segment is covered by one and only one trip. Then, a multicommodity network flow model is constructed which requires each trip to be covered exactly once. Specially, since all the trips depart from stations in the morning and arrive at stations during the night, there are no ground arcs in this time-space network (the ground arcs during the night can be eliminated by node aggregation). Side constraints for two types of maintenance are also built in the model. By constructing the one-day trips, the number of variables is reduced greatly, hence the problem becomes more solvable. However, the authors do not mention how the one-day trips are constructed and the model might cause suboptimal results in the real situation because of the way one-day trips are constructed.

In [22], another string-based model is constructed to solve the maintenance routing problem. In this model, the strings are indexed by the individual aircraft and the model ensures each aircraft will be utilized. Also, two sets of similar resource constraints (for manpower and maintenance slot) are built in this model. A branch-and-bound algorithm with modified branch-on-follow-on rule is used to solve the problem.

4 Aircraft Maintenance Routing: Extensions and Integration

As we discussed before, the aircraft maintenance routing depends on the fleet assignment model that specifies the set of flight segments to be covered by an aircraft fleet. The result produced by the aircraft maintenance routing problem is, in turn, considered as the input of the crew pairing problem. As we can see, solving these optimization problems sequentially may lead to a suboptimal solution for the entire airline operations planning problem. In the last few decades, a lot of researchers have focused on solving more than one optimization problem simultaneously which generates higher revenues and lower the cost for the airlines. In this section, we discuss several integrated scheduling problems related to the aircraft maintenance routing problem.

4.1 Aircraft Maintenance Routing Problem with Fleet Assignment Problem

The first integrated model about the aircraft maintenance routing problem appeared in [5]. Instead of integrating any maintenance routing models with other problems, a set of constraints for maintenance routing is built on the multicommodity fleet assignment model in [14]. Two types of maintenance (short maintenance and long maintenance) are considered in their work. A dummy flight arc is built for each fight segment arriving at the maintenance stations for short maintenance. The flight arriving time for the dummy flight is delayed so that there is enough time for a short maintenance. A set of maintenance arcs, namely, the *leapfrog arcs*, is built in the time-space network for long maintenance. Since long maintenance operations are only conducted during the night, a maintenance arc is only built after a flight departs and arrives at the same station at night.

Another integrated model for fleet assignment and aircraft maintenance routing problem is presented in [1]. The model is an expansion of the basic string model in Eqs. (16)-(21). The decision variable u_s will be duplicated for each fleet type $h \in H$, where H is the set of fleet types. The objective function is to maximize the total profit of the fleet assignment problem and minimize the aircraft maintenance routing cost. A branch-and-price technique is used to solve the integrated model.

4.2 Aircraft Maintenance Routing Problem with Crew Pairing Problem

In the last decade, there were plenty of research articles found to solve the integrated planning for the maintenance routing and crew pairing problems, which address the effect of *short connects*. One restriction on a valid pairing is that two sequential flights cannot be assigned to the same crew unless the time between the flights is sufficient (known as *minimum sit time*). This minimum sit time can be shortened only if the crew follows the plane turn, which is called a *short connect*. Even though the difference between the minimum turn time and the minimum sit time is small, using more short connects in planning may significantly improve the robustness of the crew scheduling. This is because during the aircraft disruption, it is possible to absorb the delay in the crew assignment if initially the crew is assigned to follow the aircraft turn.

The first research work to address the impact of short connects on the crew pairing problem appears in [15]. They demonstrate that by considering the short connects in solving the crew pairing problem, the crew cost is significantly reduced from the traditional sequential model.

In their approach, the planning problems are solved in reverse order. They first solve the crew pairing problem assuming all the short connects are valid. A set of constraints are added to the original crew pairing model to ensure that the number of the aircrafts used at any short connection period is not more than the fleet size. Then they solve a maintenance routing problem to incorporate the short connects selected in the crew pairing solution. This approach can lead to maintenance infeasibility. However, in practice they find feasible solutions for some hub-and-spoke flight networks using this approach. As long as the maintenance routing problem is feasible, the solution for the crew pairing problem is optimal for the integrated problem. This method requires no more computational effort than the traditional sequential method.

In [15] the aircraft maintenance routing problem is considered to be a feasibility problem, since the crew cost in fact dominates the cost of the integrated problem. This is reasonable because the running cost of aircraft is more or less determined at this planning stage, which takes place after the schedule design and fleet assignment stages.

A *basic integrated model* for the maintenance routing and crew pairing problems was presented in [8]. This model guarantees the maintenance feasibility. The maintenance routing cost is explicitly considered in this model. A dated planning horizon was assumed such that a set of flights may vary from day to day. The string-based maintenance routing model and set partitioning model were used for crew pairing. These models were linked by short connection constraints. In particular, if a short connect appears in a crew pairing solution, it must appear in the maintenance routing solution. The basic integrated model is formulated as follows:

$$\min \sum_{p \in P} c_p v_p + \sum_{s \in S} c_s u_s \tag{24}$$

$$\text{s.t.} \sum_{p \in P} \eta_{fp} v_p = 1 \quad \forall f \in F \tag{25}$$

$$\sum_{s \in S} \gamma_{fs} u_s = 1 \quad \forall f \in F \tag{26}$$

$$\sum_{h(s)=n} u_s + w_n^- = \sum_{t(r)=n} u_s - w_n^+ \quad \forall n \in N \tag{27}$$

$$\sum_{s \in S_O} \delta_s u_s + \sum_{n \in N_O} w_n^+ \leq G \tag{28}$$

$$\sum_{s \in S} \theta_{cs} u_s - \sum_{p \in P} \zeta_{cp} v_p \geq 0 \quad \forall c \in C \tag{29}$$

$$u_s, v_p \in \{0,1\} \quad \forall s \in S, \forall p \in P \tag{30}$$

$$w_n^+, w_n^- \geq 0 \quad \forall n \in N, \tag{31}$$

where, beside the notations defined in Sections 2.1 and 2.2, C is the set of all possible short connects, P is the set of pairings, θ_{cs} is 1 if a string s contains short connect c, 0 otherwise, ζ_{cp} is 1 if a pairing p contains short connect c, 0 otherwise.

The objective function in Eq. (24) minimizes the cost of chosen pairings and strings. The constraints in Eq. (25) are the same as the ones in the crew pairing model. The constraints in Eqs. (26)-(28) are the same as the ones in the aircraft maintenance routing model. The two models are linked by a constraint set in Eq. (29), which ensures that a short connect is selected in a crew pairing only when it appears in the maintenance routing solution. The constraints in Eqs. (30)-(31) are binary and nonnegative constraints for variables.

Here, crew pairings are constructed with all potential short connects allowed. This model results in a large-scale integer program. The model is solved using a Benders decomposition approach coupled with a heuristic branching strategy in which the maintenance routing is considered as the master problem while the crew pairing is the subproblem.

An *extended crew pairing model* was proposed in [7] to solve the integrating crew pairing and maintenance routing problem. As in [8], the two decisions for crew pairing and maintenance routing are linked by short connect constraints to ensure the crew pairing feasibility. As in [15], the model does not consider the maintenance routing cost explicitly. The decision on the maintenance routing is captured as a problem of choosing one out of all the feasible maintenance routing solutions. Formally, the model is formulated below.

$$\min \sum_{p \in P} c_p v_p \tag{32}$$

$$\text{s.t. } \sum_{p \in P} \eta_{fp} v_p = 1 \quad \forall f \in F \tag{33}$$

$$\sum_{e \in E} \tau_{ce} y_e - \sum_{p \in P} \eta_{fp} v_p \geq 0 \quad \forall c \in C \tag{34}$$

$$\sum_{s \in S} y_e = 1 \tag{35}$$

$$y_e, v_p \in \{0,1\} \quad \forall e \in E, \forall p \in P, \tag{36}$$

where E is the set of feasible maintenance routing solutions, τ_{ce} is 1 if a short connect c is included in maintenance routing solution e, 0 otherwise, y_e is 1 if a maintenance routing solution e is chosen, 0 otherwise. Without constraints in Eqs. (34)-(35), this problem is merely a crew pairing model. The constraint in Eq. (35) ensures that exactly one maintenance routing solution is chosen. The constraints in Eq. (34) ensure that a short connect is used in a pairing only when it is included in the selected maintenance routing solution. It is observed that the binary constraints for y_e can be relaxed and replaced by a set of nonnegative constraints. Hence the model becomes a mixed integer program and requires fewer integer variables than the basic integrated model used by [8]. In addition, its linear programming relaxation is tighter.

It is further observed in [7] that the collection of all the maintenance routing solutions can be reduced to a much smaller set containing only those distinct maintenance routing solutions that represent the *unique maximal short connect sets* (UMs). Given a set of maintenance solutions which contain the same set of short connects, it is observed that these maintenance solutions have the same impact on the crew pairing decisions. Hence, all the maintenance solutions which contain the same short connect set can be represented by the associated short connect set. This is referred to as *uniqueness*. If a maintenance solution has a short connect set A and another maintenance solution has a short connect set B, where $B \subset A$, then the choice of short connects to be used in crew pairings provided by A is more than B. Thus, it suffices to include only the first maintenance solution A. In other words, it suffices to include maintenance solutions representing only *maximal short connect sets*.

Two solution approaches for solving the integrated model were proposed in [7]. In the first approach, a modified string model for the aircraft maintenance routing problem is solved so that a set of UMs are generated (may not be the complete set of all the UMs). Then the extended crew pairing model is solved by including these UM sets as the maintenance solutions. This can be viewed as a restricted version of the complete problem, since only a subset of the maintenance solution variables are involved. In fact, the traditional way of airline schedule planning can be viewed as a special case, where only one maintenance solution is provided which may or may not represent a UM. Thus, a feasible solution by this approach is guaranteed to be at least as good as that found using the traditional sequential approach.

In the second approach, the problem is solved as a constrained crew pairing model. First, a crew pairing problem is solved with all potential short connects permitted. If the short connects used in the pairing solution are maintenance feasible, the optimal solution for the integrated model is obtained. Otherwise, a cut is added to eliminate the current infeasible crew pairing solution and the solving procedure continues. This cut is generated by identifying a *minimally infeasible short connect set* (MIS) of the current pairing solution. Formally, given a crew pairing solution $\bar{P}$ with the associated short connect set $\bar{C}$, for a set $\bar{C}'$ to be minimally infeasible, it must satisfy that $\bar{C}' \subseteq \bar{C}$ and any proper subset of $\bar{C}'$ is maintenance feasible while itself is not. The cut is written as below.

$$\sum_{p \in P} \sum_{c \in \bar{C}'} \eta_{cp} v_p \leq |\bar{C}'| - 1. \tag{37}$$

This cut prohibits the crew pairing solution containing $\bar{C}'$. Here, authors want maintenance infeasible short connect sets of $\bar{C}$ to be as small as possible, so that a minimal number of cuts are needed in solving the constrained crew pairing problem.

In [18], the authors extend the model in [8]. The new model includes several operational constraints for both crew pairing and aircraft maintenance routing problems. Those constraints include plane count constraints, cyclical constraints for aircraft maintenance routing problem, duty count constraints for the crew pairing problem, and a set of other operational constraints used to increase the robustness of the model. They solve the problem using the Benders decomposition technique, in which the crew pairing problem is considered as the master problem and the maintenance routing problem is considered as the subproblem. The three-phase heuristic algorithm is used to solve the complete problem.

A flight retiming model together with the aircraft maintenance routing problem and crew pairing problem was presented in [19]. In the integrated model, each flight is duplicated a number of times with in a small time interval, e.g., if the original departure time of a flight segment is 10:00 am, then we can have three possible options with the departure time 9:55 am, 10:00 am and 10:05 am, respectively. The integrated model follows the model presented in [8] and [18], with the extra constraints for duplicated flight segments. The integrated problem is also solved using Benders decomposition and the three-phase algorithm presented in [18].

5 Concluding Remarks and Discussion

In this chapter, we give a survey of recent advances in the aircraft maintenance routing problem. As discussed in this chapter, the research on the aircraft maintenance routing problem has made considerable progress in the last two decades. The most powerful methods are now capable of solving large-scale problems and researchers are tackling even more difficult problems. Those problems include an integrated maintenance routing problem with other planning problems. We believe that this integrated planning problem will still be an active research area in the next few years. Another active field related to the aircraft maintenance routing problem in recent years has been the problem to reassign the aircrafts when a schedule disruption occurs in the real situation, which is called the *aircraft recovery problem*. Several research studies have been done in the field, yet there are many great challenges for improvement of the solution approaches to this problem.

In addition to the above-mentioned four optimization problems, there are many other optimization problems arising in the airline industry. For example, when an airline carrier is in the process of replacing several of its aging aircraft, it is very important to understand the importance of identifying the optimal fleet and its impact on the airlines' profitability. Yield management (YM), whose objective is to "sell the right seat to the right passenger at the right price," has been applied as an effective planning and marketing strategy in the airline industry. YM has to deal with continuous process decisions in order to minimize the net-present-value (NPV) of future profits under many real-life constraints like financial resources, regulation, route structure, fleet maintenance, crew bases, facilities, schedule, availability, pricing, and policies. The YM practice, in turn, can be modeled as an optimization problem, which is very complex and hard to solve.

References

1. C. Barnhart, N. Boland, L. Clarke, E. Johnson, G. Nemhauser, and R. Shenoi. Flight string models for aircraft fleeting and routing. *Transportation Science*, 32(3):208–220, 1998.

2. C. Barnhart, E. Johnson, G. Nemhauser, and P. Vance. Crew scheduling. In R.W. Hall, editor, *Handbook of Transportation Science*, pages 493–521. Kluwer Scientific Publishers, 1999.

3. N. Boland, L. Clarke, and G. Nemhauser. The asymmetric traveling salesman problem with replenishment arcs. *European Journal of Operational Research*, 123:408–427, 2000.

4. G. Chartrand and O. Oellermann. *Applied and Algorithmic Graph Theory*. McGraw-Hill, 1993.

5. L. Clarke, C. A. Hane, E. Johnson, and G. Nemhauser. Maintenance and crew considerations in the fleet assignment. *Transportation Science*, 30(3):249–260, 1996.

6. L. Clarke, E. Johnson, G. Nemhauser, and Z. Zhu. The aircraft rotation problem. *Annals of Operations Research*, 69:33–46, 1997.

7. A. Cohn and C. Barnhart. Improving crew scheduling by incorporating key maintenance routing decisions. *Operations Research*, 51(3):387–396, 2003.

8. J. Cordeau, G. Stojkoviac, F. Soumis, and J. Desrosiers. Benders decomposition for simultaneous aircraft routing and crew scheduling. *Transportation Science*, 35(4):375–388, 2001.

9. M. Elf and V. Kaibel. Rotation planning for the continental service of a European airline. In W. Jager and H. Krebs, editors, *Mathematics - Key Technologies for the Future: Joint Projects between Universities and Industry*, pages 675–689. Springer, 2003.

10. A. Erdmann, A. Nolte, A. Noltemeier, and R. Schrader. Modeling and solving an airline schedule generation problem. *Annals of Operations Research*, 107:117–142, 2001.

11. FAA. *Federal Aviation Regulations*. URL: http://www.faa.gov/avr/afs, 2002.

12. M. Gamache and F. Soumis. A method for optimally solving the rostering problem. In G. Yu, editor, *Operations Research in the Airline Industry*, pages 124–157. Kluwer Academic Publishers, 1998.

13. R. Gopalan and K. Talluri. The aircraft maintenance routing problem. *Operations Research*, 46(2):260–271, 1998.

14. C. Hane, C. Barnhart, E. Johnson, R. Marsten, G. Nemhauser, and G. Sigismondi. The fleet assignment problem: Solving a large-scale integer program. *Mathematical Programming*, 70:211–232, 1995.

15. D. Klabjan, E. Johnson, G. Nemhauser, E. Gelman, and S. Ramaswamy. Airline crew scheduling with time windows and plane count constraints. *Transportation Science*, 36(3):337–348, 2002.

16. Z. Liang and W. Chaovalitwongse. Novel network based model for aircraft maintenance routing problem. Technical report, Rutgers University, Industrial & Systems Engineering Department, 2007.

17. V. Mak and N. Boland. Heuristic approaches to the asymmetric travelling salesman problem with replenishment arcs. *International Transactions in Operational Research*, 7:431–447, 2000.

18. A. Mercier, J. Cordeau, and F. Soumis. A computational study of Benders decomposition for integrated aircraft routing and crew scheduling problem. *Computer & Operations Research*, 32:1451–1476, 2005.

19. A. Mercier and F. Soumis. An integrated aircraft routing, crew scheduling and flight retiming model. *Computer & Operations Research*, 34:2251–2265, 2007.

20. R. L. Phillips, D.W. Boyd, and T.A. Grossman. An algorithm for calculating consistent itinerary flows. *Transportation Science*, 25:225–239, 1991.

21. D. Ryan and B. Foster. An integer programming approach to scheduling. In A. Wren, editor, *Computer Scheduling of Public Transport: Urban Passenger Vehicle and Crew Scheduling*, pages 269–280. North-Holland, 1981.

22. A. Sarac, R. Batta, and C. Rump. A branch-and-price approach for operational aircraft maintenance routing. *European Journal of Operational Research*, 175:1850–1869, 2006.

23. H. Sherali, E. Bish, and X. Zhu. Airline fleet assignment concepts, models and algorithms. *European Journal of Operational Research*, 172:1–30, 2006.

24. C. Sriram and A. Haghani. An optimization model for aircraft maintenance scheduling and re-assignment. *Transportation Research Part A*, 37:29–48, 2003.

25. K. Talluri. The four-day aircraft maintenance routing problem. *Transportation Science*, 32(1):43–53, 1998.

The Stochastic Vehicle Routing Problem for Minimum Unmet Demand

Zhihong Shen[1], Fernando Ordóñez[2], and Maged M. Dessouky[3]

[1] Department of Industrial and Systems Engineering
University of Southern California, Los Angeles, CA 90089
shenz@usc.edu

[2] Department of Industrial and Systems Engineering
University of Southern California, Los Angeles, CA 90089
fordon@usc.edu

[3] Department of Industrial and Systems Engineering
University of Southern California, Los Angeles, CA 90089
maged@usc.edu

Summary In this chapter, we are interested in routing vehicles to minimize unmet demand with uncertain demand and travel time parameters. Such a problem arises in situations with large demand or tight deadlines so that routes that satisfy all demand points are difficult or impossible to obtain. An important application is the distribution of medical supplies to respond to large-scale emergencies, such as natural disasters or terrorist attacks. We present a chance constrained formulation of the problem that is equivalent to a deterministic problem with modified demand and travel time parameters under mild assumptions on the distribution of stochastic parameters and relate it to a robust optimization approach. A tabu heuristic is proposed to solve this MIP and simulations are conducted to evaluate the quality of routes generated from both deterministic and chance constrained formulations. We observe that chance constrained routes can reduce the unmet demand by around 2%-6% for moderately tight deadline and total supply constraints.

1 Introduction

The classic vehicle routing problem (VRP) determines the optimal set of routes used by a fleet of vehicles to serve a given set of customers on a predefined graph; it aims at minimizing the total travel cost (proportional to the travel times or distances) and operational cost (proportional to the number of vehicles used). The stochastic VRP (SVRP) arises whenever some parameters of the VRP are random. Common examples are uncertain customers and demands and stochastic travel times or service times. In this work, we address a stochastic routing problem motivated by the problem of distributing medical supplies in large-scale emergency response.

W. Chaovalitwongse et al. (eds.), *Optimization and Logistics Challenges in the Enterprise*, Springer Optimization and Its Applications 30, DOI 10.1007/978-0-387-88617-6_13, © Springer Science+Business Media, LLC 2009

Large-scale emergencies (or major emergencies) are defined as events that overwhelm local emergency responders, which severely impact the operation of normal life, and have the potential to cause substantial casualties and property damage. Examples are natural disasters (earthquake, hurricane, flooding, etc.) and terrorist attacks, like September 11, 2001.

Careful and systematic preplanning, as well as efficient and professional execution in responding to a large-scale emergency, can save many lives. Rational policies and procedures applied to emergency response could maximize the effectiveness of the scarce resources available in relation to the overwhelming demands. A key ingredient in an effective response to an emergency is the prompt availability of necessary supplies at emergency sites. Given the challenges of delivering massive supplies in a short time period to dispersed demand areas, operations research models can play an important role in addressing and optimizing the logistical problems in this complex distribution process. Larson et al. [37, 38] conducted a detailed analysis based on well-known and recent large-scale emergencies. They emphasized the need for quantitative, model-oriented methods provided in the operations research field to evaluate and guide the operational strategies and actions in response to major emergencies.

The distinguishing characteristics of large-scale emergencies are high demand for supplies, low frequency of occurrence, and high uncertainty in many aspects (e.g., when and where they happen). In particular, the routing problem for emergency response faces the following unique challenges.

First, the initial prepositioned supplies will typically not be sufficient to cover all demands in the recommended response times. Unmet demand in an emergency situation can result in loss of life, an impact that outweighs more common VRP objectives such as the travel or operational cost. Therefore, the overriding objective in such problems is to mitigate the effect of the emergency by reducing the amount of unmet demand.

An additional aspect important in meeting the demand is the response time. Since the delivery of supplies to the population within a time frame makes an appreciable health difference, it is natural to associate a deadline with each demand, to model situations in which a late delivery leads to loss of life.

The highly unpredictable nature of large-scale emergencies leads to significant uncertainty both in demand and in travel time. For instance, at a given demand point (e.g., a neighborhood block), the quantity of required supplies (antidotes, protective equipments, medication, etc.) is often proportional to the size of the population and/or numbers of casualties. The casualty exposures or demands among "worried well" are hard to accurately predict for an overwhelming emergency occurrence. In addition, the emergency event itself may directly affect road conditions, e.g., possible congestion caused by the emergency event or destruction of the physical road network. Therefore, the travel times between points are stochastic.

We model a stochastic vehicle routing problem that minimizes unmet demand under the uncertainty in demand and travel time, with predefined service deadline and limited supply at the depot. Although it may not be possible to service all the nodes in the problem, we seek solutions that visit all the nodes, including nodes after the deadline or with an empty vehicle. Such a planned route provides a starting point for recourse action once demand and travel times are realized and there is the possibility of satisfying additional demand.

There are many ways to address parameter uncertainty in optimization problems, leading to different models and requiring different information on the uncertainty. In particular, we can obtain an answer to a routing problem under uncertainty through the solution of a representative deterministic problem by using stochastic programming and chance constrained models, by a robust optimization approach, or a Markov decision process model. The type of information available on the uncertain parameters is key in determining which model is most appropriate for a given application. In this work we ignore this application's specific concern and assume that any information that is needed is available. The chance-constrained programming relies on probabilistic information, supposed to be available, and tries to find a solution which is "optimal" in a probabilistic sense. By adopting this approach, we believe that it is not necessary to look for a solution that is always feasible, since the worst case (if there is one) is very unlikely to occur. However, robust optimization aims generally to provide the best solution feasible for all the uncertainty considered. In this work, we relate these two approaches by showing that both chance-constrained and robust optimization models require the solution of a single deterministic instance of the problem with modified uncertain parameters for our problem at hand. We therefore focus this chapter on the solution for these models of uncertainty which do not require specific solution procedures, as is the case in stochastic programming models or Markov decision models.

The rest of this chapter is organized as follows. Section 2 reviews relevant stochastic vehicle routing literature. Section 3 presents our problem formulation and the models to address uncertainty and Section 4 presents a tabu heuristic solution approach. We present some numerical experiment results in Section 5, and finally, conclude the chapter in Section 6.

2 Literature Review

In this section, we review well-known problems in the literature which are relevant to our problem at hand. This section is organized as follows: the general SVRP problem is briefly introduced and classified according to type of uncertainty, modeling methods and solution techniques. Two specific topics,

SVRP with stochastic customers and demands and SVRP with stochastic travel time and service time, follow the general discussion.

2.1 Stochastic VRPs

The vehicle routing problem is defined on a given graph $G = (V, \mathcal{A})$, where $V = \{v_1, v_2, \ldots, v_n\}$ is a set of vertices and $\mathcal{A} \subseteq \{(v_i, v_j) : i \neq j, v_i, v_j \in V\}$ is the arc set. An optimal set of routes, composed of a cyclic linkage of arcs starting and ending at the depot, is selected to serve a given set of customers at vertices. This problem was first introduced by Dantzig and Ramser in 1959 to solve a real-world application concerning the delivery of gasoline to service stations. A comprehensive overview of the vehicle routing problem can be found in [48] which discusses problem formulations, solution techniques, important variants and applications. Other general surveys on the deterministic VRP also can be found in [34]. The stochastic VRP (SVRP) introduces some element of uncertainty in the parameters within the system in question. The SVRP differs from its deterministic counterpart: several fundamental properties of the deterministic VRPs no longer hold in the stochastic case, and solution methodologies are considerably more complicated. A general review on the SVRP appeared in [23]. The stochastic vehicle routing problem can be broadly classified based on the following criteria:

- **Uncertainty in the problem:** The uncertainty can be present in different parts of the vehicle routing problem. It can be divided into VRP with stochastic customers (VRPSC), VRP with stochastic demands (VRPSD), VRP with stochastic travel time (VRPSTT) and VRP with stochastic service time (VRPSST).
- **Modeling method:** The modeling method can also be the criterion to classify SVRPs. Stochastic VRPs can be cast into a stochastic programming framework, which is further divided into chance-constrained program (CCP) and stochastic program with recourse (SPR). Another approach to model the SVRP is to view it as a Markov decision process. Some more recent modeling approaches include neurodynamic programming/reinforcement learning methodology.
- **Solution techniques:** Different solution techniques are the direct result of different modeling methods and the nature of the model. They usually broadly fall into two categories: exact methods and heuristic methods. The exact solution methods successfully applied to the SVRP include branch and bound, branch and cut, integer L-shape method and generalized dynamic programming. Under some mild assumptions, several classes of chance-constrained SVRPs can be transformed into equivalent deterministic VRPs [35, 42]. Numerous heuristics have been used to solve the SVRP, e.g., the savings algorithm, tabu search, etc. The heuristic solution techniques usually can be further divided as constructive heuristics, improvement heuristics and meta-heuristics.

2.2 SVRPs with Stochastic Customers and Demands

The VRP has stochastic demands (VRPSD) when the demands at the individual delivery (pickup) locations behave as random variables. The first proposed algorithm for the VRPSD by Tillman [46] was based on Clarke and Wright savings algorithm [15]. Another early major contribution on the VRPSD comes from Stewart and Golden [42] who applied the chance-constrained programming and recourse methods to model the problem. Later on, Dror [18] illustrated the impact that the direction of a designed route can have on the expected cost. A major contribution to the study of the VRPSD comes from Bertsimas [4, 5]. This work illustrated the a priori method with different recourse policies (reoptimization is allowed) to solve the VRPSD and derived several bounds, asymptotic results and other theoretical properties. Bertsimas and Simchi-Levi [9] surveyed the development in the VRPSD with an emphasis on the insights gained and on the algorithms proposed. Besides the conventional stochastic programming framework, a Markov decision process for single-stage and multistage stochastic models were introduced to investigate the VRPSD in [16, 17]. More recently, a reoptimization type routing policy for the VRPSD was introduced by [41].

The VRP with stochastic customers (VRPSC), in which customers with deterministic demands and a probability p_i of being present, and the VRP with stochastic customers and demands (VRPSCD), which combines the VRPSC and VRPSD, first appeared in the literature of [27, 28, 29]. Bertsimas [4] gave a more systematic analysis and presented several properties, bounds and heuristics. Gendreau et al. [22, 24] proposed the first exact solution, L-shaped method, and a meta-heuristic, tabu search, for the VRPSCD.

Another research direction of the VRP with stochastic demands or customers incorporates a dynamically changing environment, where the demands vary over time. Bertsimas and Van Ryzin [6, 7] pioneered this work and named it dynamic traveling repairman problem (DTRP). They applied queuing theory to solve it and analyzed several dispatching policies for both light and heavy traffic. Based on their results, Papastavrou et al. [40, 44] defined and analyzed a new routing policy using the branching process. The multiple scenario approach (MSA) [3] and waiting strategies [11] have also been introduced for this problem recently.

2.3 SVRPs with Stochastic Travel Time and Service Time

Compared with stochastic customers and demands, the research on the stochastic travel time and service time problem of the TSP and VRP have received less attention. The VRP with stochastic travel time (VRPSTT) describes the uncertain environment of the road traffic condition. Kao [31] first proposed heuristics based on dynamic programming and implicit enumeration for the TSP with stochastic travel time (TSPSTT). Carraway et al. [13]

used a generalized dynamic programming methodology to solve the TSP-STT. Laporte, Louveaux and Mercure [36] performed systematic research on the VRP with stochastic service and travel time (VRPSSTT). They proposed three models for the VRPSSTT: chance-constrained model, three-index recourse model, and two-index recourse model. They presented a general branch-and-cut algorithm for all three models. The VRPSSTT model was applied to a banking context and an adaptation of the savings algorithm was used in [33]. Jula et al. [30] developed a procedure to estimate the arrival time to the nodes in the presence of hard time windows. In addition, they used these estimates embedded in a dynamic programming algorithm to determine the optimal routes. Hadjiconstantinou and Roberts [26] formulated a VRP with stochastic service times to model a company who receives calls (failure reports) from customers and dispatches technicians to customer sites. They used a two-stage recourse model and a paired tree search algorithm to solve it.

3 Minimum Unmet Demand Routes

We now formulate our problem into a mixed integer programming (MIP) model. We first introduce the notation used and formulate the deterministic version of the problem. We then compare different uncertainty models for this problem.

3.1 Notation

We consider a set K of vehicles and a set D of demand nodes. We identify an additional node, node 0, as the supply node (depot) and let $C = D \cup \{\text{node } 0\}$ represent the set of all nodes. Indexed on sets K and C, we define the following *deterministic parameters*:

n: initial number of vehicles at the supply node (depot)
s :   amount of supplies at the supply node (depot)
c_k: load capacity of vehicle k
dl_i: service deadline at demand node i.

We use M as a large constant used to express nonlinear relationships through linear constraints. We also consider the following two parameters to represent the uncertain travel time and demand, respectively:

$\tau_{i,j,k}$:   time required to traverse arc(i, j) for vehicle k
ζ_i :   amount of demand needed at node i.

Finally, we define the binary and nonnegative decision variables as follows, indexed on sets K, C:
Binary:

$X_{i,j,k}$:   flow variables, equal to 1 if (i, j) is traversed by vehicle k
$S_{i,k}$:   service variables, equal to 1 if node i can be serviced by vehicle k

Nonnegative:

$Y_{i,j,k}$:   amount of commodity traversing arc(i, j) using vehicle k
U_i: amount of unsatisfied demand of commodity at node i
$T_{i,k}$:   visit time at node i of vehicle k
$\delta_{i,k}$:   delay incurred by vehicle k in servicing i.

3.2 Deterministic Model

The deterministic, minimize unmet demand problem can be expressed as follows

$$\textbf{DP}: \quad \begin{array}{c} \text{minimize} \sum_{i \in D} U_i + \kappa \sum_{i \in D, k \in K} T_{i,k} \\ \text{subject to constraints } (1) - (17), \end{array}$$

where the constraints are explained in detail below.

The objective of model DP is to minimize the weighted sum of the total unmet demands over all demand nodes and the total visit time at demand nodes of all vehicles. The κ value usually is set to be very small to make the total travel time a secondary objective compared with the unmet demand quantity. However, the travel time is a necessary term in the objective function to guide the route generation after the deadline. Since we model the routing problem in response to a large-scale emergency, the service start times (arrival times) directly associate with when the supply will be shipped and used at the dispensing sites. We would like to serve the dispensing sites as early as possible for life-saving purposes, so the arrival time is a much more important indicator of the service quality than the conventional objectives such as travel times or operational time.

We group the constraints into four parts: route feasibility constraints, time constraints, demand flow constraints and node service constraints. The following constraints (1)-(6) characterize the vehicle flows on the path and enforce the route feasibility.

$$\sum_{i \in D} \sum_{k \in K} X_{0,i,k} \leq n \tag{1}$$

$$\sum_{i \in D} \sum_{k \in K} X_{i,0,k} \leq n \tag{2}$$

$$\sum_{j \in D} X_{0,j,k} = \sum_{j \in D} X_{j,0,k} = 1 \qquad (\forall k \in K) \tag{3}$$

$$\sum_{j \in C} \sum_{k \in K} X_{i,j,k} = 1 \qquad (\forall i \in D) \tag{4}$$

$$\sum_{j \in C} \sum_{k \in K} X_{j,i,k} = 1 \qquad (\forall i \in D) \tag{5}$$

$$\sum_{j \in C} X_{i,j,k} = \sum_{j \in C} X_{j,i,k} \qquad (\forall i \in D \quad k \in K) \tag{6}$$

Constraints (1) and (2) specify that the number of vehicles to service must not exceed the available quantity ready at the supply node at the beginning of the planning horizon. The number of vehicles to service is stated by the total number of vehicles flowing from and back to the depot. Constraint (3) represents each vehicle flow from and back to the depot only once. Constraints (4) and (5) state that each demand node must be visited only once. Constraint (6) requires that all vehicles that flow into a demand point must flow out of it.

Constraints (7)-(10) guarantee schedule feasibility with respect to time considerations.

$$T_{0,k} = 0 \qquad (\forall k \in K) \tag{7}$$

$$(T_{i,k} + \tau_{i,j,k} - T_{j,k}) \le (1 - X_{i,j,k})M \qquad (\forall i, j \in C \quad k \in K) \tag{8}$$

$$0 \le T_{i,k} \le \sum_{j \in C} X_{i,j,k} M \qquad (\forall i \in D \quad k \in K) \tag{9}$$

$$0 \le T_{i,k} - \delta_{i,k} \le dl_i \sum_{j \in C} X_{i,j,k} \qquad (\forall i \in D \quad k \in K) \tag{10}$$

The fact that all vehicles leave the depot at time 0 is specified by constraint (7). Constraint (8) enforces the time continuity based on the node visiting sequence of a route. Constraint (9) sets the visit time to be zero if the vehicle does not pass a node. The variable $\delta_{i,k}$ represents the delay of the visit time if a vehicle reaches the node after the deadline and is set to zero if it arrives before the deadline in constraint (10).

This model primarily accommodates the emergency situation where late deliveries could lead to fatalities. To maximize the likelihood of saving lives, medication should be received by the affected population within the specified hours of the onset of symptoms to impact the patient survival. This is the rationale behind the preference of using a hard deadline constraint instead of the soft deadline. However, for problems where late deliveries are possible we can translate the proposed model to soft deadlines, having the penalty on the violation represent the worsening in patient condition due to late arrival.

Constraints (11)-(13) state node service constraints.

$$\delta_{i,k} \le (1 - S_{i,k})M \qquad (\forall i \in D \quad k \in K) \tag{11}$$

$$S_{i,k} \le \sum_{j \in C} X_{i,j,k} \qquad (\forall i \in D, k \in K) \tag{12}$$

$$S_{i,k}M \ge \left(\sum_{j \in C} Y_{j,i,k} - \sum_{j \in C} Y_{i,j,k} \right) \qquad (\forall i \in D, k \in K) \tag{13}$$

Binary decision variables $S_{i,k}$ are used to indicate whether a node i can be serviced by vehicle k (when it equals to 1). That is, if the vehicle k visits node

i before the deadline, then the vehicle can drop off some commodities at this node. However, the vehicle does not necessarily do it when $S_{i,k}$ equals to 1 since there might not be enough supply at the depot so the vehicle may not carry any commodities when it visits a later node in the route. We use these binary variables to keep the feasible region of this problem nonempty all the time. Constraints (4) and (5) will still enforce each node to be visited once and only once no matter before or after the deadline; however, those visits after the deadline cannot service the node any more. Constraint (11) states the deadline constraint and it can only be violated when $S_{i,k}$ equals to zero. Constraint (12) illustrates the relationship between the binary flow variables and the binary service variables. It implies the service variable can only be true when a vehicle physically passes a node. Constraint (13) requires that no commodity flows in a node after the deadline. On the other hand, there is no compulsory dropping off commodities at nodes visited before the deadline since there may not be enough supplies to meet the demand. It establishes the connection between the commodity flow and the vehicle flow.

Constraints (14)-(16) state the construction on the demand flows.

$$s - \sum_{k \in K} \left[\sum_{j \in C} Y_{0,j,k} - \sum_{j \in C} Y_{j,0,k} \right] \geq 0 \tag{14}$$

$$\sum_{k \in K} \left[\sum_{j \in C} Y_{j,i,k} - \sum_{j \in C} Y_{i,j,k} \right] + U_i - \zeta_i \geq 0 \qquad (\forall i \in D) \tag{15}$$

$$X_{i,j,k} c_k \geq Y_{i,j,k} \qquad (\forall i, j \in C, k \in K) \tag{16}$$

Constraint (14) requires the total shipment of commodity from the depot not exceeding its current supply inventory level. Constraint (15) enforces the balanced material flow requirement for the demand nodes. Constraint (16) allows the flow of commodities as long as there is sufficient vehicle capacity. It also connects the commodity flow and the vehicle flow.

$$X_{i,j,k}, S_{i,j} \text{ binary;} \quad Y_{i,j,k} \geq 0; \quad U_i \geq 0; \quad T_{i,k} \geq 0; \quad \delta_{i,k} \geq 0. \tag{17}$$

Constraint (17) states the binary and nonnegativity properties of the decision variables.

3.3 Stochastic Model

The parameters $\tau_{i,j,k}$ in constraint (8) and ζ_i in constraint (15) represent the uncertain travel time and demand parameters of our problem, respectively. If we ignore the uncertainty and replace these random quantities by representative values, such as their mean $\mu^T_{i,j,k}$ and μ^ζ_i or mode values, we can solve a deterministic problem DP to obtain a simple solution for this problem. This deterministic solution will be helpful as a benchmark to compare the quality of routes and demonstrate the merits of other more sophisticated methods we discuss next. There are two other ways to handle uncertainty that for this problem lead to the solution of a single deterministic problem DP: chance-constrained programming and robust optimization. The solution of this routing problem through other methods of representing uncertainty, such as stochastic programming and Markov decision processes require more involved solution procedures and will not be explored in this chapter.

In *chance-constrained programming* (**CCP**) we assume that the parameters $\tau_{i,j,k}$ and ζ_i are unknown at the time of planning but follow some known probability distributions. We assume they are uniformly and independently distributed. We let α_D and α_T represent the confidence level of the chance constraints defining the unmet demand at each node and the arrival time of each vehicle at each node, respectively. Thus, the constraints with stochastic parameters must hold with these given probabilities. For a given distribution on $\tau_{i,j,k}$ and ζ_i, we can rewrite constraint (8) and constraint (15) in the chance-constrained fashion with levels α_T and α_D as follows:

$$P\left\{\tau | (T_{i,k} + \tau_{i,j,k} - T_{j,k}) \leq (1 - X_{i,j,k})M\right\} \geq 1$$
$$-\alpha_T \ (\forall i, j \in C \ k \in K) \tag{18}$$

$$P\left\{\zeta | \sum_{k \in K}\left[\sum_{j \in C} Y_{j,i,k} - \sum_{j \in C} Y_{i,j,k}\right] + U_i - \zeta_i \geq 0\right\} \geq 1$$
$$-\alpha_D \ (\forall i \in D) \tag{19}$$

We call this *chance-constrained model* (**CCP** model), which is modified based on the DP model in Section 3.2, by replacing constraints (8) and (15) with constraints (18) and (19). Under some assumption of their distribution, constraint (18) and constraint (19) can be transformed to their deterministic counterpart. From this point onward in this paragraph, we use short notation τ and ζ to substitute $\tau_{i,j,k}$ and ζ_i for simplicity. For example, we assume τ and ζ follow a lognormal distribution with mean μ_τ and standard deviation σ_τ and mean μ_ζ and standard deviation σ_ζ, respectively. The logarithm $\log(\tau), \log(\zeta)$ are normally distributed as $normal(\mu'_\tau, \sigma'_\tau)$ and $normal(\mu'_\zeta, \sigma'_\zeta)$. The relationship between the parameters of lognormal distribution and normal distribution is stated as: $\mu' = \log \mu - \frac{1}{2}\sigma'^2$, $\sigma'^2 = \log(\frac{\mu^2+\sigma^2}{\mu^2})$. We let κ_T and κ_D represent the Z value for the normal distribution corresponding to the confidence level α_T and α_D and we call them "safety factors" in the later experimental results section. Therefore, the deterministic counterpart of constraint (18) and constraint (19) can be expressed as:

$$(T_{i,k} + e^{\mu'_{\tau_{i,j,k}} + \kappa_T \sigma'_{\tau_{i,j,k}}} - T_{j,k}) \leq (1 - X_{i,j,k})M \quad (\forall i,j \in C \quad k \in K) \quad (20)$$

$$\sum_{k \in K} \left[\sum_{j \in C} Y_{j,i,k} - \sum_{j \in C} Y_{i,j,k} \right] + U_i \geq e^{\mu'_{\zeta_i} + \kappa_D \sigma'_{\zeta_i}} \quad (\forall i \in D) \quad (21)$$

The *robust optimization model* assumes that the uncertain parameters $\tau_{i,j,k}$ and ζ_i are only known to belong to a given uncertainty set $\mathcal{U}$. The robust optimization approach, introduced by Ben-Tal and Nemirovski [2] and more recently extended to integer programming [8] and VRP [43], requires that the solution satisfy the constraints with uncertain parameters for all possible values in the uncertainty set $\mathcal{U}$. That is, we rewrite constraint (8) and constraint (15) as follows:

$$T_{i,k} + \tau_{i,j,k} - T_{j,k} \leq (1 - X_{i,j,k})M \quad \forall \tau_{i,j,k} \in \mathcal{U}, \quad (\forall i,j \in C \ k \in K) \quad (22)$$

$$\sum_{k \in K} \left[\sum_{j \in C} Y_{j,i,k} - \sum_{j \in C} Y_{i,j,k} \right] + U_i - \zeta_i \geq 0 \quad \forall \zeta_i \in \mathcal{U}, \quad (\forall i \in D) \quad (23)$$

These infinitely many constraints, indexed over every $\tau_{i,j,k} \in \mathcal{U}$ and $\zeta_i \in \mathcal{U}$, can be expressed by single constraints by substituting the maximum possible uncertainty parameters $\tau^*_{i,j,k} = \max_{\tau_{i,j,k} \in \mathcal{U}} \tau_{i,j,k}$ and $\zeta^*_i = \max_{\zeta_i \in \mathcal{U}} \zeta_i$. With these maximum uncertainty parameters, these robust constraints can be written as:

$$T_{i,k} + \tau^*_{i,j,k} - T_{j,k} \leq (1 - X_{i,j,k})M \quad (\forall i,j \in C \quad k \in K) \quad (24)$$

$$\sum_{k \in K} \left[\sum_{j \in C} Y_{j,i,k} - \sum_{j \in C} Y_{i,j,k} \right] + U_i - \zeta^*_i \geq 0 \quad (\forall i \in D) \quad (25)$$

As we mentioned earlier, both the chance-constrained model and the robust optimization model can be expressed as a deterministic equivalent problem, with changing demand and travel time values. Therefore these problems are just as difficult to solve as a single deterministic routing problem. The only difference between these two models of addressing uncertainty is the value of the uncertainty used, for instance, $e^{\mu'_{\tau_{i,j,k}} + \kappa_T \sigma'_{\tau_{i,j,k}}}$ versus $\tau^*_{i,j,k}$, for the time continuity constraints. Note that varying κ_T and κ_D we can represent $\tau^*_{i,j,k}$ and ζ^*_i values as well as the mean values $\mu^\tau_{i,j,k}$ and μ^ζ_i, thus in the remainder of the chapter we only consider the chance-constrained model for different confidence levels.

4 Tabu Heuristic Solution Approach

Tabu search is a local search procedure which iteratively moves from a solution to its best neighbor until some stopping criteria have been satisfied. The search keeps a tabu list which prohibits revisiting a recently explored node unless some aspiration criteria have been met to avoid cyclic movement. It allows a solution to temporarily move to a worse position to escape a local optimum. Tabu search has been successfully used in solving hard optimization problems in many fields. A more comprehensive review on this technique and its applications can be found in [25]. Tabu search was first introduced for solving VRP by Willard [49]. Later on, different groups of researchers designed a variety of neighborhood/moves and adopted some problem-specific mechanism to significantly improve the performance, see for instance [20, 39, 47, 50]. Tabu search has also been applied to major variants of VRP, e.g., VRP with time windows [45], VRP with split delivery [1], the pick up and delivery problem [10] as well as the stochastic VRP [24]. It has been shown that tabu search generally yields very good results on a set of benchmark problems and some larger instances [21].

Applying tabu search to a particular problem requires a fair amount of problem-specific knowledge. Given the success of tabu heuristic in the classic VRP and its variants, and the similar structure of our model, we believe this approach holds much promise in solving our problem. The algorithm we propose here uses some ideas from the standard VRP, and incorporates new features taking into account the unmet demand objective of our problem.

Because the problem has insufficient supplies at the depot in most scenarios, we use an *unassign node manager* list of nodes to keep track of all the nodes with unmet demand. We also use a *route manager* list of solutions to keep track of all the incomplete (may not include all the nodes) but feasible (meet both capacity and time constraints) routes. We initiate a solution with a visit-nearest-node heuristic and put all the unvisited nodes into the *unassign node manager* list.

A key element in designing a tabu heuristic is the definition of the neighborhood of a solution or equivalently the possible moves from a given solution. Beside adopting the standard 2-opt exchange move and the λ-interchange move, we implement a new DEM-move to accommodate the special needs of the potentially unmet demands. The DEM-move will insert an unassigned demand node into a current route or exchange an unassigned node with some (one to three) consecutive node(s) in a route by abiding both deadline and capacity constraints. In a λ-interchange move, with $\lambda = 3$, it includes a combination of vertex reassignments to different routes and vertex interchanges between two routes; the reassigning/interchanging segment is up to three consecutive vertices in a route.

We also maintain a *move manager* list to record all possible moves from the current solution. After a solution moves to its neighbor, instead of reconstructing the whole neighborhood of the new solution, we only update the nonoverlapping neighbors. That is, we eliminate those moves that are relevant (sharing the same route or sharing the same unassigned node, etc.) to the move that just has been executed, and generate new feasible moves that are relevant to the updated route(s) only. This will significantly reduce the computational effort of exploring the neighborhood. In the search process, we apply a random tabu duration (uniformly random generated between five to ten steps) and a "best-ever" aspiration scheme.

We stop the heuristic after a given number of nonimproving steps. After we obtain the tabu search solution, we complete the route by adopting a variation of a "next-earliest-node" heuristic to insert all unassigned nodes after the deadline of each route (keeping the before-deadline part intact). In this postprocessing heuristic, each unassigned node picks a route to locate itself after the deadline, where the summation of its arrival time and the arrival time to the next node is minimized at the time of assignment. The quality of this tabu heuristic is evaluated in the next section.

5 Experimental Results

In this section, we demonstrate how the proposed MIP model in Section 3 and the tabu heuristic solution approach in Section 4 can be used to solve our problem at hand.

The purpose of the experiments is to compare the performance of the DP and CCP routes in terms of the unmet demand through simulations. We define a *problem scenario* for every randomly generated network consisting of a depot and 50 demand nodes, with a randomly generated mean demand quantity for each demand node. We perform simulations on ten problem scenarios and average the results. Our interests are primarily focused on comparing the alternatives between short-and-risky routes and long-and-safe routes in the first set of experiments. Another experiment investigates how the value of the safety factors influences the quality of the routes.

We first describe the input parameters for our example and then we show the quality of the tabu heuristic. Finally, we demonstrate and discuss the results from the simulation experiments.

5.1 Data Generation of Input Parameters

We test our model and heuristic on ten different problem scenarios. For each scenario we uniformly generate: 50 demand nodes and one depot node in a 200 by 200 square domain; and a mean demand quantity for each demand node ranging from 5 to 15. We service this demand with a fleet of ten uncapacitated vehicles. We use the Euclidean distances between any two of the nodes and assume a symmetric complete graph topology. The mean travel time between any two nodes is proportional to the distance.

In the DP model, we use the mean value of the demand quantities and travel times as its parameters. In the CCP model, we use a lognormal distribution with the same mean value as was used in the DP model; the standard deviation is set to be proportional to the mean value for demand (20% of the mean value) and inversely proportional to the mean value of the travel time ($\sigma = \frac{UB-\mu}{100}\mu$, UB is the upper bound for the inversely proportional transformation, whose value is dependent on the graph topology, which must be greater than the longest arc in the graph). We restrict our analysis to this type of travel time distribution because we aim to compare the trade-offs between short-and-risky routes and long-and-safe routes. Note that the DP routes in our experiments tend to be short-and-risky since they do not take into account the variability, while the CCP routes tend to favor longer routes that have smaller variance. For the chance-constrained model, we set the confidence level as 95%, thus setting the values κ_D and κ_T in constraints (20)-(21) to 1.65, which we refer to as the "safety factor." Since the routes generated from the model are sensitive to both the deadline and the total supply at the depot, we vary these two parameters and observe the results. We use 70%, 80%, 90%, 100% and 120% of the *base quantity* as the available supply quantity. The *base quantity* of the total supply at the depot is defined to be the summation of the demand quantity at all the demand nodes, which is 500 on average. The deadline is set to 40%, 50%, 60%, 80%, 100% and 120% of the *base route length*. The *base route length* is defined as the average length of all the edges in the graph times 5 (50 demand nodes are served by 10 vehicles; so on average, each vehicle serves 5 demand nodes). We call one combination of the deadline and the total supply parameters a *test case*. So for each problem scenario, we have 30 *test cases*, which are identified by the different combinations of the deadline on the demand nodes (six types: 40%, 50%, 60%, 80%, 100%, 120%) and the total supply at the depot (five types: 70%, 80%, 90%, 100%, 120%). For each *test case*, both the deterministic and the CCP version of the problem are solved by the tabu heuristic.

5.2 Quality of Tabu Heuristic

To evaluate the quality of our tabu heuristic, we run this search process over ten random problem scenarios as described above. The average of the percentage of unmet demand as well as the remaining supply quantity level at the depot for the deterministic formulation is recorded for each *test case*. Note that each entry in the table is an average of ten routes. The results are presented in Table 1 and Table 2.

These results show that, in terms of minimizing the unmet demand, the six cases, which are indicated by a "*", reach the minimum. They either meet all the demands (0% unmet demand) or deliver all the supplies (zero remaining supply at the depot) over the ten problem scenarios we tested. From the first three rows of the above tables, the results are very close to the best we can do regarding to the unmet demand, since the remaining supplies

Table 1. Percentage of unmet demand over total demand for the deterministic model

	DL200 (40%)	DL250 (50%)	DL300 (60%)	DL400 (80%)	DL500 (100%)	DL600 (120%)
sup350(70%)	30.78%	30.63%	30.63%	30.61%	30.59%	30.59%
sup400(80%)	21.19%	20.75%	20.73%	20.71%	20.64%*	20.64%*
sup450(90%)	12.76%	11.12%	10.82%	10.82%	10.82%	10.78%
sup500(100%)	5.31%	2.20%	1.87%	1.77%	1.75%	1.75%
sup600(120%)	4.98%	0.20%	0.00%*	0.00%*	0.00%*	0.00%*

Table 2. Remaining supply at depot for the deterministic model

	DL200 (40%)	DL250 (50%)	DL300 (60%)	DL400 (80%)	DL500 (100%)	DL600 (120%)
sup350(70%)	0.9	0.2	0.2	0.2	0.1	0.1
sup400(80%)	2.4	0.2	0.2	0.1	0*	0*
sup450(90%)	9.7	1.5	0.1	0.1	0.1	0.1
sup500(100%)	22.1	6.5	4.7	4.4	4.3	4.2
sup600(120%)	120.3	95.9	94.8*	94.8*	94.8*	94.8*

are almost empty and the percentage of the unmet demand is near the simple lower bound given by the shortage of supplies to meet the total demand. Only the DL200-sup500 and DL200-sup600 grids give a higher percentage of unmet demand than what is due to the shortage in supplies. We suspect this is due to the tight deadline which might prevent the timely delivery even with enough supplies. A tighter lower bound with respect to a short deadline is a subject for future research. However, in general, we conclude that the tabu search method we proposed is very effective in minimizing the unmet demand, which is our primary concern in this model.

5.3 Simulation and Analysis

To evaluate the quality of the routes generated by the DP and CCP models, we observe how each type of route performs under randomly generated demand and travel time situations. We now describe this simulation process and then present two sets of results with analysis.

LP Model for Simulation

To evaluate the effectiveness of DP and CCP routes, we assume that they are planned prior to the realization of the uncertainty and have to operate in an environment where demand and travel times that occur may deviate from the ones considered in the planning phase. Thus we consider the following sequential events: first, we use the MIP problem to obtain DP and CCP routes with estimates (mean and distributions) of the uncertain demand and travel times. Then the uncertainty is realized, and these fixed routes have to meet as much of the demand that appears. Here we assume that we are able to take a recourse action on how to distribute the supplies among the vehicles to meet the realized demand. We solve a linear programming problem to determine the quantity of commodity to be carried and delivered by each vehicle based on the fixed routes. Therefore, we generate preplanned routes with the MIP model and then at the time of the emergency run a linear programming problem to determine the delivery quantity. There are several advantages to having well-established preplanned routes for training and preparation purposes.

Since most of the input parameters are known and only the commodity flows are decision variables in the simulation stage, this optimization model is a deterministic linear programming problem (DLP) with the same basic structure as the MIP problem presented in Section 3. Now the vehicle flow $X_{i,j,k}$ is fixed to the result of the MIP problem. With these variables fixed, the values of $T_{i,k}$ and $\delta_{i,k}$ are also fixed and can be calculated from the fixed routes $X_{i,j,k}$. We eliminate those demand nodes whose $T_{i,k}$ already missed the deadline and minimize the unmet demand by adjusting the commodity flow $Y_{i,j,k}$. We formally show the model as follows:

Model **DLP**:

$$\text{minimize} \sum_{i \in D} U_i + \kappa \sum_{i \in D, k \in K} T_{i,k}$$

subject to: constraints (13)-(14), (16)

$$\sum_{k \in K} \left[\sum_{j \in C} Y_{j,i,k} - \sum_{j \in C} Y_{i,j,k} \right] + U_i - d_i \geq 0 \qquad (\forall i \in D) \qquad (26)$$

$$Y_{i,j,k} \geq 0; \qquad U_i \geq 0. \qquad (27)$$

The d_i in constraint (26) is the actual demand quantity at each node after the scenario has been realized. Constraint (27) shows the nonnegativity properties of the decision variables and the DLP is a linear programming problem.

Solution and Analysis I – the Deterministic Routes versus the CCP Routes

Our first set of experiments test the efficiency and robustness of our chance-constrained model in determining preplanned routes. We compare solutions from the CCP model against routes generated by the DP model that uses the mean values of the demand quantity and travel times as problem parameters.

For every problem scenario and test case, for given deadline and total supply, we compute the DP and CCP routes with the tabu heuristic described in Section 4. We also generate 100 *instances* of the realization of the random travel time and demand quantity parameters based on the same lognormal distribution assumed by the CCP model. We evaluate how well the DP and CCP routes can meet the demand in each instance by solving the DLP problem above. We solved the linear program DLP with CPLEX 9.0 with default settings on a 3.2 GHz CPU with 2GB RAM. To summarize, in the experiments below we consider ten problem scenarios, and for each we perform 30 different test cases, varying the deadline and total supply. In each test case we evaluate the performance of the DP and CCP routes using 100 different random instances. Hence, for each *test case*, the simulated results of the unmet demand are obtained from the average over 1000 instances. The results are summarized in Table 3 and Figure 1.

The results suggest that when the deadline is very tight (between 40% to 60% of the *base route length*), the deterministic routes outperform the CCP routes. There is no benefit for utilizing a conservative routing strategy under tight deadlines since the longer less risky trips (arcs) in most generated instances are longer than the deadlines. In this case, the short-and-risky routes generated by the DP model at least have a chance in some of the generated instances of arriving on time especially for the demand points in the beginning of the route. In the situation where there is limited supply (e.g., when the total supply is only 70% of the total demand), the DP and CCP routes yield roughly the same amount of unmet demand even under a more relaxed deadline. Because of the insufficient supplies at the depot, the extended deadline cannot provide a better coverage.

Under moderate deadlines and supply quantities (when the percentage of the total supply over the total demand and the percentage of the deadline over the *base route length* both are between 80% and 120%), we can conclude that CCP routes outperform DP routes 2% to 6% in terms of the percentage of the unmet demand over the total demand, according to our simulation experiments. As the deadline and the total supply are relaxed, the percentage of the unmet demand is decreasing for both DP and CCP routes; at the same time,

Table 3. Average percentage of the unmet demand over the total demand for DP and CCP models (lognormal distribution)

	DL200 (40%) DP	DL200 (40%) CCP	DL250 (50%) DP	DL250 (50%) CCP	DL300 (60%) DP	DL300 (60%) CCP	DL400 (80%) DP	DL400 (80%) CCP	DL500 (100%) DP	DL500 (100%) CCP	DL600 (120%) DP	DL600 (120%) CCP
sup350 (70%)	31.79	35.07	31.32	31.94	31.03	30.91	30.57	30.68	30.78	30.63	31.04	30.63
sup400 (80%)	24.28	30.53	23.20	24.85	22.69	22.09	21.83	21.01	22.13	20.73	21.87	20.70
sup450 (90%)	19.67	28.75	17.48	21.00	16.35	15.63	14.59	12.16	13.98	11.21	14.45	11.00
sup500 (100%)	17.85	28.07	14.78	19.70	13.23	13.28	11.11	6.51	10.15	4.17	9.36	3.34
sup600 (120%)	17.22	28.38	14.37	19.69	11.57	13.00	10.54	5.53	8.55	2.55	7.78	1.44

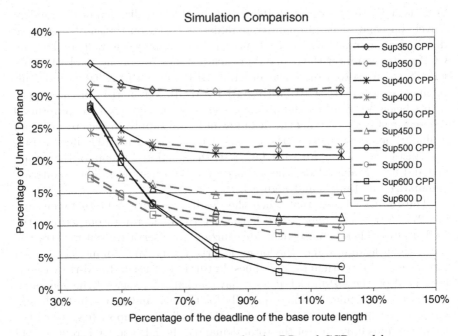

Fig. 1. Simulation comparison for the DP and CCP models

the advantage of CCP routes over DP routes gradually increases. The simulation results on the percentage of the unmet demand from the CCP routes are approaching its lower bound due to the shortage of the supply, with an increasing deadline. We believe that it is because of, under the relaxed enough deadline and supply quantities, the conservative nature of the CCP model shows its merits. It leads to a similar number of nodes in different routes in the CCP model. In contrast, the DP routes are more prone to have an uneven number of nodes between different paths leading to a higher chance of having unmet demand when they are tested over different realization *instances*.

Solution and Analysis II – the Role of the Safety Factor

Another experiment we conduct is to fix the deadline and the total supply at the depot and vary the confidence level, which has a one-to-one correspondence to the κ_T and κ_D values in constraints (20) and (21). We use the same value for both κ_T and κ_D. For one specific network topology and demand distribution (a *problem scenario*), we fix the deadline at 80% of the *base route length* and the total supply at 90% of the total demand. We generate different sets of CCP routes with different safety factors, and run the same simulation as described above. We plot the percentage of the unmet demand over the total demand in Figure 2. Since we are using a lognormal distribution, the deterministic case (mean value) lies at $e^{\mu' + \frac{\sigma'^2}{2}}$. The corresponding safety factor is bigger than 0.5 as compared with exactly 0.5 for the normal distribution. The value of the safety factor for the deterministic case in the lognormal distribution is heavily dependent on both the mean and the variance of the distribution. For the specific problem scenario we are testing, the estimated safety factor is 0.75 by averaging the variance value of all the edges in the graph in a statistical sense. It is shown as a dot on the plot. The point pinned by a square gives the best average unmet demand ratio, which is also the value we used for the CCP model for the comparison purpose in Section 5.3.2. It corresponds to the confidence level at 95% and the safety factor as 1.65. The plot in Figure 2 infers that with an increasing "contingency stuffing" (a bigger safety factor) starting from the deterministic point, it will provide better coverage. However, if we are too conservative as further increasing the safety factor after the "best CCP point," the quality of the result deteriorates. Under a given deadline, if the safety factor is too big such that all the short and risky arcs have been "stretched" to overpass the deadline, then the route planning based on those "overstretched" arcs is not very meaningful in providing any guidance since every node would miss its deadline. Hence it is crucial for planning with the "optimal" safety factor, which is neither too opportunistic nor too conservative.

As we discussed earlier, the robust optimization, which aims to optimize the worst-case scenario, also shares the same problem structure and computational complexity as the deterministic and the CCP formulations under a bounded uncertainty set or a given set of scenarios. Recently, different groups

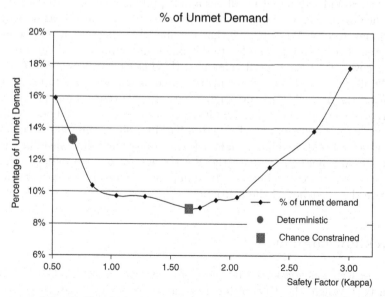

Fig. 2. Simulation comparison for different safety factors

of researchers are interested in developing more elaborated uncertainty sets in order to address the issue of overconservative in worst-case models while maintaining the computational tractability. Also several works have established the link between the chance-constrained technique and robust optimization [12, 14, 19, 32]. In our work, the only difference between the deterministic counterpart of the robust model and the CCP model are the different safety factors used by each model. In chance-constrained optimization, the constraints are enforced up to a prespecified level of probability. Robust optimization seeks a solution which satisfies all possible constraint instances. It achieves this by enforcing that the strictest version of every constraint is satisfied by substituting the largest uncertainty parameter value in each constraint. This maximal value for the uncertainty parameter has a one-to-one correspondence with a safety factor value in the chance-constrained formulation and lies somewhere on the plot in Figure 2.

6 Conclusions and Future Work

In this chapter, we model a stochastic vehicle routing problem to minimize the unmet demand with a chance-constrained programming technique, which is primarily motivated by routing medical supplies for large-scale emergency response. Unique features of this problem have been analyzed and embedded into our model. A tabu search heuristic is proposed to solve this model and its quality is evaluated by preliminary numerical experiments. Simulations are conducted to test the effectiveness of the CCP routes over the deterministic routes under different deadlines and supply quantities at the depot. The influence of the safety factor at different values is also demonstrated by the simulation results. We conclude that our model is effective in producing the preplanned routes under moderate deadlines and supplies and can provide a better coverage of the overall demand with some uncertainty present.

We are interested in the following two ongoing research directions: first, due to the nature of the large-scale emergencies, we believe it is more pragmatic to incorporate a split delivery and multidepot situation into our model to better deal with the huge amount of requests within a short time line. Second, we are interested in developing a tighter lower bound with short deadline, which does not only help to evaluate the effectiveness of our heuristic result, but could also facilitate to eliminate those infeasible routes in the earlier stage to improve the efficiency of the heuristic.

7 Acknowledgments

This research was supported by the United States Department of Homeland Security through the Center for Risk and Economic Analysis of Terrorism Events (CREATE), grant number EMW-2004-GR-0112. However, any opinions, findings, and conclusions or recommendations in this document are those of the author(s) and do not necessarily reflect views of the U.S. Department of Homeland Security. Also the authors wish to thank Ilgaz Sungur, Hongzhong Jia, Harry Bowman, Richard Larson and Terry O'Sullivan for their valuable input and comments for the improvement of this chapter.

References

1. C. Archetti, A. Hertz, and M. Speranza. A tabu search algorithm for the split delivery vehicle routing problem. *Transportation Science*, 40:64–73, 2006.
2. A. Ben-Tal and A. Nemirovski. Robust convex optimization. *Mathematics of Operations Research*, 23(4):769–805, 1998.
3. R. W. Bent and P. V. Hentenryck. Scenario-based planning for partially dynamic vehicle routing with stochastic customers. *Operations Research*, 52(6):977–987, 2004.
4. D. Bertsimas. *Probabilistic combinational optimization problems*. PhD thesis, Operation Research Center, Massachusetts Institute of Technology, Cambridge, MA, 1988.
5. D. Bertsimas. A vehicle routing problem with stochastic demand. *Operations Research*, 40(3), May 1992.
6. D. Bertsimas and G. Van Ryzin. A stochastic and dynamic vehicle routing problem in the Euclidean plane. *Operations Research*, 39(4):601–615, 1991.
7. D. Bertsimas and G. Van Ryzin. Stochastic and dynamic vehicle routing in the Euclidean plane with multiple capacitated vehicles. *Operations Research*, 41(1):60–76, 1993.
8. D. Bertsimas and M. Sim. Robust discrete optimization and network flows. *Mathematical Programming*, 98:49–71, 2003.
9. D. Bertsimas and D. Simchi-levi. A new generation of vehicle routing research: robust algorithms, addressing uncertainty. *Operations Research*, 44(2), March 1996.
10. N. Bianchessi and G. Righini. Heuristic algorithms for the vehicle routing problem with simultaneous pick-up and delivery. *Computers and Operations Research*, 34:578–594, 2006.
11. J. Branke, M. Middendorf, G. Noeth, and M. Dessouky. Waiting strategies for dynamic vehicle routing. *Transportation Science*, 39:298–312, 2005.
12. G. Calafiore and M.C. Campi. Uncertain convex programs: randomized solutions and confidence levels. *Mathematical Programming*, 102:25–46, 2005.
13. R. Carraway, T. Morin, and H. Moskowitz. Generalized dynamic programming for stochastic combinatorial optimization. *Operations Research*, 37(5):819–829, 1989.
14. X. Chen, M. Sim, and P. Sun. A robust optimization perspective on stochastic programming. *Optimization Online*, 2005. http://www.optimization-online.org/DB_HTML/2005/06/1152.html.
15. G. Clarke and J.W. Wright. Scheduling of vehicles from a central depot to a number of delivery points. *Operations Research*, 12:568–581, 1964.
16. M. Dror. Modeling vehicle routing with uncertain demands as a stochastic program: properties of the corresponding solution. *European Journal of Operational Research*, 64:432–441, 1993.

17. M. Dror, G. Laporte, and P. Trudeau. Vehicle routing with stochastic demands: properties and solution framework. *Transportation Science*, 23(3), August 1989.
18. M. Dror and P. Trudeau. Stochastic vehicle routing with modified savings algorithm. *European Journal of Operational Research*, 23:228–235, 1986.
19. E. Erdoğan and G. Iyengar. Ambiguous chance constrained problems and robust optimization. *Mathematical Programming*, 107:37–61, 2006.
20. M. Gendreau, A. Hertz, and G. Laporte. A tabu search heuristic for the vehicle routing problem. *Management Science*, 40:1276–1290, 1994.
21. M. Gendreau, G. Laporte, and J. Y. Potvin. *The Vehicle Routing Problem*, chapter Metaheuristics for the Capacitated VRP, pages 129–154. SIAM Monographs on Discrete Mathematics and Applications, SIAM Publishing, 2002.
22. M. Gendreau, G. Laporte, and R. Seguin. An exact algorithm for the vehicle routing problem with stochastic demands and customers. *Transportation Science*, 29(2), May 1995.
23. M. Gendreau, G. Laporte, and R. Seguin. Stochastic vehicle routing. *European Journal of Operational Research*, 88:3–12, 1996.
24. M. Gendreau, G. Laporte, and R. Seguin. A tabu search heuristic for the vehicle routing problem with stochastic demands and customers. *Operations Research*, 44(3), May 1996.
25. G. Glover and M. Laguna. *Tabu Search*. Kluwer, Boston, MA, 1997.
26. E. Hadjiconstantinou and D. Roberts. *The Vehicle Routing Problem*, chapter Routing under Uncertainty: an Application in the Scheduling of Field Service Engineers, pages 331–352. SIAM Monographs on Discrete Mathematics and Applications, SIAM Publishing, 2002.
27. P. Jaillet. *Stochastics in Combinatorial Optimization*, chapter Stochastic Routing Problem. World Scientific, New Jersey, 1987.
28. P. Jaillet and A. Odoni. *Vehicle Routing: Methods and Studies*, chapter The Probabilistic Vehicle Routing Problem. North-Holland, Amsterdam, 1988.
29. A. Jézéquel. Probabilistic vehicle routing problems. Master's thesis, Department of Civil Engineering, Massachusetts Institute of Technology, 1985.
30. H. Jula, M. M. Dessouky, and P. Ioannou. Truck route planning in non-stationary stochastic networks with time-windows at customer locations. *IEEE Transactions on Intelligent Transportation Systems*, 2005. to appear.
31. E. Kao. A preference order dynamic program for a stochastic travelling salesman problem. *Operations Research*, 26:1033–1045, 1978.
32. O. Klopfenstein and D. Nace. A robust approach to the chance-constrained knapsack problem. *Optimization Online*, 2006. http://www.optimization-online.org/DB_HTML/2006/03/1341.html.
33. V. Lambert, G. Laporte, and F. Louveaux. Designing collection routes through bank branches. *Computers and Operations Research*, 20:783–791, 1993.
34. G. Laporte. The vehicle routing problem: an overview of exact and approximate algorithms. *European Journal of Operational Research*, 59:345–358, 1992.
35. G. Laporte, F. Laporte, and H. Mercure. Models and exact solutions for a class of stochastic location-routing problems. *European Journal of Operational Research*, 39:71–78, 1989.
36. G. Laporte, F. Louveaux, and H. Mercure. The vehicle routing problem with stochastic travel times. *Transportation Science*, 26(3), August 1992.
37. R. C. Larson. *The McGraw-Hill Handbook of Homeland Security*, chapter Decision Models for Emergency Response Planning. The McGraw-Hill Companies, 2005.

38. R. C. Larson, M. Metzger, and M. Cahn. Emergency response for homeland security: lessons learned and the need for analysis. *Interfaces*, 2005. To appear.

39. I. H. Osman. Metastrategy simulated annealing and tabu search algorithms for the vehicle routing problem. *Annals of Operations Research*, 41:421–451, 1993.

40. J. D. Papastavrou. A stochastic and dynamic routing policy using branching processes with state dependent immigration. *European Journal of Operational Research*, 95:167–177, 1996.

41. N. Secomandi. A rollout policy for the vehicle routing problem with stochastic demands. *Operations Research*, 49(5):796–802, 2001.

42. W. Stewart and B. Golden. Stochastic vehicle routing: a comprehensive approach. *European Journal of Operational Research*, 14:371–385, 1983.

43. I. Sungur, Fernando Ordóñez, and Maged M. Dessouky. A robust optimization approach for the capacitated vehicle routing problem with demand uncertainty. Technical report, Deniel J. Epstein Department of Industrial and Systems Engineering, University of Southern California, 2006.

44. M. R. Swihart and J. D. Papastavrou. A stochastic and dynamic model for the single-vehicle pick-up and delivery problem. *European Journal of Operational Research*, 114:447–464, 1999.

45. E. D. Taillard, P. Badeau, M. Gendreau, F. Guertin, and J. Y. Potvin. A tabu search heuristic for the vehicle routing problem with soft time windows. *Transportation Science*, 31:170–186, 1997.

46. F. Tillman. The multiple terminal delivery problem with probabilistic demands. *Transportation Science*, 3:192–204, 1969.

47. P. Toth and D. Vigo. The granular tabu search (and its application to the vehicle routing problem). Technical Report OR/98/9, DEIS, Università di Bologna, Italy, 1998.

48. P. Toth and D. Vigo. *The Vehicle Routing Problem*. SIAM Monographs on Discrete Mathematics and Applications, SIAM Publishing, 2002.

49. J. A. G. Willard. Vehicle routing using r-optimal tabu search. Master's thesis, The Management School, Imperial College, London, 1989.

50. J. Xu and J. P. Kelly. A network flow-based tabu search heuristic for the vehicle routing problem. *Transportation Science*, 30:379–393, 1996.

Collaboration in Cargo Transportation

Richa Agarwal[1], Özlem Ergun[2], Lori Houghtalen[3], and
Okan Orsan Ozener[4]

[1] Department of Industrial and Systems Engineering
Georgia Institute of Technology, Atlanta, GA 30332
richaa@amazon.com (Now at Amazon.com)
[2] Department of Industrial and Systems Engineering
Georgia Institute of Technology, Atlanta, GA 30332
oergun@isye.gatech.edu
[3] Department of Industrial and Systems Engineering
Georgia Institute of Technology, Atlanta, GA 30332
lhoughtalen@babson.edu (Now at Babson College)
[4] Department of Industrial and Systems Engineering
Georgia Institute of Technology, Atlanta, GA 30332
oozener@isye.gatech.edu

Summary We discuss two forms of collaboration in cargo transportation: carrier alliances in sea and air cargo, and shipper collaborations in trucking. After discussing the current industry settings that make such collaborations beneficial, we present a set of questions that need to be answered in the collaborative setting. These questions deal with issues such as (i) what is the maximum benefit the collaboration can achieve? (ii) How should these benefits be allocated? (iii) What membership rules and regulations will increase the sustainability of a given alliance? We provide a set of models for resolving these issues and analyze the properties of the solutions obtained from them.

1 Introduction

Cooperation among companies is widely practiced today. Facing increasing pressures to improve profitability, companies are responding by seeking and implementing solutions that require strategic collaboration with external partners, because these solutions afford benefits that cannot be achieved alone. These external collaborators can be suppliers, customers, or even competitors, and the potential benefits gained by collaborating include reducing costs, decreasing lead times, and improving asset utilization and service levels. Growth in cooperation has been further encouraged by the information sharing capabilities made possible by advances in technology, the best example of course being the internet.

W. Chaovalitwongse et al. (eds.), *Optimization and Logistics Challenges in the Enterprise*, Springer Optimization and Its Applications 30, DOI 10.1007/978-0-387-88617-6_14, © Springer Science+Business Media, LLC 2009

Collaboration among companies occurs across a variety of levels and business functions. There are numerous examples of collaboration among *buyers*, among *sellers*, and among buyers and sellers. Collaborations among buyers or among sellers are examples of *horizontal* collaboration, which are collaborations composed of companies with similar characteristics. One of the earliest examples of collaboration among buyers is a group purchasing organization, or GPO. In a GPO, companies who are potential competitors in the end market collaborate to procure goods. GPOs help members lower their purchasing costs by utilizing volume discounts from sellers. Among the many industries where GPOs exist are the health care industry (Vha.com), the manufacturing industry (Mfrmall.com), and the automotive industry (Covisint.com). There have been numerous studies of such organizations; we refer the reader to [14], [23], and [5] for three examples. In another example, discussed later in this chapter, shippers work together for joint procurement of transportation services (Nistevo.com). In such a scenario, the shippers exploit the operational synergies among their shipments to get discounts from truckload carriers by reducing the carriers' expected empty truck movements. On the other hand, sellers collaborate to take advantage of synergies among their operations, by increasing asset utilization and improving service levels. For example, collaboration among cargo carriers (both in sea and air), as explored further in this chapter, increases the profitability of cargo space. The sharing of seat capacity through the use of code-sharing among airlines is another demonstration of collaboration among sellers [8], [64].

Vertical collaboration, such as a collaboration *between* buyers and sellers, occurs across different levels of the supply chain. For example, suppliers and manufacturers often collaborate to increase forecasting accuracy, reduce inventory, and achieve efficiencies by coordinating supply chains through co-managed processes (cpfr.org). Vendor managed inventory (VMI), where the manufacturer (or seller) is responsible for maintaining the distributors' (or buyers') inventory level, is one of the most successful examples of collaboration among participants at different levels of the supply chain. There are many studies that explore supply chain collaboration from different angles. Thomas and Griffin [63] provide a review of literature on the coordination between two or more stages of the supply chain.

Clearly, collaborations help reveal synergies among members to achieve benefits. However, they pose many challenges as well. Many of the traditional optimization algorithms that work well when there is only one decision maker become ineffective in a collaborative setting, where there are multiple decision makers. This is because individual decision makers, though often working towards a common objective, are guided by their own self-interests. While a centralized approach is desirable in order to achieve maximum gain from collaboration, the impact of collaboration must also be evaluated from the perspective of the individual participants. Therefore in a collaborative setting one seeks a solution that is both centrally and individually attractive. Furthermore, the involvement of multiple decision makers means that there must be a well-defined mechanism to regulate the benefits achieved from collaboration and to ensure the sustainability of the collaborative system.

In this chapter, we discuss opportunities and insights gained from a collaborative approach to logistics by considering three applications in cargo routing: the ocean liner shipping industry, the air cargo industry, and the trucking industry. The first two applications focus on collaboration among carriers (sellers) and the last application focuses on collaboration among shippers (buyers). Each application aims to develop membership criteria to ensure that the benefits of collaboration are achieved. The problems are approached from the common perspective of how to model, analyze, and operate a realistic collaboration among cargo carriers or shippers. Through the study of collaboration in cargo routing we will address the following key questions related to collaborative logistics:

- How does one assess the maximum potential benefit from collaborating? In many cases, this can be determined by finding the centralized system optimal solution. However, it is difficult to obtain such a solution for applications in which the underlying computational problem is NP hard.
- How should a membership mechanism be formed, and what are the desired properties that such a mechanism should possess? For logistics applications, this involves issues related to the design of the service network and utilization of assets, such as the allocation of ship capacity among collaborating carriers. We design mechanisms that facilitate voluntary participation, enable good utilization of assets, and discourage members from colluding to form subcoalitions.
- How should the benefits achieved by collaborating be allocated among the members in a "fair" way? In the cargo routing setting we investigate what constitutes a "fair" allocation and how such an allocation may be achieved in the context of day-to-day operations.
- Are there insights to be gained from the solution procedure itself? Through the study of how to design membership and allocation mechanisms, we intend to gain intuition about which collaborations should form and why, about ways to evaluate potential new partners, or about how the offering of new products or services may impact the collaboration.

These questions also appear in a wide variety of other, sometimes seemingly unrelated, fields where a number of players interact with varying degrees of collaboration. Examples include internet routing, auctions, and telecommunications. The mathematical tools and insights most appropriate to understand these problems are obtained from mathematical programming and game theory. Operations research tools are often used to design efficient algorithms for the underlying optimization problem and concepts from mathematical economics and game theory are commonly united to design membership mechanisms and algorithms to allocate benefits. Game theory is used to determine what an appropriate distribution of costs or benefits among members in a collaboration should be, and linear programming machinery, especially ideas from duality theory such as primal dual methods, is heavily used in the literature to develop benefit allocation algorithms [21], [30], [43].

The rest of the chapter is organized as follows. In the next section, we present some of the key game theoretic ideas that are relevant to this chapter. In Section 3 we present issues related to collaboration among carriers in the sea cargo industry and the air cargo industry. In Section 4 we present collaboration among shippers in the trucking industry. Finally, we summarize our discussion in Section 5.

2 Game Theory Basics

Games are broadly classified into two categories - noncooperative games, in which any type of collusion among participants is forbidden, and cooperative games, where participants are allowed to work together. Assuming the existence of a transferrable utility (money, for example), cooperative games can be further categorized into cooperative games with transferable payoffs, where participants are allowed to transfer utility among each other, and cooperative games with nontransferable payoffs, in which side payments between participants are forbidden.

It is an interesting research issue to question the right framework for studying collaborations in transportation and logistics problems. Voss and Shi [65] provide related insights for studying alliance formation among sea cargo carriers. In the airline environment, a noncooperative setting is appropriate for analyzing decisions such as pricing of seats and determination of fare classes, as legal and logistical difficulties as well as competitive advantage considerations prevent airlines from merging their revenue management systems [8], [41]. However, decisions regarding capacity are amenable to analysis from the cooperative perspective. For example, in passenger airlines, code sharing requires cooperation among participating carriers to determine on which flights to code share, how many seats to offer to partner carriers, and how to share revenue across segments flown by a passenger.

In this chapter we focus on applications of cooperative game theory in cargo routing. The increasing trend towards collaboration among businesses drives the need to study cooperative transportation games. Collaboration is further fostered as governments are taking steps to relax antitrust regulations in order to stimulate economic growth. For example, carriers in the sea cargo industry receive antitrust immunity. Similarly, the Federal Aviation Administration deregulated all-cargo domestic carriers in 1997 [69].

In the next section we present some key game theoretic concepts. It is first necessary, however, to introduce some notation. A collaborative transportation or logistics game is defined by a set $\mathcal{N} = \{1, 2, \cdots n\}$ of different companies or organizations, henceforth referred to as players, and a *characteristic function* $v : 2^{\mathcal{N}} \rightarrow \mathbb{R}$, associating a value $v(S)$ with every subset $S \subset \mathcal{N}$. When players collaborate to increase their overall benefits, $v(S)$ is interpreted as the total benefit the members of S achieve. When players collaborate to reduce their overall costs, $v(S)$ is interpreted as the total cost the members of S incur by cooperating. It is generally assumed that $v(\emptyset) = 0$. A subset of players is generally called a coalition, and the entire set $\mathcal{N}$ is referred to as the *grand coalition*. When the problem or game is considered for the grand coalition, it is referred to as the *system* or *centralized* problem. The optimal solution $v(\mathcal{N})$ to the system problem is the best solution that can be achieved by the grand coalition. A payoff allocation vector is denoted by $x = \{x^1, \ldots, x^n\}$, where x^k represents the payoff allocated to player k.

2.1 Allocation Methods

In this section we focus on methods of allocating benefits among players. One of the most prominent and widely accepted notion of fair allocation of benefits is the "core" of the game. The core of a collaborative game is analogous to the Nash equilibrium in noncooperative game theory. An allocation of benefits is said to be in the core if (1) the sum of payoffs over all players equals their maximum attainable profit, and (2) no subset of players can collude and obtain a better payoff for its members. Mathematically, a payoff vector $x = \{x^1, \ldots, x^n\}$ is said to be in the core if:

$$\sum_{k \in \mathcal{N}} x^k = v(\mathcal{N}) \tag{1}$$

$$\sum_{k \in S} x^k \geq v(S) \quad \forall S \subset \mathcal{N}. \tag{2}$$

When $v(S)$ represents the total costs (rather than benefits) incurred by a coalition S, an allocation in the core is obtained by reversing the sign of the inequality in Equation (2). The first condition is referred to as the *budget balance* condition and the second condition is referred to as the *stability* condition. Intuitively, an allocation in the core implies that the grand coalition is not threatened by any of its subcoalitions. Thus, the core represents a very strong type of stability. Unfortunately, the core of a game is frequently empty. Shapley, however, proved that if the characteristic function of a game is supermodular (i.e., $v(S) + v(T) \leq v(S \cup T) + v(S \cap T) \ \forall S, T \subset \mathcal{N}$), then the core is nonempty [56].

When the core of a game is nonempty, it is likely that multiple core allocations exist. In such a case, some allocations may be particularly desirable with respect to others in the core. One such well-studied allocation is the *nucleolus*. The nucleolus is, intuitively, the "center" of the core, although the nucleolus can be defined for games with empty cores as well. Let the excess of a coalition $\emptyset \neq S \neq \mathcal{N}$ with respect to an allocation $x \in \mathbb{R}^n$ be defined as $e(S, x) = \sum_{k \in S} x^k - v(S)$. We say that the excess of S with respect to x reflects the "attitude" of coalition S towards x. That is, if $e(S, x) > e(T, x)$, then allocation x is more acceptable to coalition S as compared to coalition

T. Given an allocation $x \in \mathbb{R}^n$ define the *excess vector* $\theta(x)$ as the $2^n - 2$ dimensional vector whose components are the excesses $e(S, x)$ arranged in non-decreasing order. The nucleolus is then the unique allocation $x^* \in \mathbb{R}^n$ that lexicographically maximizes the excess vector $\theta(x)$.

Another well-studied allocation method is the *Shapley value*, which is defined for each player as the weighted average of the player's marginal contribution to each subset of the collaboration. The marginal contribution of player k to any coalition S which does not include player k is:

$$\Delta^k(S) = v(S \cup \{k\}) - v(S)$$

Then the Shapley allocation for player k is given by:

$$\varphi^k(\mathcal{N}, v) = \sum_{S \subseteq \mathcal{N} \setminus \{k\}} \frac{|S|! \, |\mathcal{N} \setminus (S \cup \{k\})|!}{|\mathcal{N}|!} \Delta^k(S). \tag{3}$$

A unique Shapely allocation exists for all games and the Shapley value is a budget balanced cost allocation method [40]. However, the Shapely allocation is not guaranteed to be in the core. We refer the reader to [44] for a detailed discussion.

In general, we can characterize some desirable properties of any allocation. Stability and budget balance, discussed previously, are obviously highly desirable. Other desirable properties include *cross monotonicity* and *positive benefits*. Cross monotonicity suggests that the payoff of any player should not decrease as the number of players in the collaboration grows. Moulin [40] shows that a cross monotonic cost allocation which is budget balanced is also guaranteed to be stable (and therefore in the core). An allocation possessing the positive benefits property ensures that each player receives a strictly positive benefit from collaborating. We refer the reader to [67] for a thorough review of basic cost allocation methods.

Algorithms to allocate the benefits of a collaboration among its members are available in literature for various logistics games. Sanchez-Soriano et al. [55] and Sanchez-Soriano [54] study the core of transportation games. Hamers [25] and Granot et al. [22] study delivery games associated with the Chinese postman problem. Göthe-Lundgren and Värbrand [35] and Engevall et al. [17] study the core and the nucleolus of the vehicle routing game with homogeneous and heterogeneous fleets, respectively. Allocation algorithms for many network-related games are also available in the literature. Some examples include the network design game [33], the assignment game [57], and the facility location game [21]. Finally, studies of the bin packing and knapsack games, which are classic combinatorial games, are also prevalent [15].

2.2 Mechanism Design

Mechanism design is another important notion from game theory that is relevant to this chapter. It is a subfield of game theory and microeconomics and it aims to study how privately known preferences of many players in a game can be aggregated towards a "social choice." In many situations it is not enough to obtain only an overall allocation of benefits among the players with desired properties, say an allocation in the core, but also to design algorithms that motivate players to obtain these allocations. The goal of mechanism design is to design such algorithms and protocols that perform well when individual players behave selfishly.

Formally, the mechanism design problem concerns a set of players $\mathcal{N}$ and a set $\mathcal{O}$ of possible outcomes for these players. Each player k has a preference relation for different outcomes $\tilde{o}$ in the set $\mathcal{O}$, denoted by $v^k(\tilde{o})$ to quantify his valuation of outcome $\tilde{o}$. The goal of the mechanism designer is to design an algorithm that chooses an outcome $\tilde{o} \in \mathcal{O}$ and an n-tuple of side payments $\{s^1, \cdots, s^n\}$ such that the total payment, x^k, to player k is $x^k = v^k(\tilde{o}) + s^k$. (The total payment x^k is what each individual player aims to optimize.) Intuitively, a mechanism solves a given problem by providing side payments to players to assure that the required output occurs, even when players selfishly choose their strategies so as to maximize their own profits. This field has received widespread attention recently and has been used successfully to develop algorithms for interconnected collections of computers, such as on the internet, and for task scheduling problems. For a detailed discussion of mechanism design we refer the reader to [42].

3 Carrier Alliances

As deregulation affected all modes of cargo transportation, carriers started forming alliances by sharing capacity on assets. In this section, we discuss large-scale transportation networks that operate as an alliance among carriers by focusing on two examples: liner shipping and combination air cargo carriers. In both of these businesses the cargo is transported on regularly scheduled service routes operated by the carriers' assets (ships and airplanes, respectively), whether the asset is full or not. This involves higher fixed costs and administrative overhead as compared to the type of services that wait until the asset is fully loaded; for example, tramp shipping. Also, owning the assets (especially ships and airplanes) involves a large capital investment (usually millions of US dollars) and the cost of idling these assets runs in the tens of thousands of dollars per day. Thus carriers collaborate primarily to share infrastructure setup and capital costs.

In the following presentation of carrier alliances, we assume that we are given a set of carriers and a set of nodes, representing airports or ports. Each carrier has a fleet of assets and a set of cargo to be delivered. Delivering a unit of the cargo earns a given revenue. To form an alliance, carriers bring their fleets into a pool and operate them together. The alliance may have to jointly resolve all or a subset of the following operational and tactical questions: (i) What is the right service network; that is, on which routes should the assets operate? (ii) How should the assets be allocated for servicing the chosen network? (iii) How should each asset's capacity be allocated among the alliance members? (iv) Which cargo should be accepted or rejected for servicing and which *path(s)* should be used to deliver the selected cargo?

Given the specific characteristics of the industry one considers, the subset of the questions that are jointly decided by the alliance changes. For example, in the liner shipping industry questions (i)-(iii) are resolved jointly. In the combination air-cargo industry only question (iii) is considered jointly since the service network design and asset allocation decisions for each carrier are generally driven by the carrier's passenger transportation business. Question (iv) is generally decided by the individual carriers given the resolutions for (i)-(iii). However, for the alliance to be able to perform as successfully as possible as a combined system, the decision-making process for resolving (i)-(iii) should take into account the fact that at the end, each carrier will be making some individual decisions selfishly. In this section, we propose a general methodology for resolving (i)-(iv) for carrier alliances.

In both the liner shipping and air cargo industries, alliances function as (at least partially) decentralized systems, as carriers make their own routing decisions. Thus, unlike in a centralized setting where a central controller makes all the routing decisions and then assigns payments to the carriers, it is often not possible to assign only an overall payment to a carrier. Furthermore, even if it were possible to determine an overall payment to a carrier, a carrier might be interested in evaluating the worth of his own resources (demand and the capacity on the edges of the network) in the alliance. Thus we need algorithms and protocols that not only provide payments to the carriers in an alliance such that the overall allocation has desirable properties, but are realistically implementable in the sense that they guide the members of the alliance to achieve these payments in the absence of a centralized controller.

With this in mind, we design an allocation mechanism to distribute the benefits and costs among the members of an alliance. Note that individual carriers working in an alliance cannot be assumed to accept the solution suggested by the centralized optimization algorithm, but follow their own self-interests. Hence, for forming sustainable alliances, the task is not only to obtain a good solution, but also to provide algorithms to share the benefits and costs of an alliance in such a way that all carriers are motivated to collaborate. Sharing benefits and costs generally translates into exchanging capacity on assets among the carriers. Thus, one way to regulate capacity exchanges among the carriers is to assign suitable *capacity exchange costs* so that the carrier who owns the capacity on a ship or plane is motivated to sell the capacity to a carrier who can utilize it to deliver cargo. To help the overall alliance achieve its maximum potential revenue, we provide incentives to the carriers to pursue the solution suggested by the centralized optimization model. Because the benefits obtained from the centralized solution directly are often not enough to motivate individual carriers to behave in the best interest of the alliance, we provide side payments, via the capacity exchange costs, to the carriers so that they are motivated to "play along." But first, we overview the liner shipping and combination air-cargo industries and survey the relevant literature.

3.1 The Liner Shipping Industry

Sea cargo is the freight carried by ships and it includes anything traveling by sea other than mail, persons and personal baggage. According to the American Association of Port Authorities, in the United States, which is the largest trading nation in the world for both imports and exports, accounting for nearly 20% of world trade, sea cargo is responsible for moving over 99% (by volume) of the international cargo. *Liner shipping* is a mode of global shipping and it mainly involves carrying cargo stored in containers on predetermined, regularly scheduled service routes, regardless of whether the ship is full. Thus the number of ships required for a given liner service route is determined principally by the frequency required on the service route, the distance traveled by a ship on the route and the speed of the ship. For example, a weekly liner service between New York and Hamburg may require four ships to maintain the necessary frequency. An emerging trend in liner shipping is the use of *transshipment ports*. A transshipment port is a port where cargo is transferred from an inbound ship to an outbound ship for further transportation. Transshipment services provide carriers with additional routing options, reduce transit times, and act as a facilitator of international trade. For example, the Hutchinson terminal in Freeport, Bahamas has become a major transshipment port between the Eastern Coast of the United States, the Gulf of Mexico, the Caribbean, South America, and trade lanes to European, Mediterranean, far Eastern and Australian destinations.

Since 1990 when Sea-Land and Maersk introduced the alliance system and began sharing capacity on ships in the Atlantic and Pacific oceans, mergers have become increasingly common in liner shipping. In the mid 1990s an estimated 60% of the total global liner capacity was accounted for by alliances. Liner shipping alliances are most common on long deep sea routes since alliances provide carriers with an opportunity to maintain good frequency on long routes. For example, maintaining weekly frequency on an Asia-North America route requires up to eight ships, which is an expensive capital commitment for an individual carrier. However, if two or more carriers pool their ships to operate this route, then they only need to commit, on an average, less than four ships each.

In the liner shipping alliance problem, carriers bring their fleets into a pool to operate them together and seek answers for the tactical and operational level questions described previously. The service network is designed by creating the ship routes, i.e., the sequence of port visits by a given fleet and the assignment of ships to these routes. Ships move in cycles, referred to as *service routes*, from one port to another following the same port rotation for the entire planning horizon. The determination of service routes must take into account a minimum frequency, which is desired in order to achieve high market share. Further, carriers decide which cargo to accept or reject for servicing and which path(s) to use to deliver the selected cargo. The cargo is allowed to be transshipped; that is, it is allowed to travel on ships on multiple service routes before reaching its final destination. Once a set of service routes is decided, members of the alliance assign their ships for operating the chosen routes and allocate each ship's capacity among the alliance members.

Relevant Liner Shipping Literature

There are two streams of liner shipping literature relevant to our work - service network design and alliance formation among carriers.

The network design problem in liner shipping comes in various flavors. Tailor-made models for specific problems with specialized constraints and objectives are available in the literature. Rana and Vickson [49] provide a nonlinear integer program to maximize total profit by finding an optimal sequence of ports to visit for each container ship and an optimal number of cargo units to be transported between each pair of ports by each ship for a special network structure in which loading and unloading of cargo is allowed only at specific ports. Fagerholt [20] considers the liner shipping problem in a special network where all cargo is transported from a set of production ports to a single depot. They impose a weekly frequency constraint on the operated routes; however, the feasible routes for their particular problem have a maximum route time of only one week. Thus on any of the feasible routes a single ship can maintain a weekly frequency. Perakis [48] provides a review of linear and integer programming models that consider the deployment of a fleet of liner ships, with different ship types, on a set of predetermined routes with targeted service frequencies to minimize operating and layup costs. Agarwal and Ergun [3] develop a mixed integer linear program to simultaneously schedule the ships and route the cargo. It considers a weekly frequency constraint on the operated routes and allows the cargo to be transshipped on multiple routes. Also, it provides various LP/IP based algorithms to solve the integrated model efficiently. For a comprehensive review of literature on ship scheduling and cargo routing we recommend [50] for the work done before 1983, [51] for the decade 1982-1992 and [11] for the last decade.

The allocation of benefits and costs in order to satisfy all the members of an alliance is an intriguing research topic, and very little is available in literature on the systematic mathematical study of alliances in liner shipping. Midoro and Pitto [36], Slack et al. [60], Ryoo and Thanopoulou [52], Song and Panayides [61] discuss the importance of strategic alliances in liner shipping. In particular, [36] studies the factors that led to the advent of strategic alliances among liner carriers 30 years ago, and the changes in the industry in the 1990s (for example, the increase in demand due to globalization) that made the previous alliances inadequate and called for a new generation of strategic partnerships. It suggests that differentiation in the contribution of each member, depending on their core competencies, can lead to alliances that deliver more than the sum of individual contributions. Also, alliances with fewer members or ones that are led by a dominant partner are more likely to succeed. Slack et al. [60] provides industry data to support the claim that alliances lead to intensification in service frequency and an increase in ship size. It points out that as a result of alliances, carriers are becoming more similar (with similar service routes, serving the same markets and employing comparable ships) and although individual carriers who form alliances serve more ports than before, the total number of ports served by the overall industry remains remarkably constant. Ryoo and Thanopoulou [52] study the progression of collaborations from consortia, which are route-based forms of cooperation, to alliances, which cooperate on a global level, among Asian carriers. They argue that the reasons behind this trend are the flexibilities and synergies provided by alliances in global perspective. Song and Panayides [61] make use of cooperative game theory and provide a quantitative study to analyze liner shipping alliances by considering two small examples (three ports and two to three carriers). They explicitly write all the core inequalities to analyze the alliance and allocate the revenue among the members in ratio of their shipping capacity. However, they do not provide any framework for dealing with larger instances for which it becomes harder to explicitly write all the core inequalities. Finally, [59] investigates liner shipping alliances and in particular the influence of these alliances on the performance of port operations.

3.2 The Air Cargo Industry

Air cargo is the freight (excluding mail and passenger baggage) carried by aircraft, and in this chapter we consider cargo transported using passenger aircraft. Carriers who transport passengers and cargo on the same aircraft are referred to as *combination carriers*. In the US, approximately 40% of air cargo is transported on passenger aircraft, while in Asia the passenger and cargo business is even more integrated, with up to 60% of cargo being transported in the belly of passenger aircraft [69]. As passengers generate more revenue than cargo, flight networks (and alliances) for combination carriers are typically driven by passenger markets. Increasingly, however, carriers are taking steps to improve the profitability of their cargo business. In contrast to the passenger industry and ocean liner shipping industry where alliances are well established, air cargo alliances are considerably younger; the first air cargo alliance, composed of Aeromexico, Air France, Delta, and Korean Air, was formed in 2000. It is an interesting observation that carriers compatible for passenger alliances may not be compatible for a cargo alliance, due to differences in flow patterns: passengers typically complete a round trip, resulting in balanced flow, while cargo flow follows unbalanced trade patterns [69].

Similar to code sharing in the passenger setting, carriers collaborate in the cargo setting by sharing capacity. Decisions about how much space to assign to each member and how to distribute revenue among the members are both applicable. A key difference between passengers and cargo, however, is the relative insensitivity of cargo to route selection. Because of this flexibility, the decision of how to route cargo becomes a relevant factor in considering collaborations among air cargo carriers, and is an important justification for the independent analysis of collaboration in the air cargo industry.

In this chapter, to study the benefits of collaboration among air cargo carriers, we study a system in which we assume that carriers accept or reject loads for transport. Origins and destinations for load demand correspond to airports. Because we assume that the flight schedule is motivated by the passenger industry and is therefore fixed, we do not consider costs incurred by operating the network. As in the liner shipping industry, carriers form alliances by integrating their networks.

Relevant Airline Alliance Literature

There is very little available in the literature relating to air cargo alliances, which is likely due to the fact that the first air cargo alliance was formed in 2000. Most literature concerning air cargo is related to dedicated cargo carriers, cargo operations, or the relationship between the cargo and passenger industries. The network design problem is addressed by [24] and [34], but these works focus on dedicated cargo carriers, in contrast to the assumed context of this chapter, which is the cargo business of combination carriers. The problem of short-term capacity planning for air cargo space is studied in [10]. Differences between the cargo revenue management problem and the passenger yield management problem are discussed in [32], as well as complexities in developing additional models to facilitate cargo revenue management.

Analysis of airline alliances in the passenger industry is more prevalent; significant work has been done to evaluate the impact of alliances on the passenger industry. For example, [47] investigates the impact of international alliances on the passenger market by comparing alliances composed of airlines with complementary and parallel networks; it is predicted that an alliance that joins complementary networks will be more profitable. In response to a concern that alliances would lead to a situation where major carriers would have a monopoly, [39] finds instead that alliances have merely allowed carriers to preserve, not increase, their narrow profit margins through an increase in load factors and productivity. Bamberger et al. [7] in fact finds that consumers benefit from the formation of passenger alliances; in the two domestic alliances that were studied fares decreased on the markets impacted by the alliance, in part due to increased competition from rivals competing with the alliance. The notion of "virtual" code sharing, in which an itinerary involving a single operating carrier is marketed by a code share partner, is the subject of [29]. It is demonstrated that virtual code share itineraries are less expensive than itineraries operated and marketed by the same carrier, leading the authors to suggest that virtual code sharing is increasingly being used to further segment the passenger market. An empirical analysis of alliance revenue management mechanisms is conducted in [66]; a free sale scheme and several dynamic trading schemes are analyzed to determine their effect on the equilibrium behavior of the alliance.

Finally, there is limited research available on the impact that an alliance in one industry (air cargo or passenger) can have on the other. Morrell and Pilon [38] studied a passenger alliance between KLM and Northwest and found that, ultimately, the effects on cargo service were positive. From the other perspective, [68] investigates the effect of an air cargo alliance on the passenger market. Using an oligopoly model with four passenger airlines, it finds that when two carriers integrate their cargo service, the outputs of the partnering carriers are increased in both the cargo and passenger markets.

3.3 Problem Definition

Because our focus is on developing a methodology to manage interactions among carriers in an alliance, we make some simplifying assumptions to improve tractability. First, we assume that both cargo demand and cargo capacity are deterministic and measured in single-dimension units. We also permit the splitting of loads so that the flow of loads through the network can be modeled using a standard multicommodity flow model. For details regarding assumptions specific to the liner shipping and air cargo applications, we refer the reader to [3] and [26], respectively.

Let V denote the set of nodes (airports or ports) in the network and $\mathcal{N} = \{1, 2, \ldots, n\}$ denote the set of carriers. Each carrier $k \in \mathcal{N}$ operates a fleet of assets N_k, and each asset in carrier k's fleet has some amount of capacity. (All assets in the set N_k do not necessarily have the same capacity.) For a carrier k, each demand is characterized by its origin (o)-destination (d) pair, the maximum demand that can arise, denoted by $D^{(o,d,k)}$, and the revenue obtained by satisfying one unit of demand, denoted by $R^{(o,d,k)}$; (o, d, k) is used to identify a demand from o to d for carrier k and the demand set of carrier k is identified by Θ_k. In an alliance formed by pooling assets $N = \sum_k N_k$ and consolidating demand $\Theta = \cup_k \Theta_k$, the carriers face some combination of the following problems:

1. Together they need to design their service network. For this, they need to design a set of service routes $(\overline{C} = \{C_1, \cdots, C_r\})$ to operate to utilize their assets.
2. Together they also need to decide a set of cargo $(\overline{\Theta} \subset \Theta)$ to deliver, as well as the paths to use to deliver the selected cargo.
3. The members of an alliance need to decide how to realize the service routes in $\overline{C}$. For example, they need to decide the number of assets that each carrier should assign to the service routes in $\overline{C}$.
4. Each carrier k needs to compute the valuation, v^k, of the solution given by $(\overline{C}, \overline{\Theta})$, depending on the cost incurred by him and the revenue generated by delivering his demands.

In reality, the carriers only pool their assets to form an alliance. That is, after the asset allocations are decided each carrier k individually decides which subset of the cargo in θ_k to deliver on which paths. There is no guarantee that these individual cargo accept-reject and routing decisions made by each carrier will together achieve the maximum system profit. Moreover, there is no guarantee that carriers will be satisfied by their valuations v^k alone. Hence the alliance needs to decide on an n-tuple of side payments, $\{s_1, \cdots, s_n\}$, for its members such that the total payment, $x_k = v_k + s_k$, to carrier k is optimal for him. With the help of side payments, carriers are motivated to make accept-reject decisions and route cargo as prescribed in the system solution. The range of feasible side payments is impacted by the choice of the model, henceforth referred to as the behavioral model, used to represent a single carrier's behavior within the alliance. The goal of studying these models is to analyze how side payments can impact the routing decisions of an individual carrier, and to use this analysis to guide the computation of side-payments. Note that as carriers work in collaboration with others, the routing decisions or the behavioral model of an individual carrier should also account for the flow of other carriers in the alliance. As discussed later, the choice of behavioral model impacts the total payment to a carrier in the alliance as well as the overall performance of the mechanism. For example, depending on the behavioral model chosen, side payments must be selected to ensure that they do not create additional complications such as a secondary market for capacity.

3.4 Solution Strategy

Before providing the details, we first present an outline of our solution strategy. Each carrier wants to design a service network which maximizes his profit. However, since he is working in collaboration with other carriers, a network that generates maximum overall revenue for all carriers is selected. Clearly, such a network can be obtained by combining the individual carriers into one large pseudocarrier. Often this involves solving a hard problem for which only a near optimal solution can be computed. As we are interested in finding the best solution for the alliance that can be computed, we refer to the optimal (or the best available) centralized solution by $opt(\mathcal{N})$, and the total revenue obtained by this solution as $v(\mathcal{N})$, i.e., the value of the grand coalition. This section describes the network design problem and the problem of assigning resources to service routes in $opt(\mathcal{N})$. Next, the valuation of solution $opt(\mathcal{N})$ is calculated for each carrier by calculating the revenue generated by him and the costs incurred by him. However, the valuation obtained from solution $opt(\mathcal{N})$ alone is not guaranteed to provide enough motivation for a carrier to act according to the solution $opt(\mathcal{N})$, so side payments are provided to the carriers. Two possible models are presented to compute side payments, reflecting different behavioral models of the carriers. Side payments are constructed from a collection of capacity exchange costs, which are computed using inverse optimization techniques. We also discuss implications relating to behavioral model selection. Next, we provide the details of our solution strategy.

Network Design

As we have previously discussed, for combination carriers in the airline industry the service network is determined by the passenger business. This network is then utilized to route cargo. The centralized solution for the air cargo problem can thus be obtained by integrating the passenger network of individual carriers and finding an optimal set and routing of cargo loads. Houghtalen et al. [27] present a multicommodity flow model to study this problem in detail.

As opposed to the combination carriers in the air cargo industry, sea cargo carriers do in fact design at least a part of their network as an alliance. To model the centralized network design problem we replace individual carriers by a fictitious carrier with a fleet equal to the combined fleet of all the carriers, $N = \sum_k N_k$ and demand equal to the combined demand of all the carriers, $\Theta = \cup_k \Theta_k$. Next, a set of service routes for the fleet of ships in N is determined to satisfy some or all of the demands in Θ. Note that the set of service routes operated by the ships determine which paths can be selected to deliver the cargo. The cargo delivered, and the paths chosen to deliver the cargo, determine the revenue that can be generated, and hence determines the profitability of the service network. These two problems are highly interdependent and [3] presents a mixed integer programming model to solve both problems in an integrated fashion. They prove that the problem is NP hard and develop various heuristic and LP-based algorithms to solve it efficiently.

Once the centralized solution $opt(\mathcal{N})$ is determined, carriers need to realize the schedule determined by $opt(\mathcal{N})$. For this they first need to determine which carrier(s) should assign ship(s) to each of the operated routes. Assigning a ship on a route incurs various costs, such as the cost of operating the ship as well as port visit costs. Since each carrier pays for maintaining and operating his assets on the collaborative routes, the assignment of ships to the operated routes influences the cost incurred by a carrier in the collaborative solution. Depending on the flow of demands (o, d, k) in $opt(\mathcal{N})$, the utility of assigning a ship to a service route is different for different carriers. The problem of assigning ships to the operated service routes can be modeled as a generalized assignment problem, once the utility of assigning a ship to an operated route is determined for the carriers. Agarwal and Ergun [2] study different models for computing the utility function. Also they study exact as well as various heuristic methods to solve the assignment problem. For an individual carrier, the choice of utility function and solution strategy used to solve the assignment problem influences the assignment of his ships to the operated routes.

Valuation and Side Payments

For a carrier, the valuation of the centralized solution $opt(\mathcal{N})$ is determined by calculating the revenue earned by him and the costs incurred by him. Let $(\overline{C}, \overline{\Theta})$ denote the sets of optimal service routes and delivered demand, respectively, that comprise the solution $opt(\mathcal{N})$. The revenue earned by a carrier k is calculated by summing over the revenue generated by satisfying demand (o, d, k) such that $(o, d, k) \in \overline{\Theta} \cap \Theta_k$. The costs incurred by a carrier k are due to maintaining and operating his assets on the collaborative routes. Let $\overline{f_e^{(o,d,k)}}$ denote the flow for demand (o, d, k) on edge e according to the solution $opt(\mathcal{N})$. ($\overline{f}$ is therefore the system optimal flow.) Then the valuation of solution $opt(\mathcal{N})$ for a sea cargo carrier is given by:

$$v^k(opt(\mathcal{N})) = \sum_{(o,d,k')\in\Theta_k} R^{(o,d,k')} \overline{f_{(d,o,k')}^{(o,d,k')}} - Cost\ of\ operating\ routes. \quad (4)$$

In the air cargo context, where we assume a fixed network, we ignore the costs of operating the service routes. The valuation of $opt(\mathcal{N})$ for carrier k is therefore given by:

$$v^k(opt(\mathcal{N})) = \sum_{(o,d,k')\in\Theta_k} R^{(o,d,k')} \overline{f_{(d,o,k')}^{(o,d,k')}}. \quad (5)$$

Recalling that we assume carriers will act according to their own interests, we need to model such selfish behavior of individual carriers within the alliance. We will present two different behavioral models; in both models we assume that given the collaborative network the carriers solve their cargo routing problems individually. Note that it is in the best interest of the collaboration that the carriers make their cargo routing decisions as in $\overline{f}$. However, $\overline{f}$ requires carriers to share capacity on the assets. We facilitate this by allowing a carrier to charge other carriers for using his asset on an edge e, whenever he has an asset assigned to edge e. We refer to this payment as the capacity exchange cost on edge e and denote it by $cost_e$; the capacity exchange costs provide side payments to the carriers. The total profit for a carrier in the alliance is then obtained by adding these side payments to the valuation (4) or (5).

In the network (designed by the alliance in the case of sea cargo and assumed given in the context of air cargo), capacity on an edge belongs to carriers. In fact, in the sea cargo context several carriers may own the capacity on a single edge. However, to simplify the presentation we assume that capacity on an edge belongs to a unique owner. (For the case in which there are multiple owners on a single edge, we refer the reader to [2].) Let E_k denote the set of edges owned/operated by carrier k. Let Cap_e denote the capacity on an edge. The $f_e^{(o,d,k)}$ denotes the flow of demand (o, d, k) on edge e. We denote by $f_{(d,o,k)}^{(o,d,k)}$ the flow on a fictitious edge from d to o for commodity (o, d, k). Note that the maximum flow on the fictitious edge $f_{(d,o,k)}^{(o,d,k)}$ is bounded by $D^{(o,d,k)}$, the maximum demand for (o, d, k).

We now present two behavioral models to capture the individual behavior of a carrier in the alliance. These models are not literal descriptions of the actual optimization problem solved by an individual carrier. Rather, the goal of these models is to determine suitable capacity exchange prices by approximating how capacity exchange prices impact the behavior of individual carriers. Both the models discussed here are static in the sense that they assume all demand information is available when the participants make their decisions. Dynamic demand models for passenger airlines have been considered by [66]. The choice of model depends on the amount of information available to individual carriers and often there are legal barriers to such information sharing. However, as mentioned earlier, deregulation has affected all modes of cargo transportation, thereby easing alliance formation among carriers from a legal perspective. The first model below ($M1^k$) assumes that all carriers have complete information regarding the capacities on the edges in the network and the demands of their partners. Though the capacities on the edges are generally known to all the carriers, the assumption of full information regarding the exact demand distribution of the partners is unrealistic in situations when partners are also competitors, cannibalizing each other's demand. However, carriers often have at least rough estimates of the demand distribution of their partners. In the second model for the behavior of an individual carrier within the alliance ($M2^k$), the capacity on each edge is restricted to a fixed amount (computed according to a formula explained later) depending on the characteristics of the centralized solution. Though both the models make certain strong assumptions they lead to interesting results as explained in Section 3.5 and Section 3.6.

Behavioral Model 1 In this model we assume that an individual carrier can modify the flow of other carriers in addition to his own. However, in practice, an individual carrier can only make decisions regarding his own flow. In this sense Model 1 is a conservative approach, since the maximum revenue that carrier k can obtain by behaving selfishly will always be less than the optimal value of $M1^k$ below. For this model, the mathematical formulation representing the optimization problem solved by carrier k in the alliance is given by:

$$(M1^k) : \max \sum_{(o,d,k')\in\Theta_k} f_{(d,o,k')}^{(o,d,k')} R^{(o,d,k')}$$

$$+ \sum_{e\in E_k} cost_e \left(\sum_{(o,d,k')\notin\Theta_k} f_e^{(o,d,k')} \right) - \sum_{e\notin E_k} cost_e \left(\sum_{(o,d,k')\in\Theta_k} f_e^{(o,d,k')} \right) \quad (6)$$

subject to

$$\sum_{e\in\delta^-(v)} f_e^{(o,d,k')} - \sum_{e\in\delta^+(v)} f_e^{(o,d,k')} \le 0 \quad \forall v \in V, \forall (o,d,k') \in \Theta \quad (7)$$

$$\sum_{(o,d,k')\in\Theta} f_e^{(o,d,k')} \le Cap_e \quad \forall e \in E \quad (8)$$

$$f_{(d,o,k')}^{(o,d,k')} \le D^{(o,d,k')} \quad \forall (o,d,k') \in \Theta \quad (9)$$

$$f_e^{(o,d,k')} \ge 0 \quad \forall e \in E, \forall (o,d,k') \in \Theta. \quad (10)$$

The objective function (6) consists of three terms. The first term maximizes the revenue generated by satisfying demand. The second term computes the cost paid to carrier k by other carriers for using capacity on an edge owned by carrier k. Similarly, the third term represents the cost paid by carrier k to other carriers for using capacity on their edges. Constraints (7)-(10) are network flow constraints. Constraints (7) are the flow balance constraint at every vertex v and for every demand triplet where $\delta^-(v)$ and $\delta^+(v)$ denote the incoming and outgoing edges of vertex v, respectively. Constraints (8) are the capacity constraints on the edges in the network, while (9) are the capacity constraints on the fictitious edges. Constraints (10) are the nonnegativity constraints on the flow variables.

Behavioral Model 2 Alternatively, we can reflect the assumption that carriers cannot control the flow of other carriers in the alliance. Instead, the flow of other carriers is acknowledged in the model through the enforcement of capacity restrictions for each carrier on each edge in the network. That is, we assume that the alliance centrally controls both available capacity and exchange prices on each edge. In this model we denote by $\overline{Cap_e^k}$ the capacity on edge e allocated to carrier k. For every edge e, $\overline{Cap_e^k} = \sum\limits_{(o,d,k) \in \Theta_k} \overline{f}_e^{(o,d,k)} +$ $extra_e$, where $extra_e$ is the amount of extra capacity on edge e with respect to the optimal flow $\overline{f}$. Note that this capacity allocation ensures that $\overline{f}$ is feasible. The mathematical formulation to model carrier k's behavior in the alliance under this alternative assumption is as follows:

$$(M2^k) : \max \sum_{(o,d,k') \in \Theta_k} f_{(d,o,k')}^{(o,d,k')} R^{(o,d,k')} - \sum_{e \notin E_k} cost_e \left(\sum_{(o,d,k') \in \Theta_k} f_e^{(o,d,k')} \right) \quad (11)$$

subject to

$$\sum_{e \in \delta^-(v)} f_e^{(o,d,k')} - \sum_{e \in \delta^+(v)} f_e^{(o,d,k')} \leq 0 \quad \forall v \in V, \forall (o,d,k') \in \Theta_k \quad (12)$$

$$\sum_{(o,d,k') \in \Theta_k} f_e^{(o,d,k')} \leq \overline{Cap_e^k} \quad \forall e \in E \quad (13)$$

$$f_{(d,o,k')}^{(o,d,k')} \leq D^{(o,d,k')} \quad \forall (o,d,k') \in \Theta_k \quad (14)$$

$$f_e^{(o,d,k')} \geq 0 \quad \forall e \in E, \forall (o,d,k') \in \Theta_k. \quad (15)$$

The objective function (11) is equal to the revenue earned by delivering loads minus the total leg costs paid to other carriers. Because it does not account for payments received, the solution of $M2^k$ is a lower bound on the actual profit earned by k in the collaboration. Constraints (12) are the flow balance constraints at every vertex v and for every demand triplet that belongs to carrier k; note that in this model we consider only the demands for carrier k, rather than all demand, as in Model 1. Constraints (13) are the capacity constraints on the edges of the network, which depend on the amount of capacity allocated to carrier k on each edge in the network. Constraints (14) are the demand constraints (for carrier k's demands only) and (15) are the nonnegativity constraints on the flow variables.

Calculating Side Payments In both models, decision variables for individual carriers are flow variables for a given vector *cost*. However, our aim is to identify the *cost* vector such that the flow suggested by the centralized solution $(\overline{f})$ is also an optimal decision for each carrier k. Thus in both the models we know the value of the flow variables that we would like to attain in the optimal solution, and we wish to determine the parameter $(cost)$ that will make these flow values optimal. This fits well into the *inverse optimization* framework. In a typical optimization problem, one identifies the value of the variables given the values of model parameters (cost coefficients, vector of right-hand sides), whereas in an inverse optimization problem one identifies the values of model parameters that make a given feasible solution optimal. Inverse problems have been studied in a variety of fields, such as portfolio optimization [12] and traffic equilibrium [13]. We refer the reader to [4] for a detailed discussion on a unified framework for studying inverse optimization problems for various network flow problems.

Our goal is to determine the *cost* vector such that the flow $\overline{f}$ is optimal for all individual carrier problems. Inverse optimization techniques have been used by [2] to obtain a *cost* vector in the context of sea cargo and by [27] in the context of air cargo. The approach employed in both papers is similar in the way they obtain the inverse problem to find the suitable *cost* vector. They consider the optimization problem for each individual carrier k in the alliance, and construct the inverse problem corresponding to that carrier. Then they combine all the individual inverse problems into one problem to obtain the common *cost* vector that satisfies all the individual inverse problems. The difference between the two papers arises from the models they consider to evaluate an individual carrier's behavior. More specifically, [2] uses model $M1$ and [27] uses model $M2$ to model individual carrier behavior.

Let $INVP1$ denote the inverse problem obtained by considering model $M1$ and $INVP2$ denote the inverse problem obtained by considering model $M2$. Let $\overline{cost^{M1}}$ and $\overline{cost^{M2}}$ denote the cost vector obtained by solving $INVP1$ and $INVP2$, respectively. Then for model Mi, the overall payment to carrier k is given by:

$$x_{Mi}^k = v^k(opt(\mathcal{N})) + s_{Mi}^k \tag{16}$$

where the vector of side payments $\{s_{Mi}^1, s_{Mi}^2, \cdots, s_{Mi}^n\}$ is such that,

$$s_{Mi}^k = \sum_{e \in E_k} \overline{cost_e^{Mi}} \left(\sum_{(o,d,k') \notin \Theta_k} f_e^{(o,d,k')} \right) - \sum_{e \notin E_k} \overline{cost_e^{Mi}} \left(\sum_{(o,d,k') \in \Theta_k} f_e^{(o,d,k')} \right). \tag{17}$$

Note that the vector of payoffs $\{x_{Mi}^1, x_{Mi}^2, \cdots, x_{Mi}^n\}$ is such that $\sum_{k \in \mathcal{N}} x^k = v(\mathcal{N})$. This is easy to see, since once a feasible solution is found for the inverse problem ($INVP1$ or $INVP2$), the flow in the network is the same as the flow of optimal solution $opt(\mathcal{N})$. Also note that some of the side payments may be negative and the vector of side payments $\{s^1_{Mi}, s^2_{Mi}, \cdots, s^n_{Mi}\}$ is such that $\sum_{k \in \mathcal{N}} s^k_{Mi} = 0$.

3.5 Results

In this section we present some results that have been derived from the two models previously introduced. For supporting proofs and analysis, we refer the reader to [1], [2] and [27]. First, we state a theorem to indicate that a *cost* vector which makes $\overline{f}$ optimal for all the individual carrier problems can indeed be found, irrespective of the model considered to capture individual carrier behavior.

Theorem 1. *The inverse problems $INVP1$ and $INVP2$ are feasible.*

To prohibit carriers from colluding to form subcoalitions, we in fact want to have a cost vector such that for any subset $S \subset \mathcal{N}$, $\overline{f}$ is an optimal solution for the corresponding problem $M1$ or $M2$. However, there are an exponential number of such subsets. Therefore, including the corresponding constraint for each subset in the inverse problem will cause the inverse problem to become exponential in size. Theorem 2 guarantees that it is sufficient to consider only single carrier problems in $INVP1$ to determine a suitable cost vector.

Theorem 2. *The inverse problem in $INVP1$ identifies a cost vector such that $\overline{f}$ is optimal for any subset $S \subset \mathcal{N}$ of carriers.*

It is reasonable to assume that a subset of carriers would seek a higher payoff in the alliance as compared to the revenue that they can generate on their own. Recall that for a subset $S \subset \mathcal{N}$, $v(S)$ denotes the maximum revenue that the carriers in set S can obtain when operating on their own. Then for subset S, we can write a rationality constraint as:

$$v(S) \leq \sum_k x^k. \tag{18}$$

To verify if the overall payment x^k_{Mi} allocated by our mechanism is in the core we need to check if Equation (18) is satisfied for every subset $S \subset \mathcal{N}$. Also, we need to compute $v(S)$ for every subset S. Note that if the underlying network is given, as is the case with air cargo alliances, $v(S)$ is computed easily by solving a multicommodity flow problem. Behavioral model (M1) ensures that in this case, the payoff made by the mechanism, (16), satisfies rationality constraints (18) for all the subsets $S \subset \mathcal{N}$ (please refer to [1] for the proof). In other words, the payoff made by the mechanism is in the core. However, computation of $v(S)$ is not always easy. For sea cargo alliances an NP-hard problem (including the network design component) needs to be solved to compute $v(S)$. In this case, it becomes hard to directly compare $v(S)$ with the payoff allocated by the mechanism to the members of set S and neither of the models can guarantee that the rationality constraints can be satisfied for all the subsets. In closely linked networks in transportation, for example, liner shipping and air cargo, there are typically three to four participants. For a four player alliance, 2^4 hard problems need to be solved to compute $v(S)$ for every subset $S \subset \mathcal{N}$. It might still be possible to compute $v(S)$ for every subset S in such scenarios. However, in cases when it is computationally expensive to compute $v(S)$ for all subsets of carriers, the inverse optimization model can be enhanced by adding limited rationality constraints. Such limited rationality constraints for individual carriers and each pair of two carriers can be written as (19) and (20), respectively.

$$v(\{k\}) \geq x_k \text{ for each } k \in \mathcal{N}. \tag{19}$$

$$v(\{k,i\}) \geq x_k + x_i \text{ for } k, i \in \mathcal{N}. \tag{20}$$

3.6 Comparison of Models

In this section we present a small example to compare models $M1$ and $M2$, and then discuss some advantages and disadvantages associated with each model.

Consider the network illustrated in Figure 1. In this example, a time-expanded network with two locations and two time periods is depicted; nodes 1 and 2 represent the same location, one time period apart, as do nodes 3 and 4. There are four demands in the system, each of unit size, described in Table 1. For example, demand $(1, 3, A)$ represents a load of carrier A with

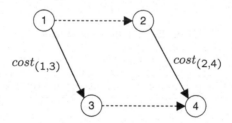

Fig. 1. System network

Table 1. Demands table

Demand	revenue $(R^{(o,d,k)})$
$(1,3,A)$	2
$(2,4,A)$	2
$(1,4,B)$	6
$(2,4,C)$	3

ready time and origin location corresponding to node 1, and delivery deadline
and destination corresponding to node 3. The revenue $R^{(1,3,A)}$ associated with
this demand is 2. From the set of demands we see that we have a three-carrier
example. There are two capacitated edges in the system (edges $(1,3)$ and
$(2,4)$) that represent service routes operated by carrier A. Let the capacity on
each of these edges be one. Edges $(1,2)$ and $(3,4)$ are uncapacitated fictitious
ground edges representing the ability of a cargo load to remain in a location
over time.

The system optimal flow $\overline{f}$ is depicted in Figure 2. We see that $\overline{f}_{(1,3)}^{(1,4,B)} =$
$1, \overline{f}_{(3,4)}^{(1,4,B)} = 1$, and $\overline{f}_{(2,4)}^{(2,4,C)} = 1$, while the demands $(1,3,A)$ and $(2,4,A)$ are
unfulfilled.

Under Model 1 (and therefore using $INVP1$), a feasible solution for the
$cost$ vector is $cost_{(1,3)} = 2$, $cost_{(2,4)} = 2$. The resulting side payments and
overall allocations are listed in Table 2. We note that this is in fact a core
solution. Furthermore, we find that using Model 1, the maximum feasible
value for $cost_{(1,3)}$ and $cost_{(2,4)}$ is 3 and their minimum feasible value is 2.

Under Model 2 (using $INVP2$), we find that $0 \leq cost_{(1,3)} \leq 6$ and $0 \leq$
$cost_{(2,4)} \leq 3$. Therefore, the solution obtained under Model 1 is certainly
feasible for Model 2, and would result in an identical allocation to that given
in Table 2. However, the set of feasible $cost$ vectors is larger. For example, the

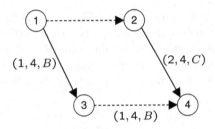

Fig. 2. System optimal flow

Table 2. Allocations under $cost_{(1,3)} = 2, cost_{(2,4)} = 2$

Carrier	Valuation of $opt(\mathcal{N})$	Side payments	Total allocation
A	0	4	4
B	6	-2	4
C	3	-2	1

Table 3. Allocations under $cost_{(1,3)} = 5, cost_{(2,4)} = 2$

Carrier	Valuation of $opt(\mathcal{N})$	Side payments	Total allocation
A	0	7	7
B	6	-5	1
C	3	-2	1

$cost$ vector $cost_{(1,3)} = 5$, $cost_{(2,4)} = 2$, is feasible for $INVP2$ and also results in a core allocation. The resulting side payments and overall allocations for this solution are listed in Table 3.

Flexibility and Fairness

In general, we see that the set of feasible solutions to $INVP2$ is larger than the set of feasible solutions to $INVP1$, affording more flexibility in the choice of edge costs. We now turn our focus to introducing an objective function to the inverse problems, with the goal of selecting a feasible solution such that the resulting allocation will be perceived as fair by the carriers. Note that in this discussion we do not consider the core of a problem to be synonymous with a fair allocation; we allow the possibility for solutions outside the core to be considered fair, as well as solutions within the core to have varying degrees of fairness. Put another way, not all core solutions are equally fair.

A simple definition of fairness might be the following: every carrier in the collaboration should receive equal benefit from collaborating. While clearly this would not always be perceived as fair by all participating carriers, the simple *equal benefits* rule is easy to implement and is interesting to explore as a base case. In order to achieve fairness according to the equal benefits rule, we include the rationality constraints with a slack variable $slack_k \geq 0$ for every carrier k. Using the objective function $\min \sum_k (slack_k)^2$ we will obtain a *cost* vector that achieves an allocation in which the benefit from collaborating is distributed as equally as possible among carriers.

We may instead choose to define fairness based on well-studied cooperative game theory allocation methods, namely the Shapley value or the nucleolus. While the Shapley allocation is computationally difficult to find, assuming we are able to compute it we can then consider it a target allocation. Let $\overline{x} = \{\overline{x}^k : \overline{x}^k = \text{Shapley allocation for carrier } k\}$. Then we add the following constraints to the inverse problem:

$$\sum_{(o,d,k)\in\Theta_k} \overline{f}_{(d,o)}^{(o,d,k)} R^{(o,d,k)} + \sum_{e\in E_k} cost_e \Big(\sum_{(o,d,k)\notin\Theta_k} \overline{f}_e^{(o,d,k)} \Big)$$

$$- \sum_{e\notin E_k} cost_e \Big(\sum_{(o,d,k)\in\Theta_k} \overline{f}_e^{(o,d,k)} \Big) + slack_k \geq \overline{x}_k \ \ \forall k \in \mathcal{N} \quad (21)$$

where $slack_k$ is an unrestricted slack variable associated with each of the $\mathcal{N}$ constraints. Again using the objective function $\min \sum_k (slack_k)^2$ we obtain a set of edge costs that gets as close to the Shapely allocation as possible. Similarly, if we choose to define fairness based on the nucleolus, we simply let $\overline{x} = \{\overline{x}^k : \overline{x}^k = \text{nucleolus allocation for carrier } k\}$. In general, the inverse problems provide the flexibility to implement a variety of fairness measures with minor changes. Note that $INVP2$ provides a larger set of feasible solutions to draw from to implement different fairness measures.

Secondary Market for Capacity

If we examine the solution $cost_{(1,3)} = 5$, $cost_{(2,4)} = 2$ more closely, we find that it gives rise to a secondary market for capacity on edge $(2, 4)$. A capacity exchange in the secondary market will occur if carrier B realizes that a unit of capacity on leg $(2, 4)$ can be bought from carrier C for the price of 4, instead of buying capacity on $(1, 3)$ directly from carrier A for the price of 5. The resulting allocations actually realized by the carriers are given in Table 4. Note that the revenue earned deviates from the valuation of $opt(\mathcal{N})$ for carrier C, because carrier C cannot fulfill his demand, as he is selling his capacity to carrier B instead. The total system profit resulting from this solution is 8 (from delivering $(1, 3, A)$ and $(1, 4, B)$), which is less than the profit of 9 obtained by $\bar{f}$. Clearly the existence of a secondary market for capacity is detrimental for the profit of the overall collaboration. We can take steps to restrict the solution space of $INVP2$ in order to prevent a secondary market; this is discussed further in [27]. However, we find that a strength of using behavioral model 1 is that $INVP1$ leads to solutions for which a secondary market for capacity will not exist. We refer the reader to [26] for a proof and further discussion.

It can be easily seen that if carriers participating in a collaboration were required to pay for all capacity up front, we would effectively be preventing a secondary market. This is because the existence of a secondary market depends on carriers having the opportunity to exchange information after edge costs have been fixed. However, in a more realistic pay-as-you-go system (that is, carriers pay the exchange cost for capacity as the capacity is used, rather than up front), we must be aware of the impact a secondary market can have on our system. Traditionally, secondary markets have been studied in the context of markets for used goods, such as automobiles and books. Our interpretation here is slightly different, since the capacity is being sold as "unused" in the secondary market, but the existence of the secondary market causes similar complications with stability in the primary market.

Table 4. Allocations after exchanges in the secondary market

Carrier	Revenue earned	Capacity sold/ purchased	Side payments	Total allocation
A	0	(2,4) sold	2	2
B	6	(2,4) purchased from C	-4	2
C	2	(2,4) purchased from A, sold to B	-2+4	4

4 Shipper Collaboration

In this section, we will focus on collaboration among buyers in the **trucking industry**.

Trucking is the backbone of US freight movement. According to the American Trucking Association (ATA), the trucking industry hauled 68.9% of the total volume of freight transported in the United States in 2005, which generated $623 billion in revenue. In spite of the size of the industry - there are 573,000 carriers operating in the US - a major portion of the trucking companies are small businesses. According to the ATA, 86% of these companies operate six or fewer trucks, while 96% operate 28 or fewer trucks, resulting in a highly fragmented industry structure, intense competition, and low profit margins.

Meanwhile, a high number of shippers also increases the inefficiencies in truck transportation. Most of the individual shippers do not have enough volume of shipments or load balance among locations to plan their shipments in a cost-effective manner. In such a setting, collaboration among shippers for exploiting operational synergies that might exist among their shipments becomes a necessity. For example, the "less-than-truckload" (LTL) operations might be considered as an implicit collaboration among the shippers with relatively small freight. In LTL transportation a carrier collects freight from various shippers, bundles the freight at a breakbulk terminal, transfers it to another breakbulk terminal, where loads are split, and then finally delivers each load to its consignee. Although LTL terminal operations take more time than truckload (TL) operations and are costlier in general, LTL services still offer cost savings to the shippers. These cost savings are generated by sharing the trailer capacity between two breakbulk terminals, which is usually the long-haul portion of the route the shipments have to travel.

Collaborative opportunities among the shippers are not restricted to the LTL setting. Shippers with a high volume of freight can form alliances to reduce the imbalance of loads among different locations, resulting in a decrease in empty hauling miles. Instead of sharing the truck capacity along a single origin-destination movement, they share it over a cycle that starts and ends at the same location. As the complementarity of the shipments increases (that is, they form continuous routes requiring fewer empty truck movements), the overall cost of transportation decreases.

In a traditional TL logistics market, a shipper with multiple truckload shipments, each between a specific origin and a destination, negotiates prices with the carrier(s). A truckload shipment between an origin and a destination is referred to as a *lane*. The negotiated prices include the anticipated empty truck movement costs, *asset repositioning* or *deadheading* costs, required for handling all the shipments from the shipper. The inefficiencies due to empty asset repositioning are reflected in the lane prices given by the carrier. For instance, the carrier can charge a percentage of the lane cost as the anticipated asset repositioning cost along a lane. According to the estimates of the ATA, asset repositioning movements of TL carriers correspond to approximately 35 million empty miles monthly [6].

A collaboration for procurement of truckload transportation services is established when shippers work together to minimize their total transportation costs by better utilizing the truck capacity of a carrier. The possibility of finding efficient continuous routes increases when multiple shippers pool their lane sets. On the other hand, the total asset repositioning costs incurred by a carrier depend on the entire set of lanes served by the carrier. In the absence of complementary lanes in the carrier's network, the shipper is held responsible for the expected deadheading cost associated with its lane. However, when the shippers come together and identify a set of complimentary lanes, they might be able to provide continuous moves, reducing the expected repositioning for the carrier and in turn allowing them to negotiate better rates with the carrier.

Nistevo, Transplace, and One Network Enterprises are examples of collaborative logistics networks that enable shippers and carriers to manage their transportation activities. As an example, by managing 75% of its freight through the Nistevo Network, Land O'Lakes expects a 10% reduction in product shipment costs (according to www.nistevo.com). Similarly, other companies such as General Mills and Georgia-Pacific achieved considerable savings in their transportation expenses by being members of such networks.

The collaborative setting we consider in this section is relevant for companies that regularly send truckload shipments, use dedicated fleets for such shipments, and in which the trucks are expected to return back to the initial location periodically. In such a setting the shippers in the collaboration are faced with the following two questions: (i) What is the best set of cycles that cover the joint lane set? (ii) Once prices are negotiated for the above set of cycles, how should the total cost be allocated?

The question of how to best group the lane sets of the shippers into cycles (or continuous moves) can be resolved by solving a large-scale and possibly NP-hard optimization problem. We expect that the total cost obtained from the carrier decreases when the shippers pool their lanes. Hence a mechanism must be devised to distribute the gains (or equivalently, costs) among the members of the collaboration. Otherwise, the collaboration is faced with the risk of collapsing. In current practice, collaborative networks allocate benefits proportionally to the stand-alone cost of the participating shippers' lanes. We define the "stand-alone cost" of a lane as the cost of covering the lane plus the expected asset repositioning cost along the lane. First, the total stand-alone costs of the participating lanes of all shippers in the collaborative transportation network are summed. The difference between the total stand-alone cost and the total cost of the collaborative tours is the savings to be distributed among the participants. These savings are then distributed proportionally based on the percentage of the stand-alone cost each shipper has contributed to the collaborative transportation network.

Consider the example in Figure 3. There are three shippers, shipper A with a lane from node 1 to node 2, and shippers B and C with one lane each from node 2 to node 1. Suppose that the cost of covering a lane with a full

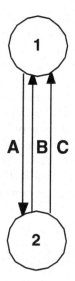

Fig. 3. A shippers' network

or empty truckload is equal to 1. The total cost of covering the lanes in this network is 4. A cost allocation method which allocates costs proportional to stand-alone costs allocates a cost of $\frac{4}{3}$ to each shipper. Unfortunately, this cost allocation is not stable since shippers A and B (equivalently A and C) are better off collaborating on their own with a total cost of 2. The only stable cost allocation for this example allocates $(0, 2, 2)$ to shippers A, B and C, respectively. However, allocating a cost of 0 to A makes A a *free rider*, and allocating a cost of 2 to B and C will give them no positive benefits resulting from collaborating. This simple example suggests that basic cost allocation methods may not be suitable in our setting and one must identify the set of desirable properties and design a cost allocation method accordingly. In the following sections, we discuss how to design cost allocation methods for a shippers' collaboration under single or multiple carrier settings. We refer the reader to [45] and [46] for a more complete discussion on collaboration in transportation procurement.

4.1 The Core of the Shippers' Collaboration Problem

In the shippers' collaboration problem, we consider a group of shippers each with a set of lanes that needs to be served. The collaboration's goal is to minimize the total cost of transportation such that the demand of each shipper in the collaboration is satisfied. Before the problem description, we first summarize our assumptions about the problem setting. First, we assume that every shipper is accepted to be a member of the collaboration, that is, each shipment request has to be served. The cost of collaborating among the members of the collaboration is assumed to be negligible and a shipper may decide to take out a subset of its lanes from the collaboration if it is profitable to do so. Finally, we assume that there are no side constraints (like time windows, driver restrictions, etc.).

We will refer to the centralized asset repositioning cost minimization problem as the *lane covering problem* (LCP). The LCP is defined on a complete directed Euclidian graph $G = (V, E)$ where V is the set of nodes $\{1, \ldots, n\}$, E is the set of edges, and $L \subseteq E$ is the lane set requiring service. Let c_{ij} be the cost of covering arc (i, j) with a truckload and let the asset repositioning cost associated with this arc be θc_{ij}, where $0 < \theta \leq 1$ is the "deadhead coefficient." The LCP is the problem of finding the minimum cost set of cycles traversing the arcs in L.

$$(P): \quad z_L(r) = \min \sum_{(i,j) \in L} c_{ij} x_{ij} + \theta \sum_{(i,j) \in E} c_{ij} z_{ij} \tag{22}$$

subject to

$$\sum_{j \in V} x_{ij} - \sum_{j \in V} x_{ji} + \sum_{j \in V} z_{ij} - \sum_{j \in V} z_{ji} = 0 \quad \forall i \in V \tag{23}$$

$$x_{ij} \geq r_{ij} \quad \forall (i, j) \in L \tag{24}$$

$$z_{ij} \geq 0 \quad \forall (i, j) \in E \tag{25}$$

$$x_{ij}, z_{ij} \in \mathbb{Z}. \tag{26}$$

In (P), x_{ij} represents the number of times lane $(i, j) \in L$ is covered with a full truckload and z_{ij} represents the number of times arc $(i, j) \in E$ is traversed as a deadhead. We let r_{ij} be the number of times lane $(i, j) \in L$ is required to be covered with a truckload and r be the vector of r_{ij}'s. Ergun et al. [19] showed that the LCP can be solved in polynomial time since solving its linear relaxation is sufficient to find an integer solution. The LCP with side constraints such as temporal considerations and driver restrictions is NP-hard and has been studied by [18] and [19].

In this setting, our goal is to allocate the total cost of the collaboration among all the lanes served. Therefore, the characteristic function of the shipper collaboration problem assigns $S \subset L$ the optimal objective function value of the LCP over lane set S. Hence, it is reasonable to assume that the lanes in L are the players for this game. The total allocated cost to a shipper is then the sum of the allocated costs to the lanes that belong to the shipper.

Owen [43] proves that for LP games a cost allocation in the core can be computed from an optimal solution to the dual of the linear program. Following this result, the optimal dual solutions of LCP yield cost allocations in the core.

Let I_{ij} be the dual variables associated with constraints (24) and y_i be the dual variables associated with constraints (23), then the dual of the LP relaxation of P is as follows:

$$D: \quad d_L(r) = \max \sum_{(i,j) \in L} r_{ij} I_{ij} \tag{27}$$

subject to

$$I_{ij} + y_i - y_j = c_{ij} \quad \forall (i,j) \in L \tag{28}$$
$$y_i - y_j \leq \theta c_{ij} \quad \forall (i,j) \in E \tag{29}$$
$$I_{ij} \geq 0 \quad \forall (i,j) \in L \tag{30}$$

Let the allocated cost, $x^{(i,j)}$, for covering lane $(i,j) \in L$ be equal to the corresponding optimal dual variable, I_{ij}^*. Then the vector x gives a cost allocation in the core. Furthermore, [45] extends the results of [16], [31], and [53] and shows that the core of the transferable payoffs shippers' collaboration problem is completely characterized by the set of optimal dual solutions.

This dual-based cost allocation method can be used in practice to allocate the costs to the participating shippers of a collaborative network. The collaborative networks already have the information about the demand of the shippers and the prices of the carrier(s). Besides this information, the dual-based method requires an optimizer interface such as CPLEX or XPRESS and completes within a few minutes for sufficiently large instances. Hence, this dual-based method (or its variants) is a practical method to allocate costs for a shippers' collaboration.

There may be multiple cost allocations in the core for an instance of the shipper collaboration problem. Among all the solutions in the core, the nucleolus is the one that lexicographically maximizes the minimal gain. Therefore, unless an implicit technique is devised, in order to compute the nucleolus one must compute the gains for all subsets of the collaboration, of which there is an exponential number. Hence, although the nucleolus may be a preferred cost allocation in general, for a regular instance of the shipper collaboration problem computing the nucleolus may take a considerable amount of time due to the size of the instance. However, alternative approaches may be employed to find the "most preferred" solution in the core.

4.2 Alternative Cost Allocation Methods

Next, we consider cost allocations that are not in the core in general but have additional properties that are desirable in the shipper collaboration game context.

We start with a well-known cost allocation method - the Shapley value. It is a budget balanced yet not necessarily stable cost allocation method. Consider again the example in Figure 3. If only two complementary lanes are present (i.e., the lanes of shippers A and B), then their Shapley value is equal to 1. When all three lanes are present, the Shapley values become $1 - \frac{\theta}{3}$ for shipper A and $1 + \frac{2\theta}{3}$ for shippers B and C. Since $x^A + x^B = 2 + \frac{\theta}{3} > 2$; Shapley value is not stable, and hence is not in the core for this simple instance. Like the nucleolus, unless an implicit technique is found, calculating the Shapley value requires an exponential effort, and hence it is an impractical cost allocation method for the shippers' collaboration problem.

Next, we consider cost allocation methods with the cross monotonicity property. Cross monotonicity ensures that the cost allocated to a player does not increase as the collaboration expands. Cross monotonicity is a relevant property in the shipper collaboration game context, since the shipment requirements for the shippers may change over time and there may be some new entrants to the collaboration. Therefore, the set of collaborating shippers naturally has a dynamic structure.

The cost function for the shippers' collaboration game is subadditive, which guarantees that when a new member joins the collaborative network, the overall benefit will be nonnegative. However, there may be some shippers whose cost allocations increase with the expansion of the collaboration, and hence these shippers may be opposed to new members. Hence, cross monotonic cost allocation methods are preferable in our setting. Unfortunately, cost allocations in the core are not cross monotonic for the shipper collaboration problem. To see this, consider the example in Figure 4. Suppose that the cost of covering all the lanes with a truckload is 1 and the deadhead coefficient is θ. If only shippers A and B are present, then the total cost of covering

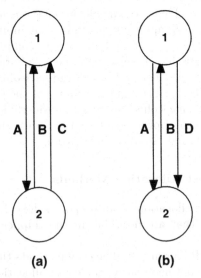

(a)   **(b)**

Fig. 4. Two shippers' networks

their lanes would be 2. In that case, any cost allocation that allocates a value between $(1 - \theta, 1 + \theta)$ to each lane and allocates the total network cost of 2 among the lanes is a cost allocation in the core.

If shipper C enters the collaboration, then the only cost allocation in the core allocates $(1 - \theta, 1 + \theta, 1 + \theta)$ to shippers A, B and C, respectively. If shipper D enters the collaboration instead of C, then the only cost allocation in the core allocates $(1 + \theta, 1 - \theta, 1 + \theta)$ to shippers A, B and D, respectively. Therefore, regardless of the initial cost allocations, either the cost allocation of shipper A or the cost allocation of shipper B increases with the newcomer. Hence, there does not exist a cross monotonic cost allocation in the core. For a more detailed discussion of this proof, see [45].

A cross monotonic and stable cost allocation can guarantee to recover at most $\frac{1}{1+\theta}$ of the total cost of the collaboration. Consider the example in Figure 5, which is very similar to the previous example, except we increase the number of lanes from node 2 to node 1. In the previous example, a cross monotonic cost allocation scheme can allocate at most 1 to each lane. Therefore, as the number of lanes, n, from node 2 to node 1 approaches infinity, the total cost of the network approaches $n \times (1 + \theta)$, whereas the total allocated cost approaches n. Thus, any cross monotonic cost allocation can recover at most $\frac{1}{1+\theta}$ for this instance. This result is not surprising, provided that similar results are reported in the literature for various problems such as the edge, vertex and set cover problems as well as the metric facility location problem [28]. As stated before, the cross monotonicity property is a useful property, but the existence of such bounds restricts its use in practice.

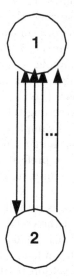

Fig. 5. Infinite example

Next, we consider the "minimum liability restriction" setting where every lane pays at least its original truckload lane cost and only the asset repositioning costs are distributed among the members of the collaboration. Therefore, there will be no *free riders* (shippers with zero allocated costs) in this setting. However, the core of the shippers' collaboration game may be empty in this setting. Consider the basic example with three shippers A, B and C. The total cost of the collaboration is $3 + \theta$ and the allocated cost to each lane is at least 1. Regardless of the way we allocate the remaining cost (θ), the sum of the cost allocations to A and B (or equivalently A and C) will be greater than 2, which is the stand-alone cost of the subset A and B (or equivalently A and C). Therefore, the core is empty for this instance.

Ozener and Ergun [45] design cost allocation methods for the shippers' collaboration game with minimum liability restriction by relaxing budget balance or stability conditions. It is concluded that when the budget balance property is relaxed, it is not possible to devise cost allocations with a reasonable bound on the fraction of the budget that is guaranteed to be collected. However, [45] suggests that relaxing the stability condition in a limited way may lead to implementable cost allocation schemes for the shipper collaboration problem due to asymmetric information exchange, limited rationality, and the cost of collaborating.

4.3 Multicarrier Setting

Next, we consider a collaboration among shippers that procures transportation services from multiple carriers with different characteristics such as regional operations, preferred routes, pre-existing networks, and fleet structures. In the basic multicarrier setting, the carriers have similar cost structures, so only the pre-existing networks of the carriers are different. The price charged for a lane varies across carriers because the value of the lane in the carriers' networks differs. For example, a carrier with many lanes from New York to Detroit in its pre-existing network will charge a lower price for lanes from Detroit to New York compared to its competitors.

We extend this setting by extending the differentiation of the carriers to nonidentical cost structures. For example, a carrier which mainly operates between metropolitan areas may charge a higher cost for the loads between rural areas. Other notions such as preferred routes, varied fleet structures, and specialized operational procedures may also cause the carriers to have dissimilar cost structures.

Collaboration among shippers is established in a manner similar to the single carrier setting. The goal of the collaboration is still to minimize asset repositioning costs by better utilizing carriers' truck capacity. However, in the multicarrier setting, both the individual shippers and the collaboration as a whole have more than one option to choose from for their transportation requirements. Also, the collaboration (hence the shippers individually) is allowed to make contracts with several carriers. Otherwise, the collaboration will try to identify the carrier that offers the lowest cost for the overall lane network of the collaboration and make contracts with that carrier. Hence the problem reduces to multiple independent single carrier cost allocation problems. We also assume that even though the carriers may have different cost structures, each carrier's cost figures satisfy the triangle inequality. Note that some special cases such as the existence of a limited number of carriers, identical cost structures, etc., may suggest relatively easier optimization methods for solving the system problem.

The formulation of the multicarrier lane covering problem (MCLCP) is similar to the LCP for the single carrier problem, where we only add a *carrier index* to the variables in the former case. We obtain the system optimal solution for the collaboration by solving the MCLCP. In the two-carrier lane covering problem, there exists a feasible integral optimal solution to the linear programming relaxation of the MCLCP[46]. However, it is not true in general. Ozener and Ergun [46] prove the NP-hardness of the problem (by a reduction from the 3-SAT problem) when the number of carriers is greater than two.

After finding the system optimal solution to MCLCP, the next step is to develop a procedure for allocating the costs among the members of the collaboration. In the single carrier problem, we use a dual-based approach to find cost allocations in the core of the shipper collaboration problem. Although the integrality gap may be nonzero for the MCLCP, we can still use the dual of the MCLCP to characterize the core of the problem. Ozener and Ergun [46] show that the core of the multicarrier shipper collaboration problem is nonempty if and only if the integrality gap is zero.

There are numerous issues one can consider in the multicarrier shipper collaboration problem such as the pricing decision of carriers, collaboration among carriers, different types of contractual agreements between shippers and carriers, carrier preferences or commitments to shippers, etc. For instance, to serve the shippers' lanes, the carriers may enter a "lane auction" and bid prices on the lanes according to their cost structures and existing networks. Also, information about competitors' networks and cost structures and willingness of the shippers to give away the lanes with specified prices affect the pricing decisions of the carriers. Auction mechanisms in transportation procurement have already been studied to some extent in the literature, primarily in the setting of one shipper and multiple carriers ([9], [37], [58], and [62]). After the auction process is finalized and the final networks of the carriers are formed, there may be further opportunities of collaboration among carriers, either by exchanging lanes or by subcontracting lanes with a specified price. Also, various contracting schemes may be investigated to maximize the total gain of the carriers and the shippers and to allocate these gains among all the participants in a fair manner. Finally, conditions under which a shipper commits to a specific carrier may be evaluated to achieve or approximate a centralized optimal solution in a decentralized system with selfish players.

5 Conclusions

In this chapter, we discussed two types of collaboration practices in the cargo transportation industry: air and sea cargo carrier alliances for better asset utilization on long routes and shipper collaborations for truckload transportation procurement. We presented issues that are relevant to managing the collaboration structures effectively so that the collaborations are sustainable.

Our main concerns were to develop implementable mechanisms to allocate gains from the collaboration in a stable manner and to design membership rules so that the continuity of the collaboration is ensured. We developed several models and analyzed the properties of the solutions that are generated by these models.

6 Acknowledgments

Ozlem Ergun was partially supported by NSF ITR Grant 0427446.

References

1. R. Agarwal and O. Ergun. Mechanism design for a multicommodity flow game in service network alliances. Operations Research Letters, 36(5): 520–524, 2008.
2. R. Agarwal and O. Ergun. Network design and allocation mechanisms for carrier alliances in liner shipping. Under revision for Operations Research, 2008.
3. R. Agarwal and O. Ergun. Ship scheduling and network design for cargo routing in liner shipping. *Transportation Science*, 42(2):175–196, 2008.
4. R. K. Ahuja and J. B. Orlin. Inverse optimization. *Operations Research*, 49(5):771–783, 2001.
5. K. S. Anand and R. Aron. Group buying on the web: A comparison of price-discovery mechanisms. *Management Science*, 49(11):1546–1562, 2003.
6. ATA. Trucking activity report. Technical report, American Trucking Associations, Arlington, VA, 2005. http://www.trucking.org.
7. G. E. Bamberger, D. W. Carlton, and L. R. Neumann. An empirical investigation of the competitive effects of domestic airline alliances. *Journal of Law and Economics*, XLVII, April 2004.
8. A. Boyd. Airline alliances. *OR/MS Today*, 25(5):28–31, 1998.
9. C. Caplice and Y. Sheffi. Optimization-based procurement for transportation. *Journal of Business Logistics*, 24(2):109–128, 2003.
10. E.-P. Chew, H.-C. Huang, E. L. Johnson, G. L. Nemhauser, J. S. Sokol, and C.-H. Leong. Short-term booking of air cargo space. *European Journal of Operational Research*, 174:1979–1990, 2006.
11. M. Christiansen, K. Fagerholt, and D. Ronen. Ship routing and scheduling: Status and perspectives. *Transportation Science*, 38(1):1–18, 2004.
12. R. Dembo and D. Rosen. The practice of portfolio replication: A practical overview of forward and inverse problems. *Annals of Operations Research*, 85(0):267–284, 1999.

13. R. Dial. Minimal revenue congestion pricing Part II: An efficient algorithm for the general case. *Transportation Research B*, 34:645–665, 2000.
14. R. J. Dolan. Quantity discounts: Managerial issues and research opportunities. *Marketing Science*, 6(1):1–22, 1987.
15. M. Dror. Cost allocation: The traveling salesman, bin packing, and the knapsack. *Applied Mathematics and Computation*, 35:191–207, 1990.
16. R. Engelbrecht-Wiggans and D. Granot. On market prices in linear production games. *Mathematical Programming*, 32(3):366–370, 1985.
17. S. Engevall, M. Gothe-Lundgren, and P. Varbrand. The heterogeneous vehicle-routing game. *Transportation Science*, 38(1):71–85, 2004.
18. O. Ergun, G. Kuyzu, and M. Savelsbergh. Reducing truckload transportation costs through collaboration. *Transportation Science*, 41(2):206–221, 2007.
19. O. Ergun, G. Kuyzu, and M. Savelsbergh. The shipper collaboration problem. *Computers and Operations Research*, 34(6):1551–1560, 2007.
20. K. Fagerholt. Optimal fleet design in a ship routing problem. *International Transactions in Operational Research*, 6:453–464, 1999.
21. M. Goemans and M. Skutella. Cooperative facility location games. *Annual ACM-SIAM Symposium on Discrete Algorithms*, 2000.
22. D. Granot, H. Hamers, and S. Tijs. On some balanced, totally balanced and submodular delivery games. *Mathematical Programming*, 86(2):355–366, 1999.
23. H. Gurnani. A study of quantity discount pricing models with different ordering structures: Order coordination, order consolidation, and multi-tier ordering hierarchy. *International Journal of Production Economics*, 72(3):203–225, 2001.
24. R. W. Hall. Configuration of an overnight package air network. *Transportation Research A*, 23(2):139–149, 1989.
25. H. Hamers. On the concavity of delivery games. *European Journal of Operational Research*, 99(2):445–458, 1997.
26. L. Houghtalen. *Designing Allocation Mechanisms for Carrier Alliances*. Ph.D. thesis, Georgia Institute of Technology, 2007.
27. L. Houghtalen, O. Ergun, and J. Sokol. Designing mechanisms for the management of carrier alliances. Working paper, 2008.
28. N. Immorlica, M. Mahdian, and V. S. Mirrokni. Limitations of cross-monotonic cost sharing schemes. In *Proceedings of 16th ACM Symposium on Discrete Algorithms*, pages 602–611, 2005.
29. H. Ito and D. Lee. Domestic codesharing, alliances, and airfares in the U.S. airline industry. *The Journal of Law and Economics*, 50:355–380, 2007.
30. K. Jain and V. V. Vazirani. Applications of approximation algorithms to cooperative games. In *Proceedings of 33rd ACM Symposium on Theory of Computing*, pages 364–372, 2001.
31. E. Kalai and E. Zemel. Generalized network problems yielding totally balanced games. *Operations Research*, 30(5):998–1008, 1982.
32. R. Kasilingam. Air cargo revenue managment: Characteristics and complexities. *European Journal of Operational Research*, 96:36–44, 1996.
33. M. Kubo and H. Kasugai. On the core of the network design game. *Journal of Operational Research Society of Japan*, 35:250–255, 1992.
34. M. J. Kuby and R. G. Gray. The network design problem with stopovers and feeders: The case of Federal Express. *Transportation Research A*, 27:1–12, 1993.
35. K. J. M. Gothe-Lundgren and P. Varbrand. On the nucleolus of the basic vehicle routing game. *Mathematical Programming*, 72:83–100, 1996.

36. R. Midoro and A. Pitto. A critical evaluation of strategic alliances in liner shipping. *Maritime Policy and Management*, 27(1):31–40, 2000.

37. E. W. Moore, J. M. Warmke, and L. R. Gorban. The indispensable role of management science in centralizing freight operations at Reynolds Metals Company. *Interfaces*, 21(1):107–129, 1991.

38. P. Morrell and R. Pilon. KLM and Northwest: A survey of the impact of a passenger alliance on cargo service characteristics. *Journal of Air Transport Management*, 5:153–160, 1999.

39. S. Morrish and R. Hamilton. Airline alliances–who benefits? *Journal of Air Transport Management*, 8:401–407, 2002.

40. H. Moulin. *Cooperative Microeconomics: A Game-Theoretic Introduction*. Princeton University Press, Princeton, NJ, 1995.

41. S. Netessine and R. A. Shumsky. Revenue management games: Horizontal and vertical competition. *Management Science*, 51(5):813–831, 2005.

42. N. Nisan and A. Ronen. Algorithmic mechanism design. *Games and Economic Behavior*, 35:166–196, 2001.

43. G. Owen. On the core of linear production games. *Mathematical Programming*, 9(3):358–370, 1975.

44. G. Owen. Indices of power. In *Game Theory*, pages 261–266. Academic Press, 2001.

45. O. O. Ozener and O. Ergun. Allocating costs in a collaborative transportation procurement network. *Transportation Science*, 42(2):146–165, 2008.

46. O. O. Ozener and O. Ergun. Collaborative opportunities in a multicarrier transportation procurement network. Working paper, 2008.

47. J.-H. Park, A. Zhang, and Y. Zhang. Analytical models of international alliances in the airline industry. *Transportation Research*, 35B:865–886, 2001.

48. A. N. Perakis. Fleet operations optimization and fleet deployment. In C. T. Grammenos, editor, *The Handbook of Maritime Economics and Business*, pages 580–597. Lloyd's of London, 2002.

49. K. Rana and R. G. Vickson. Routing container ships using Lagrangean relaxation and decomposition. *Transportation Science*, 25(3):201–214, 1991.

50. D. Ronen. Cargo ships routing and scheduling: Survey of models and problems. *European Journal of Operational Research*, 12:119–126, 1983.

51. D. Ronen. Ship scheduling: The last decade. *European Journal of Operational Research*, 71(3):325–333, 1993.

52. D. K. Ryoo and H. A. Thanopoulou. Liner alliances in the globalization era: A strategic tool for Asian container carriers. *Maritime Policy and Management*, 26(4):349–367, 1999.

53. D. Samet and E. Zemel. On the core and dual set of linear-programming games. *Mathematics of Operations Research*, 9(2):309–316, 1984.

54. J. Sánchez-Soriano. Pairwise solutions and the core of transportation situations. *European Journal of Operational Research*, 175:101–110, 2006.

55. J. Sánchez-Soriano, M. A. Lopez, and I. Garcia-Juardo. On the core of transportation games. *Mathematical Social Science*, 41:215–225, 2001.

56. L. Shapley. Cores of convex games. *International Journal of Game Theory*, 1:11–26, 1971.

57. L. Shapley and M. Shubik. The assignment game I: the core. *International Journal of Game Theory*, 1:111–130, 1972.

58. Y. Sheffi. Combinatorial auctions in the procurement of transportation services. *Interfaces*, 34(4):245–252, 2004.

59. X. Shi and S. Voss. Container terminal operations under the influence of shipping alliances. In K. Bichou, M. Bell, and A. Evans, editors, *Risk Management in Port Operations, Logistics and Supply Chain Security.* Informa, 2007.

60. B. Slack, C. Comtois, and R. McCalla. Strategic alliances in the container shipping industry: A global perspective. *Maritime Policy and Management,* 29(1):65–76, 2002.

61. D. W. Song and P. M. Panayides. A conceptual application of cooperative game theory to liner shipping strategic alliances. *Maritime Policy and Management,* 29(3):285–301, 2002.

62. J. J. Song and A. Regan. Combinatorial auctions for transportation service procurement - the carrier perspective. In *Freight Policy, Economics, and Logistics; Truck Transportation,* Transportation Research Record, pages 40–46. 2003.

63. D. J. Thomas and P. M. Griffin. Coordinated supply chain management. *European Journal of Operational Research,* 94(1):1–15, 1996.

64. B. Vinod. Alliance revenue management. *Journal of Revenue and Pricing Management,* 4(1):66–82, 2005.

65. S. Voss and X. Shi. Non-cooperative games in liner shipping strategic alliances. 2006.

66. C. Wright, H. Groenevelt, and R. Shumsky. Dynamic revenue management in airline alliances. *Working paper,* 2006.

67. H. P. Young. *Cost Allocation: Methods, Principles, Applications.* North-Holland, Amsterdam, The Netherlands, 1985.

68. A. Zhang, Y. V. Hui, and L. Leung. Air cargo alliances and competition in passenger markets. *Transportation Research,* 40E:83–100, 2004.

69. A. Zhang and Y. Zhang. Issues on liberalization of air cargo services in international aviation. *Journal of Air Transport Management,* 8:275–287, 2002.

Communication Models for a Cooperative Network of Autonomous Agents

Ashwin Arulselvan[1], Clayton W. Commander[2], Michael J. Hirsch[3], and Panos M. Pardalos[4]

[1] Department of Industrial and Systems Engineering
University of Florida, Gainesville, FL 32611
ashwin@ufl.edu
[2] Air Force Research Laboratory
Munitions Directorate, Eglin AFB, FL 32542
clayton.commander@eglin.af.mil
[3] Network Centric Systems
Raytheon Inc., St. Petersburg, FL 33710
michael_j_hirsch@Raytheon.com
[4] Department of Industrial and Systems Engineering
University of Florida, Gainesville, FL 32611
pardalos@ufl.edu

Summary We consider the problem of maximizing the total connectivity for a set of wireless agents in a mobile ad hoc network. That is, given a set of wireless units each having a start point and a destination point, our goal is to determine a set of routes for the units which maximizes the overall connection time between them. Known as the COOPERATIVE COMMUNICATION PROBLEM IN MOBILE AD HOC NETWORKS (CCPM), this problem has several military applications including coordination of rescue groups, path planning for unmanned air vehicles, and geographical exploration and target recognition. The CCPM is $\mathcal{NP}$-hard; therefore heuristic development has been the major focus of research. In this work, we survey the CCPM examining first some early combinatorial formulations and solution techniques. Then we introduce new continuous formulations and compare the results of several case studies. By removing the underlying graph structure, we are able to create a more realistic model of the problem as supported by the numerical evidence.

1 Introduction

Research in the area of cooperative networks has surged in recent years [5, 6, 17, 22]. This particular branch of telecommunications is leading the way for future technologies and the development of new network organizations [25]. In particular, so-called *mobile ad hoc networks* (MANETs) are at the forefront of the work in autonomous cooperative networks [10]. MANETs are composed of a set of loosely coupled agents which communicate via a shared radio channel

W. Chaovalitwongse et al. (eds.), *Optimization and Logistics Challenges in the Enterprise*, Springer Optimization and Its Applications 30, DOI 10.1007/978-0-387-88617-6_15, © Springer Science+Business Media, LLC 2009

with other agents within a specified range. The unique feature of MANETs that separates them from traditional cellular networks is the fact that the topology of MANETs is dynamic. That is, with each movement of the agents, a new topology is established.

The lack of a pre-established infrastructure is an attractive feature of MANETs. They are particularly useful in situations where communication is required, but no fixed telecommunication system exists. MANETs are also helpful when a set of mobile users need to be in constant contact with each other. Specific examples include combat search and rescue teams, and medical teams. In the wake of disasters such as the terrorist attacks of September 11, 2001, and Hurricane Katrina, the nation saw firsthand that communication among the emergency responders was critical to the success of the rescue operations.

Most cooperative networks require coordination among the group of users in order to accomplish the objective. The coordination of the system usually depends on communication being guaranteed amongst the agents. In typical ad hoc networks, bandwidth and communication time are very limited resources. Therefore, we see that the lack of a central command center for MANETs, while appealing from a distributed perspective, does lead to several problems in terms of routing, communication, and path planning [4, 8]. Perhaps the most important among these, and the focus of this chapter, is the study of communication models in the network. In particular, we study the problem of coordinating a set of wireless agents involved in a task that requires them to travel from a source location to a destination. The objective is to determine the paths, or trajectories for the agents which maximizes the connectivity between them subject to constraints on the initial and final configurations, and several limitations on the movements of the agents [23]. This problem is known as the cooperative communication problem in mobile ad hoc networks (CCPM), and is known to be $\mathcal{NP}$-hard [16]. In the next section, we review the currently existing work on the CCPM, which is primarily focused on heuristics for the problem posed as a discrete optimization problem.

The organization of this chapter is as follows. In Section 2, we provide the problem formulation for the CCPM. Then in Section 3 we present a review of the previous work in the area of communication in cooperative networks. In Section 4, we derive two mixed integer formulations for the CCPM using the combinatorial problem as a guide. We provide some preliminary numerical results in Section 5 and discuss conclusions and directions of future research in Section 6.

2 Problem Formulation

Consider an undirected graph $G = (V, E)$, where $V = \{v_1, v_2, \ldots, v_n\}$ represents the set of available positions for the wireless agents. Each node in V is assumed to be connected only to nodes that can be reached in one time step.

Let U represent the set of agents, $S = \{s_1, s_2, \ldots, s_{|U|}\} \subseteq V$ the set of initial positions, and $D = \{d_1, d_2, \ldots, d_{|U|}\} \subseteq V$ the set of destination positions for the agents. Furthermore, let $N(v) \subseteq 2^V$, for $v \in V$, represent the set of neighbors, or nodes, which are adjacent to node v. Given a time horizon T, the objective of the problem is to determine a set of routes for the agents, such that each agent $u_i \in U$ starts at source node s_i and finishes at its respective destination node d_i after at most T units of time [10].

For each agent $u \in U$, the function $p_t : U \to V$ returns the position of the agent at time $t \in \{1, 2, \ldots, T\}$. Then at each time instant t, an agent $u \in U$ can either remain in its current location, i.e., $p_{t-1}(u)$, or move to a node in $N(p_{t-1}(u))$.

We can represent a route for an agent $u \in U$ as a path $\mathcal{P} = \{v_1, v_2, \ldots, v_k\} \subseteq V$ where $v_1 = s_u$, $v_k = d_u$, and, for $i \in \{2, \ldots, k\}$, $v_i \in N(v_{i-1}) \cup \{v_i\}$. Finally, if $\{\mathcal{P}_i\}_{i=1}^{|U|}$ is the set of trajectories for the agents, we are given a corresponding vector $\mathcal{L}$ such that $\mathcal{L}_i$ is a threshold on the size of path $\mathcal{P}_i$. This value is typically determined by fuel or battery life constraints on the wireless agents.

We assume that the agents have omnidirectional antennas and that two agents in the network are connected if the distance between them is less than some radius $r \in \mathbb{R}$. The particular value of r is determined by the capabilities of the wireless equipment, such as the antenna strength and power amplifier. More specifically, let $\delta : V \times V \to \mathbb{R}$ represent the Euclidean distance between a pair of nodes in the graph. Then, we can define a function $c : V \times V \to \{0, 1\}$ such that

$$c(p_t(u_i), p_t(u_j)) = \begin{cases} 1, & \text{if } \delta(p_t(u_i), p_t(u_j)) \leq r \\ 0, & \text{otherwise.} \end{cases} \qquad (1)$$

With this, we can define the CCPM as the following optimization problem as given by Commander et al. [10]:

$$\max \sum_{t=1}^{T} \sum_{u,v \in U} c(p_t(u), p_t(v)) \qquad (2)$$

$$\text{s.t.} \sum_{j=2}^{n_i} \delta(v_{j-1}, v_j) \leq \mathcal{L}_i, \ \forall \ \mathcal{P}_i = \{v_1, v_2, \ldots, v_{n_i}\} \qquad (3)$$

$$p_1(u) = s_u \quad \forall \ u \in U \qquad (4)$$

$$p_T(u) = d_u \quad \forall \ u \in U, \qquad (5)$$

where constraint (3) ensures that the length of each path $\mathcal{P}_i$ is less than or equal to its maximum allowed length $\mathcal{L}_i$.

It has been determined [23] that the problem described above is $\mathcal{NP}$-hard. This can be shown by a reduction from the well-known 3SAT problem. Moreover, it is $\mathcal{NP}$-hard even to find an optimal solution for one stage of the problem at a given time t. To see this, consider an algorithm that maximizes the number of connections at time t, by defining the positions for members of the network. Run this algorithm for different sets U^i, with $|U^i| = i$ and i varying from 1 to T. Then the algorithm stops when the number of connections is less then $\binom{i}{2}$, and the value returned is $i-1$; clearly, the algorithm described above computes the value of the maximum clique on the underlying unit graph [9]. However, computing optimal solutions for the maximum clique problem on a unit graph is known to be $\mathcal{NP}$-hard [7].

Due to the computational complexity of the problem, real-world instances cannot be solved exactly. Therefore, we turn our attention to the design and implementation of efficient heuristics to solve large-scale instances within reasonable computing times. In the following section, we review the recent work in this area and describe the implementation of the first advanced metaheuristic for the CCPM based on the greedy randomized adaptive search procedure [26].

3 Previous Work

Since the introduction of the CCPM by Oliveira and Pardalos [23], heuristic design has been a major focus [9, 10]. In [10], the authors introduced a construction heuristic for the CCPM based on shortest paths [1]. The goal was to provide a way to quickly calculate sets of feasible trajectories for the agents. Pseudocode for this algorithm is provided in Figure 1. The procedure takes as input an instance of the CCPM consisting of the graph G, the set of agents U, source nodes S, destination nodes D, and a maximum travel time T. The total number of connections (c) represents the value of the objective function from Equation (2) and is initialized to zero. The set of trajectories for the agents (solution) is initialized to the empty set. In line 3, we compute

```
procedure ShortestPath(G, U, S, D, T)
1    c ← 0
2    solution ← ∅
3    Compute all shortest paths SP(sᵢ,dᵢ) for each pair (sᵢ,dᵢ) ∈ S × D
4    for i = 1 to |U| do
5        Pᵢ ← SP(sᵢ, dᵢ )
6        if lengthof Pᵢ > T then
7            return ∅
8        else
9            solution ← solution ∪ Pᵢ
10           c ← c + new connections generated by Pᵢ
11       end
12   end
13   return (c,solution)
end procedure ShortestPath
```

Fig. 1. Pseudocode for the shortest-path construction heuristic

the shortest source-destination path for each agent using the Floyd-Warshal algorithm [15, 27]. For each agent $i \in U$, the corresponding shortest path is assigned as the trajectory $\mathcal{P}_i$ for the agent. The trajectory is feasible if agent i is able to reach its destination d_i in at most T time units. Any agent which reaches its target location in less than T time steps will remain there until all other agents reach their respective destinations. If any path is infeasible, the algorithm terminates. Otherwise, the number of connections is updated and the process repeats until all agents have been considered.

The aforementioned algorithm provides feasible solutions for instances of the CCPM in $\mathcal{O}(|V|^3)$ time. However, the trajectories calculated are not guaranteed to be locally optimal, let alone globally optimal. Therefore, a local neighborhood search enhancement was applied. In general, a local search method receives a feasible solution as input and returns a solution that is guaranteed to be optimal with respect to the given neighborhood structure. Let S be the set of feasible solutions for an instance Π of the CCPM. Then for some $s \in S$ the neighborhood of s, denoted $\mathcal{N}(s)$, can be defined as the set of all solutions $s' \in S$ that differ from s in exactly one route. Notice that the number of feasible paths between any source-destination pair is exponential, and could lead to unreasonable computation times. Therefore instead of exhaustively searching the entire neighborhood the authors probe only $|U|$ neighbors at each iteration (one for each source-destination pair). Also, because of the exponential size of the neighborhood, the maximum number of iterations performed was limited to a constant MaxIter.

Pseudocode for the local improvement heuristic can be seen in Figure 2. Let f represent the objective function for the CCPM as given in Equation (2) above. New routes are computed using a randomized version of the standard

```
procedure HillClimb(solution)
1    c ← f(solution)
2    while solution not locally optimal and iter < MaxIter do
3        for i = 1 to |U| do
4            solution ← solution\{𝒫ᵢ}
5            𝒫ᵢ′ ← DFS(sᵢ, dᵢ)
6            c' ← f(solution ∪ 𝒫ᵢ′)
7            if length of 𝒫ᵢ′ < T and c' > c then
8                c ← c'
9                iter ← 0
10           else
11               Restore path 𝒫ᵢ
12           end
13       end for
14       iter ← iter+1
15   end while
16   return (solution)
end procedure HillClimb
```

Fig. 2. Pseudocode for the hill climbing intensification procedure

depth-first search (DFS) [1]. As mentioned in [10], at each step of the randomized DFS, the node selected to explore is uniformly chosen among the available children of the current node. Randomization helps to find a route that may improve the solution, while avoiding being trapped at a local optimum after only a few iterations. The local search is a standard hill-climbing method. Beginning with the feasible solution from the shortest path constructor, the local search begins computing new trajectories for the agents by using the randomized DFS to explore the neighborhood as described above. The method iterates over all the agents and repeats a total of MaxIter iterations after which the current best solution is deemed locally optimal and returned.

The two aforementioned heuristics can be combined into a one-pass heuristic for the CCPM as shown in Figure 3. Using this method, locally optimal solutions can be efficiently computed in $O(\max\{n^3, kTu^2m\})$ time, where T is the time horizon, $u = |U|$, $n = |V|$, $m = |E|$, and $k = $ MaxIter is the maximum number of iterations allowed in the local search phase.

We now describe the implementation of a more advanced randomized multistart heuristic for the CCPM based on the greedy randomized adaptive search procedure (GRASP) [26] framework. GRASP is a two-phase metaheuristic for combinatorial optimization that aims to find very good solutions through the controlled use of random sampling, greedy selection, and local search. GRASP has been used extensively in the last decade on numerous optimization problems and produces excellent results in practice [14]. Let F be the set of feasible solutions for the a problem Π, where each solution $S \in F$ is composed of k discrete components $a_1, \ldots, a_k$. GRASP constructs a sequence $\{S\}_i$ of solutions for Π, such that each S_i is feasible for Π. At the end, the algorithm returns the best solution found.

Pseudocode for the GRASP is provided in Figure 4. Notice that each GRASP solution is built in two stages, called *greedy randomized construction* and *intensification* phases. The construction phase receives as parameter an instance of the problem, a ranking function $g : A(S) \rightarrow R$ (where $A(S)$ is the domain of feasible components $a_1, \ldots, a_k$ for a partial solution S), and a constant $0 < \alpha < 1$. It starts with an empty partial solution S. Assuming that $|A(S)| = l$, the algorithm creates a list of the best ranked αl components in $A(S)$, and returns a uniformly chosen element x from this list. The current

```
procedure OnePass(G, U, S, D, T)
1    solution ← ShortestPath(G, U, S, D, T)
2    solution ← HillClimb(solution)
3    return (solution)
end procedure OnePass
```

Fig. 3. Pseudocode for the one-pass heuristic

```
procedure GRASP(MaxIter)
1    X* ← ∅
2    for i = 1 to MaxIter do
3        X ← ConstructionSolution(G, g, X)
4        X ← LocalSearch(X, MaxIterLS)
5        if f(X) ≥ f(X*) then
6            X* ← X
7        end
8    end
9    return X*
end procedure GRASP
```

Fig. 4. GRASP for maximization

partial solution is augmented to include x, and the procedure is repeated until the solution is feasible, i.e., until $S \in F$.

The intensification phase consists of the implementation of a hill-climbing procedure. Given a solution $S \in F$, let $N(S)$ be the set of solutions that can found from S by changing one of the components $a \in S$. Then, $N(S)$ is called a neighborhood of S. The improvement algorithm consists of finding, at each step, the element S^* such that

$$S^* = \underset{S' \in N(S)}{\operatorname{argmax}} f(S'),$$

where $f : F \to R$ is the objective function of the problem. At the end of each step, we assign $S \leftarrow S^*$ if $f(S) > f(S^*)$. The algorithm will converge to a local optimum, in which case the procedure above will generate a solution S^* such that $f(S^*) \geq f(S)$ for each $S \in N(S^*)$.

To apply GRASP to the CCPM, we need to specify the set A, the greedy function g, the parameter α, and the neighborhood $N(S)$, for $S \in F$. The components of each solution S are feasible moves of a member of the ad hoc network from a node v to a node $w \in N(v) \cup \{v\}$. The complete solution is constructed according to the following procedure. Start with a random $u \in U$ and find the shortest path P from s_u to d_u. If the total distance of P is greater than D_u, then the instance is clearly infeasible, and the algorithm ends. Otherwise, the algorithm considers each feasible move. A feasible move connects the final node of a subpath P_v, for $v \in U \setminus \{u\}$, to another node w, such that the shortest path from w to d_v has distance at most $D_v - \sum_{e \in P_v} \operatorname{dist}(e)$. The set of all feasible moves in a solution is defined as $A(S)$.

The greedy function g returns for each move in $A(S)$ the number of additional connections created by that move. As described above, the construction procedure will rank the elements of $A(S)$ according to g, and return one of the best $\alpha \cdot |A(S)|$ elements. This is repeated until a complete solution for the problem is obtained.

The improvement phase is defined by the perturbation function, which consists of selecting a wireless agent $u \in U$ and rerouting it, i.e., finding a complete path using the procedure described above for each time step 1 to T. The set of all perturbations of a solution S is its neighborhood $N(S)$. At each step, all elements $u \in U$ are tested, and the procedure stops when no such element u that improves the current solution can be found [9].

4 Continuous Formulations

In this section, we present continuous formulations of the CCPM [2]. These formulations will provide a more realistic scenario than the discrete formulation provided above, in that movement is not restricted to a discrete set of positions. We will assume that the agents are operating in a battlespace $\mathcal{Q} \subseteq \mathbb{R}^d$, where $\mathcal{Q}$ is a compact, convex set with unit volume and the Euclidean norm $||\cdot||_2$ in $\mathbb{R}^d$. For our purposes, we are going to consider the planar case, i.e., $d = 2$, with the understanding that extensions to higher dimensions are easily achieved. Suppose there are M wireless agents in the ad hoc network. The M agents are assumed to be omnidirectional and are modeled as point masses. We allow the agents to move freely within $\mathcal{Q}$ at some bounded velocity.

4.1 Formulation 1: A Continuous Analog of CCPM-D

In order to derive a continuous formulation, we need to to define an objective function that is consistent with that of the discrete formulation. Let R_{ij} be the communication constant for agents i and j. That is, R_{ij} is the radius of communication for the two agents. One possible objective is to maximize the *heaviside function*, defined as

$$H_1\left[R_{ij} - ||\mathbf{x}(t)^i - \mathbf{x}(t)^j||_2\right] = \begin{cases} 1, & \text{if } ||\mathbf{x}(t)^i - \mathbf{x}(t)^j||_2 \leq R_{ij} \\ 0, & \text{if } ||\mathbf{x}(t)^i - \mathbf{x}(t)^j||_2 > R_{ij}. \end{cases} \quad (6)$$

A graphical representation of H_1 is displayed in Figure 5. While this function will work as an objective, it is very extreme in the sense that there is a large jump from perfect communication at distances less than or equal to R_{ij} to no communication as soon as the distance becomes larger than R_{ij}. A more desirable function is one that approximates H_1 but degrades in a continuous fashion from perfect to no communication.

We consider two alternatives to H_1. The first is a piecewise continuous, linear function defined by

$$H_2\left[R_{ij} - ||\mathbf{x}_t^i - \mathbf{x}_t^j||_2\right] = \begin{cases} 1, & \text{if } ||\mathbf{x}_t^i - \mathbf{x}_t^j||_2 \leq R_{ij} \\ 2 - \frac{||\mathbf{x}_t^i - \mathbf{x}_t^j||_2}{R_{ij}}, & \text{if } R_{ij} < ||\mathbf{x}_t^i - \mathbf{x}_t^j||_2 \leq 2R_{ij} \\ 0, & \text{if } ||\mathbf{x}_t^i - \mathbf{x}_t^j||_2 \geq 2R_{ij}. \end{cases} \quad (7)$$

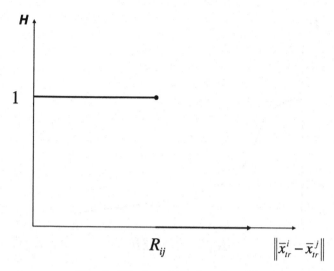

Fig. 5. The heaviside function, H_1

This function, whose graph is provided in Figure 6, has a value equal to 1 if agents i and j are within the communication radius R_{ij} of one another. The function then decreases constantly until the agents have distance $2R_{ij}$, at which time they are unable to communicate.

The third and final objective function we will consider is a continuously differentiable decreasing function of the distance between agents i and j. This function, displayed in Figure 7, is defined by

$$H_3\left[||\mathbf{x}_t^i - \mathbf{x}_t^j||_2, R_{ij}\right] = e^{-\left(\frac{||\mathbf{x}_t^i - \mathbf{x}_t^j||_2}{R_{ij}}\right)^2}. \tag{8}$$

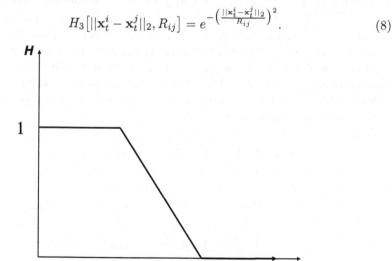

Fig. 6. H_2, continuous approximation to H_1

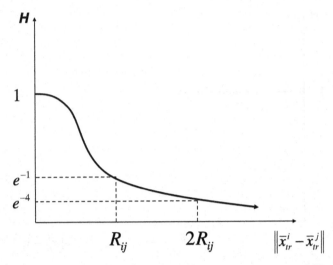

Fig. 7. H_3, continuously differentiable approximation of H_1

This is perhaps the best approximation of H_1 in that it can be interpreted as the probability of agents i and j directly communicating as a function of the distance between them.

Now that we have found a suitable objective function we can define the remaining parameters and constraints of the problem. Let $\mathbf{x}^i(t)$ be the position of agent i at time t. Similarly, let $\mathbf{v}^i(t)$ be the velocity of agent i at time t. The relationship between velocity and position is the standard one, given by $\mathbf{v}^i(t) = \frac{dx^i(t)}{dt}$. In order to formulate the continuous time analog of the CCPM, we must constrain the maximum speed of each agent. This is the continuous time analog of the constraints on the maximum distance traveled in the discrete formulation, between any two time steps. If $s_i \in \mathbb{R}^2$ is the starting position of agent i, and $d_i \in \mathbb{R}^2$ is the destination point of agent i, then we can formulate the continuous cooperative communication problem on mobile ad hoc networks (CCPM-C) as follows:

$$\max \int_0^T \sum_{i<j} H_3\big[||\mathbf{x}^i(t) - \mathbf{x}^j(t)||_2, R_{ij}\big] \tag{9}$$

s.t.

$$\mathbf{x}^i(0) = s_i, \ \forall \ i = 1, \dots, M \tag{10}$$

$$\mathbf{x}^i(T) = d_i, \ \forall \ i = 1, \dots, M \tag{11}$$

$$||\mathbf{v}^i(t)|| \leq V_i, \ \forall \ i = 1, \dots, M, t \in [0, T] \tag{12}$$

$$\mathbf{x}^i(t) \in \mathbb{R}^2, \ \forall \ i = 1, \dots, M, t \in [0, T]. \tag{13}$$

The above formulation provides a set of trajectories along which the agents can be routed in order to ensure that the communication between them is maximized. Clearly, if the agents remain in a tightly coupled formation, then communication will be maximized; however, unlike the discrete version of the problem in which the vertices of a graph had to be traversed, in a continuous setting this problem is relatively easy and worse yet, not very interesting. Alternatively, consider a set of UAVs involved in a search-and-rescue or reconnaissance mission. For obvious reasons, missions of this sort generally require the UAVs to traverse a large portion of the battlespace before arriving at their destinations. In this case, the above formulation is not helpful. With this in mind, we move on to develop a second continuous formulation which not only maximizes the communication between the agents, but also maximizes the coverage of predefined regions of the battlespace.

4.2 Formulation 2: A Continuous Multiobjective Formulation

In the following paragraphs, we derive a second continuous formulation which guarantees that certain locations will be visited by the UAVs as they traverse the battlespace from their sources to their respective destinations. Previous work on target visitation problems appear in [3]. Once again, we are considering a set of M UAVs. We keep the assumption that the ith UAV starts at a position $s_i = (s_{ix}, s_{iy})$, at time 0, and ends at position $d_i = (d_{ix}, d_{iy})$, at time T. The ith UAV, at time $t \in [0, T]$, has position $\mathbf{x}^i(t) = (x_i(t), y_i(t))$. Assume that the following holds:

$$\mathbf{x}^i(t) \in [x_{low}, x_{high}] \times [y_{low}, y_{high}] \; \forall \; i = 1, \ldots, M, \; t \in [0, T].$$

Furthermore, assume that there exists J positions in the domain, each of which must be visited by at least one UAV in the time interval $[0, T]$. These positions are given by $Q_j = (\bar{x}_j, \bar{y}_j)$, for each $j = 1, \ldots, J$. Lastly, the ith UAV has a minimum and maximum speed given by ϵ_i^{min} and ϵ_i^{max} for each $i = 1, \ldots, M$, respectively.

In order to implement a solution technique in a digital computer, we make use of the $\mathcal{L}_1$-norm as a measure of the distance between two points and discretize the time domain into ρ equal time steps, $\Delta t = T/(\rho - 1)$. Let $t_k = k\Delta t$, for each $k = 0, \ldots, \rho - 1$. Thus the position of the ith UAV at time step k is given by $\mathbf{x}^i(t_k) = (x_i(t_k), y_i(t_k))$, for each $i = 1, \ldots, M$, and for each $k = 0, \ldots, \rho - 1$.

Then the problem, which is denoted as **CCPM-C**, can be written as:

$$\min \sum_{i_1 < i_2} \sum_{k=0}^{\rho-1} \left[|x_{i_1}(t_k) - x_{i_2}(t_k)| + |y_{i_1}(t_k) - y_{i_2}(t_k)| \right] \tag{14}$$

s.t.

$$x_i(0) = s_{ix}, \quad y_i(0) = s_{iy}, \; \forall \; i = 1, \ldots, M \tag{15}$$

$$x_i(T) = d_{ix}, \quad y_i(T) = d_{iy}, \; \forall \; i = 1, \ldots, M \tag{16}$$

$$\epsilon_i^{\min} \leq \frac{1}{\Delta t} \left[|x_i(t_k) - x_i(t_{k-1})| + |y_i(t_k) - y_i(t_{k-1})| \right] \ \forall \ i,$$
$$\forall \ k = 1, \ldots, \rho - 1 \tag{17}$$

$$\epsilon_i^{\max} \geq \frac{1}{\Delta t} \left[|x_i(t_k) - x_i(t_{k-1})| + |y_i(t_k) - y_i(t_{k-1})| \right] \ \forall \ i,$$
$$\forall \ k = 1, \ldots, \rho - 1 \tag{18}$$

$$\beta_{ijk} \left[|x_i(t_k) - \bar{x}_j| + |y_i(t_k) - \bar{y}_j| \right] = 0 \ \forall \ i, \ \forall \ j \in J, \forall \ k \tag{19}$$

$$\sum_{i=1}^{M} \sum_{k=0}^{\rho-1} \beta_{ijk} \geq 1 \ \forall \ j \tag{20}$$

$$x_{low} \leq x_i(t_k) \leq x_{high} \quad y_{low} \leq y_i(t_k) \leq y_{high} \ \forall \ i, \ \forall \ k \tag{21}$$

$$\beta_{ijk} \in \{0, 1\} \quad \forall \ i, \ \forall \ j, \ \forall \ k. \tag{22}$$

Theorem 1 *The above formulation for **CCPM-C** is correct.*

Proof. Clearly, the objective function minimizes the pairwise distances between the agents. Thus as the distance between the agents decreases, the communication increases. Constraints (15) and (16), respectively, specify the starting points and the destination points for the agents. Together, constraints (17) and (18) bound the speed of the agents. Next (19) and (20) ensure that at least one agent colocates with the set of points in the domain which must be visited. More specifically, (19) implies that for all points $j \in J$, there must be a time when an agent occupies position j. In the constraint, this is accomplished by ensuring the distance between the visiting agent and point j is 0. Constraint (20) implies that at least one agent must visit each point $j \in J$. Finally, constraints (21) and (22) define the domain of the decision variables.

We can linearize the mixed integer programming formulation in (14)-(22) as follows. To begin with, replace the objective function (14) with

$$\sum_{i_1 < i_2} \sum_{k=0}^{\rho-1} \hat{x}_{i_1 i_2 k} + \hat{y}_{i_1 i_2 k} \tag{23}$$

with the additional constraints

$$\hat{x}_{i_1 i_2 k} \geq x_{i_1}(t_k) - x_{i_2}(t_k) \ \forall \ i_1, i_2 = 1, \ldots, M, i_1 < i_2,$$
$$\forall \ k = 0, \ldots, \rho - 1 \tag{24}$$

$$\hat{x}_{i_1 i_2 k} \geq -\left[x_{i_1}(t_k) - x_{i_2}(t_k) \right] \ \forall \ i_1, i_2 = 1, \ldots, M, i_1 < i_2,$$
$$\forall \ k = 0, \ldots, \rho - 1 \tag{25}$$

$$\hat{y}_{i_1 i_2 k} \geq y_{i_1}(t_k) - y_{i_2}(t_k) \ \forall \ i_1, i_2 = 1, \ldots, M, i_1 < i_2,$$
$$\forall \ k = 0, \ldots, \rho - 1 \tag{26}$$

$$\hat{y}_{i_1 i_2 k} \geq -\left[y_{i_1}(t_k) - y_{i_2}(t_k) \right] \ \forall \ i_1, i_2 = 1, \ldots, M, i_1 < i_2,$$
$$\forall \ k = 0, \ldots, \rho - 1. \tag{27}$$

Next, we replace (17) and (18) with

$$\epsilon_i^{\min} \leq \alpha_{ik} + \bar{\alpha}_{ik} \leq \epsilon_i^{\max} \tag{28}$$

adding the constraints

$$\alpha_{ik} \geq x_i(t_k) - x_i(t_{k-1}) \ \forall \ i = 1, \ldots, M, \ \forall \ k = 0, \ldots, \rho - 1 \tag{29}$$

$$\alpha_{ik} \geq -\big[x_i(t_k) - x_i(t_{k-1})\big] \ \forall \ i = 1, \ldots, M, \ \forall \ k = 0, \ldots, \rho - 1 \tag{30}$$

$$\bar{\alpha}_{ik} \geq y_i(t_k) - y_i(t_{k-1}) \ \forall \ i = 1, \ldots, M, \ \forall \ k = 0, \ldots, \rho - 1 \tag{31}$$

$$\bar{\alpha}_{ik} \geq -\big[y_i(t_k) - y_i(t_{k-1})\big] \ \forall \ i = 1, \ldots, M, \ \forall \ k = 0, \ldots, \rho - 1. \tag{32}$$

Finally, we replace (19) with

$$\phi_{ijk} + \bar{\phi}_{ijk} = 0 \tag{33}$$

and add the constraints

$$\theta_{ijk} \geq x_i(t_k) - \bar{x}_j, \ \forall \ i, \ \forall \ j, \ \forall k \tag{34}$$

$$\theta_{ijk} \geq -\big[x_i(t_k) - \bar{x}_j\big], \ \forall \ i, \ \forall \ j, \ \forall \ k \tag{35}$$

$$\bar{\theta}_{ijk} \geq y_i(t_k) - \bar{y}_j, \ \forall \ i \ \forall \ j \ \forall k \tag{36}$$

$$\bar{\theta}_{ijk} \geq -\big[y_i(t_k) - \bar{y}_j\big], \ \forall \ i, \ \forall \ j, \ \forall \ k \tag{37}$$

$$\phi_{ijk} \leq \beta_{ijk}(x_{high} - x_{low}), \ \forall \ i, \ \forall \ j, \ \forall \ k \tag{38}$$

$$\phi_{ijk} \geq 0, \ \forall \ i, \ \forall \ j, \ \forall \ k \tag{39}$$

$$\phi_{ijk} \leq \theta_{ijk}, \ \forall \ i, \ \forall \ j, \ \forall \ k \tag{40}$$

$$\phi_{ijk} \geq \theta_{ijk} - \big[1 - \beta_{ijk}\big]\big[x_{high} - x_{low}\big], \ \forall \ i, \ \forall \ j, \ \forall \ k \tag{41}$$

$$\bar{\phi}_{ijk} \leq \beta_{ijk}(y_{high} - y_{low}), \ \forall \ i, \ \forall \ j, \ \forall \ k \tag{42}$$

$$\bar{\phi}_{ijk} \geq 0, \ \forall \ i, \ \forall \ j, \ \forall \ k \tag{43}$$

$$\bar{\phi}_{ijk} \leq \bar{\theta}_{ijk}, \ \forall \ i, \ \forall \ j, \ \forall \ k \tag{44}$$

$$\bar{\phi}_{ijk} \geq \bar{\theta}_{ijk} - \big[1 - \beta_{ijk}\big]\big[y_{high} - y_{low}\big], \ \forall \ i, \ \forall \ j, \ \forall \ k. \tag{45}$$

Thus the problem becomes one of minimizing (23) subject to the constraints (15), (16), (20)-(22), (24)-(45). The resulting formulation is a mixed integer linear program (MILP) and can be solved using a number of commercial software packages. In the following section, we present some preliminary results from one such package, as well as providing a discussion of the experiments.

5 Case Studies

We have implemented the MILP formulation of the CCPM using the CPLEX^{TM} optimization suite from ILOG [13]. CPLEX contains an implementation of the simplex method [18], and uses a branch-and-bound algorithm [28] together with advanced cutting-plane techniques [21, 24]. The instances were tested on grids of size 10. The set of coordinates, J, to be visited were generated uniformly at random. Three sets of coordinates were generated and each visited by three different sets of UAVs, numbering 5, 7, and 10. The y-coordinates of the starting and ending positions were also randomly generated using a uniform distribution and the x-coordinates were assumed to be 0 and 10, respectively. The scenarios were solved making use of the MILP formulation derived above. The optimal solutions were obtained for the instances with five UAVs. The instances with seven UAVs and ten UAVs were stopped at optimality tolerances of 10% and 25%, respectively. A time frame of 10 units was provided as an input and the minimum and maximum speed of the UAVs were 1 and 2 units, respectively.

We have provided three graphical representations of the trajectories of the agents from each problem set. Figures 8 shows the paths traversed in one of the scenarios containing 5 agents. The movements of the agents are from left to right in the figure. The points which must be visited are denoted as stars. We see that from their starting points, the agents tend to converge into a tight formation. Notice that UAV2 separates from the group around point $(7, 6)$ in order to visit three "must visit" points and arrive at its destination. The remaining agents travel together until they must diverge to reach their

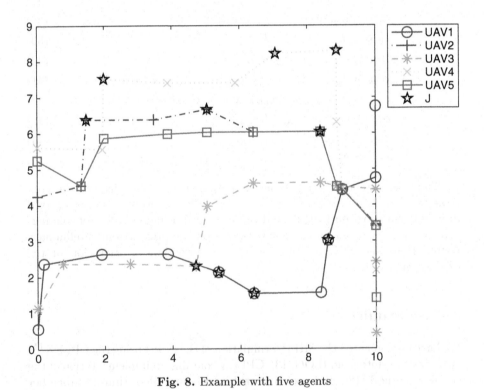

Fig. 8. Example with five agents

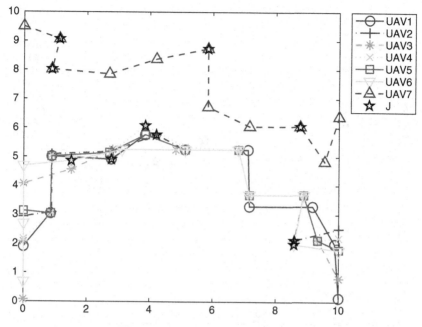

Fig. 9. Example with seven agents

destinations. In Figure 9 we have an example scenario containing seven agents. As before, the agents quickly converge to a tightly coupled formation with agents leaving the group only to visit the points in J or arrive at their destinations. A ten-agent scenario is depicted in Figure 10 and similar behaviors of the agents can be observed.

Figures 11-16 show how the paths of the agents change (from the ten-agent scenario in Figure 10) as agents are randomly removed from the scenario. As expected, the remaining agents are forced to spread out in order to ensure visiting the J target points. The results presented are very promising. Indeed, the agents exhibit the exact behavior we would expect to see given the nature of the CCPM. As communication strength is inversely proportional to distance, the convergence to a common path clearly indicates that the UAVs are attempting to maximize the communication amongst the group. Consider the scenario in Figure 8. Notice that midway through the mission a very clear clustering effect can be seen as two distinct groups of agents make their way towards the "must visit" points and ultimately their destinations.

6 Conclusion

In this chapter, we provided a review of recent work in the area of cooperative communication in mobile ad hoc networks. While inherently a problem of path planning, our formulations incorporated communication as a measure of the

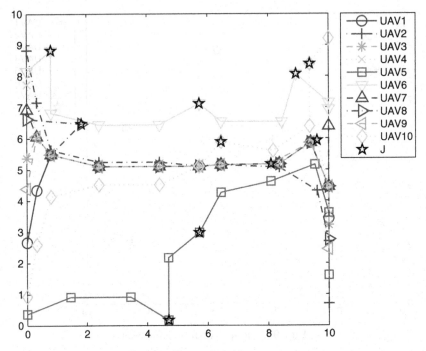

Fig. 10. Example with ten agents

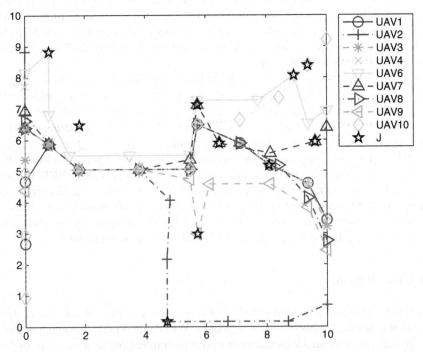

Fig. 11. Example derived from ten-agent example, with one agent removed

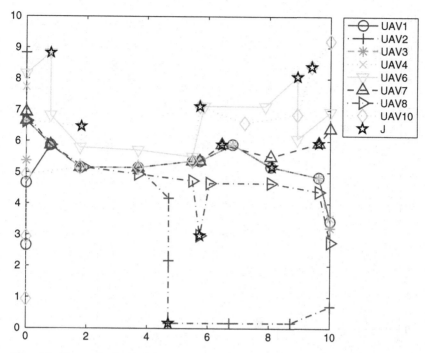

Fig. 12. Example derived from ten-agent example, with two agents removed

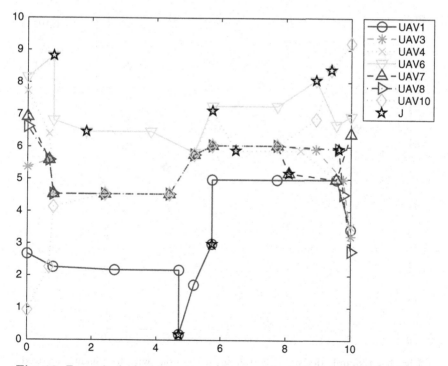

Fig. 13. Example derived from ten-agent example, with three agents removed

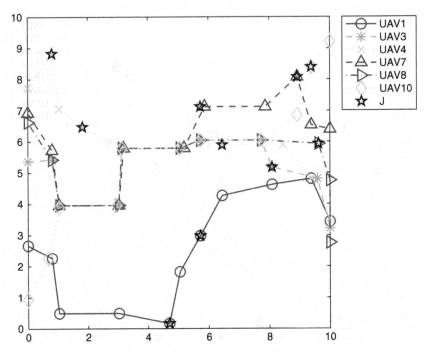

Fig. 14. Example derived from ten-agent example, with four agents removed

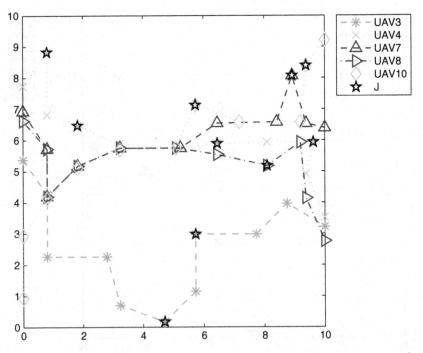

Fig. 15. Example derived from ten-agent example, with five agents removed

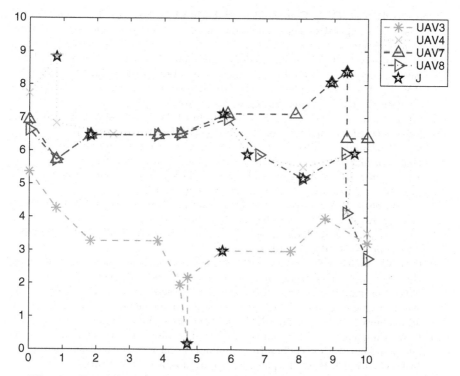

Fig. 16. Example derived from ten-agent example, with six agents removed

fitness of a given solution. We presented some discrete versions of the problem and derived two continuous formulations, the first time this has been considered. The advantage of the new models is that they ensure that a specified amount of the battlespace is explored by the agents. This addition is important in real-world applications particularly in the areas of surveillance, reconnaissance, and rescue operations. The preliminary numerical results demonstrate the effectiveness of the proposed models.

Due to the inherent complexity of the problem, future research will focus on continuous heuristic techniques for the newly proposed models, similar to those found in [19, 20]. Percentile risk constraints will be incorporated into the formulation. Commonly applied in financial applications, risk measures such as value-at-risk (VaR) and conditional value-at-risk have proven to be effective tools for military applications as well [8, 11, 12]. We also plan to provide some theoretical results regarding feasibility of problem instances.

References

1. R.K. Ahuja, T.L. Magnanti, and J.B. Orlin. *Network Flows: Theory, Algorithms, and Applications.* Prentice-Hall, 1993.

430   A. Arulselvan et al.

2. A. Arulselvan, C.W. Commander, M.J. Hirsch, P.M. Pardalos, and M.G.C. Resende. Cooperative communication in ad-hoc networks. In *National Fire Control Symposium Proceedings*, 2007.
3. A. Arulselvan, C.W. Commander, and P.M. Pardalos. A random keys based genetic algorithm for the target visitation problem. In M.J. Hirsch, P.M. Pardalos, R. Murphey, and D. Grundel, editors, *Advances in Cooperative Control and Optimization*. Springer, Inc., 2007.
4. S.I. Butenko, X. Cheng, C.A.S. Oliveira, and P.M. Pardalos. A new algorithm for connected dominating sets in ad hoc networks. In S. Butenko, R. Murphey, and P. Pardalos, editors, *Recent Developments in Cooperative Control and Optimization*, pages 61–73. Kluwer Academic Publishers, 2003.
5. S.I. Butenko, R.A. Murphey, and P.M. Pardalos, editors. *Cooperative Control: Models, Applications, and Algorithms*. Springer, 2003.
6. S.I. Butenko, R.A. Murphey, and P.M. Pardalos, editors. *Recent Developments in Cooperative Control and Optimization*. Springer, 2004.
7. B.N. Clark, C.J. Colbourn, and D.S. Johnson. Unit disk graphs. *Discrete Mathematics*, 86:165–177, 1990.
8. C.W. Commander. *Optimization Problems in Telecommunications with Military Applications*. Ph.D. dissertation, University of Florida, August 2007.
9. C.W. Commander, C.A.S. Oliveira, P.M. Pardalos, and M.G.C. Resende. A greedy randomized algorithm for the cooperative communication problem on ad hoc networks. In *8th INFORMS Telecommunications Conference*, 2006.
10. C.W. Commander, C.A.S. Oliveira, P.M. Pardalos, and M.G.C. Resende. A one-pass heuristic for cooperative communication in mobile ad hoc networks. In D.A. Grundel, R.A. Murphey, P.M. Pardalos, and O.A. Prokopyev, editors, *Cooperative Systems: Control and Optimization*. Springer, 2007.
11. C.W. Commander, P.M. Pardalos, V. Ryabchenko, S. Uryasev, and G. Zrazhevsky. The wireless network jamming problem. *Journal of Combinatorial Optimization*, to appear, 2007.
12. C.W. Commander, P.M. Pardalos, O. Shylo, and S. Uryasev. Recent advances in eavesdropping and jamming communication networks. In P.M. Pardalos D.A. Grundel, R.A. Murphey and O.A. Prokopyev, editors, *6th International Conference on Cooperative Control and Optimization*. World Scientific, 2006.
13. ILOG CPLEX. http://www.ilog.com/products/cplex, Accessed October 2006.
14. P. Festa and M.G.C. Resende. GRASP: An annotated bibliography. In C. Ribeiro and P.Hansen, editors, *Essays and Surveys in Metaheuristics*, pages 325–367. Kluwer Academic Publishers, 2002.
15. R.W. Floyd. Algorithm 97 (shortest path). *Communications of the ACM*, 5(6):345, 1962.
16. M.R. Garey and D.S. Johnson. *Computers and Intractability: A Guide to the Theory of NP-Completeness*. W.H. Freeman and Company, 1979.
17. D.A. Grundel, R.A. Murphey, and P.M. Pardalos, editors. *Theory and Algorithms for Cooperative Systems*. World Scientific, 2004.
18. F.S. Hillier and G.J. Lieberman. *Introduction to Operations Research*. McGraw Hill, 2001.
19. M.J. Hirsch. *GRASP-based Heuristics for Continuous Global Optimization Problems*. Ph.D. dissertation, University of Florida, December 2006.
20. M.J. Hirsch, C.N. Meneses, P.M. Pardalos, and M.G.C. Resende. Global optimization by continuous GRASP. *Optimization Letters*, 1(2):201–212, 2007.

21. R. Horst, P.M. Pardalos, and N.V. Thoai. *Introduction to Global Optimization*, volume 3 of *Nonconvex Optimization and its Applications*. Kluwer Academic Publishers, 1995.

22. R.A. Murphey and P.M. Pardalos, editors. *Cooperative Control and Optimization*. Springer, 2002.

23. C.A.S. Oliveira and P.M. Pardalos. An optimization approach for cooperative communication in ad hoc networks. Technical report, School of Industrial Engineering and Management, Oklahoma State University, 2005.

24. C.A.S. Oliveira, P.M. Pardalos, and T.M. Querido. A combinatorial algorithm for message scheduling on controller area networks. *International Journal of Operations Research*, 1(1/2):160–171, 2005.

25. M.G.C. Resende and P.M. Pardalos. *Handbook of Optimization in Telecommunications*. Springer, 2006.

26. M.G.C. Resende and C.C. Ribeiro. Greedy randomized adaptive search procedures. In F. Glover and G. Kochenberger, editors, *Handbook of Metaheuristics*, pages 219–249. Kluwer Academic Publishers, 2003.

27. S. Warshall. A theorem on boolean matrices. *Journal of the ACM*, 9(1):11–12, 1962.

28. L. Wolsey. *Integer Programming*. Wiley, 1998.